Advanced Structured Material

Volume 148

Common engineering materials reach in many applications their limits and new developments are required to fulfil increasing demands on engineering materials. The performance of materials can be increased by combining different materials to achieve better properties than a single constituent or by shaping the material or constituents in a specific structure. The interaction between material and structure may arise on different length scales, such as micro-, meso- or macroscale, and offers possible applications in quite diverse fields.

This book series addresses the fundamental relationship between materials and their structure on the overall properties (e.g. mechanical, thermal, chemical or magnetic etc.) and applications.

The topics of *Advanced Structured Materials* include but are not limited to

- classical fibre-reinforced composites (e.g. glass, carbon or Aramid reinforced plastics)
- metal matrix composites (MMCs)
- micro porous composites
- micro channel materials
- multilayered materials
- cellular materials (e.g., metallic or polymer foams, sponges, hollow sphere structures)
- porous materials
- truss structures
- nanocomposite materials
- biomaterials
- nanoporous metals
- concrete
- coated materials
- smart materials

Advanced Structured Materials is indexed in Google Scholar and Scopus.

More information about this series at http://www.springer.com/series/8611

Muhamad Husaini Abu Bakar ·
Mohd Nurhidayat Zahelem · Andreas Öchsner
Editors

Progress in Engineering Technology III

Springer

Editors
Muhamad Husaini Abu Bakar
Malaysian Spanish Institute
Universiti Kuala Lumpur
Kulim, Kedah, Malaysia

Mohd Nurhidayat Zahelem
Malaysian Spanish Institute
Universiti Kuala Lumpur
Kulim, Kedah, Malaysia

Andreas Öchsner
Faculty of Mechanical Engineering
Esslingen University of Applied Sciences
Esslingen am Neckar, Baden-Württemberg
Germany

ISSN 1869-8433 ISSN 1869-8441 (electronic)
Advanced Structured Materials
ISBN 978-3-030-67752-7 ISBN 978-3-030-67750-3 (eBook)
https://doi.org/10.1007/978-3-030-67750-3

This Springer imprint is published by the registered company Springer Nature Switzerland AG
The registered company address is: Gewerbestrasse 11, 6330 Cham, Switzerland

Preface

This book contains the selected and peer-reviewed manuscripts that were presented in the Conferences on Multidisciplinary Engineering and Technology (COMET 2019), held at the University Kuala Lumpur Malaysian Spanish Institute (UniKL MSI), Kedah, Malaysia from September 18 to 19, 2019. The aim of COMET 2019 was to present current and ongoing research being carried out in the field of mechanical, manufacturing, electrical, and electronics for engineering and technology. Besides, this book also contains manuscripts from the System Engineering and Energy Laboratory (SEELAB) research cluster, UniKL which is actively doing research mainly focused on artificial intelligence, internet of things, metal–air batteries, advanced battery materials, and energy material modelling fields. This volume is the fourth edition of the progress in engineering technology, Advanced Structured Materials which provides in-depth ongoing research activities among academia of UniKL MSI. Lastly, it is hoped to foster cooperation among organizations and research in the covered fields.

Kulim, Malaysia — Muhamad Husaini Abu Bakar
Kulim, Malaysia — Mohd Nurhidayat Zahelem
Esslingen am Neckar, Germany — Andreas Öchsner

Contents

Chapter 1
Vibration Response of Magneto-Rheological Elastomer Sandwich Plates

Muhammad Hafize Zaini, Mohd Nurhidayat Zahelem, Faizatul Azwa Zamri, and Muhamad Husaini Abu Bakar

Abstract In recent years, the research on magneto-rheological elastomer sandwich structures has increased due to the adjustable properties by varying the magnetic field. However, the study is mainly focused on the theoretical analysis of the magneto-rheological elastomers MRE sandwich plates. In this works, the numerical and experimental investigation on the vibration response of MRE sandwich plates has been conducted. The composition of the iron particles and silicon rubber varied with 30/70, 40/60, 50/50, 60/40, and 70/30% by its weight percentage, respectively. According to the finite element analysis, the magnetic field produced was 0.6, 0.9, and 1.3 T with 2, 4, and 6 mm of the gap between the magnet and MRE composites. The natural frequency and stiffness of the MRE sandwich structured plates was 19.68–23.97 Hz, and 152.6–23.97 N/m, respectively for 0.6, 0.9, and 1.3 T of the applied magnetic field. It is concluded that by increasing the magnetic fields, the stiffness of the material can be changed. This work will contribute to the understanding of the properties of the MRE in the sandwich structures application.

Keywords Magneto-rheological elastomers · Iron particle · Silicon rubber · MRE composites sandwich structured plate

---

M. H. Zaini
YDI Synergy Sdn. Bhd, Bandar Bukit Puchong, 47120 Puchong, Selangor, Malaysia
e-mail: Muhammad.hafize@ydi.my

M. N. Zahelem (✉)
Mechanical Section, Universiti Kuala Lumpur Malaysian Spanish Institute, Kulim Hi-Tech Park, 09000 Kulim, Kedah, Malaysia
e-mail: mnurhidayat@unikl.edu.my

F. A. Zamri · M. H. Abu Bakar
System Engineering and Energy Laboratory, Universiti Kuala Lumpur Malaysian Spanish Institute, Kulim Hi-Tech Park, 09000 Kulim, Kedah, Malaysia
e-mail: faizatul.zamri08@s.unikl.edu.my

M. H. Abu Bakar
e-mail: muhamadhusaini@unikl.edu.my

M. H. Abu Bakar et al. (eds.), *Progress in Engineering Technology III*, Advanced Structured Materials 148, https://doi.org/10.1007/978-3-030-67750-3_1

1.1 Introduction

Smart materials of magneto-rheological fluids (MRF) and magneto-rheological elastomers (MRE) belong to a family of rheological materials that undergo rheological changes under the application of magnetic fields (Rajhan et al. 2015; Liu and Xu 2019; Babu and Vasudevan 2016). The application of MRF is limited due to the accumulation of iron particles in the absence of a magnetic field, and its high cost resulting in the fact that MRE offer a better solution (Babu and Vasudevan 2016).

MR elastomers consist of three essential components which are mainly the polarized magnetic particles, an elastomer/rubber matrix, and additives (usually silicon oil). The constituents are assorted together to form a compound with a large density whereat the magnetic particle are randomly spread or pre-arranged in a low-density matrix (Rajhan et al. 2015). MRE samples are fabricated using different weight percentages of carbonyl iron particles, silicon rubber, and silicon oil (Chen et al. 2007).

Sandwich structures can provide several benefits such as enhanced strength-to-weight ratio which is essential in several engineering applications. In recent years, the interest on MRF/MRE-based sandwich structures has increased significantly (Selvaraj and Ramamoorthy 2020). These interests are mainly due to the ability of the MRE sandwich structure to control their stiffness and damping by varying the applied magnetic fields (Kolekar et al. 2019; Eshaghi et al. 2016; Ismail et al. 2014; Kallio 2005). However, most research (Babu and Vasudevan 2016; Yeh 2013; Aguib et al. 2014) has focused on the analysis of the MRE sandwich plate structures.

Therefore, numerical and experimental investigation was carried out to investigate the vibration response and its stiffness for different compositions of iron particle and silicon rubber in the magneto-rheological sandwich plates subjected to the magnetic field. The natural frequency and stiffness observed have significantly increased with the increased iron particle content for a magnetic field applied as 0.3, 0.6, and 1.3 T, respectively.

1.2 Methodology

The MRE sandwich structural plate consists of MRE composites, and fiberglass as core and skins, respectively. The MRE composites were fabricated using iron particles of a size of 10 μm and silicon rubber according to the composition ratio as shown in Table 1.1. A total 2.0 g of silicon oil was used as adhesive to bind the composition. Meanwhile, the fiberglass was fabricated using a tissue mat, resin, and catalyst. A total of five pieces of the sample of the MRE composite and 10 pieces of the fiberglass were fabricated to produce five samples of the MRE sandwich structural plates in this work.

Table 1.1 Composition ratio of the MRE samples

Iron particles (%)	Silicone rubber (%)
30	70
40	60
50	50
60	40
70	30

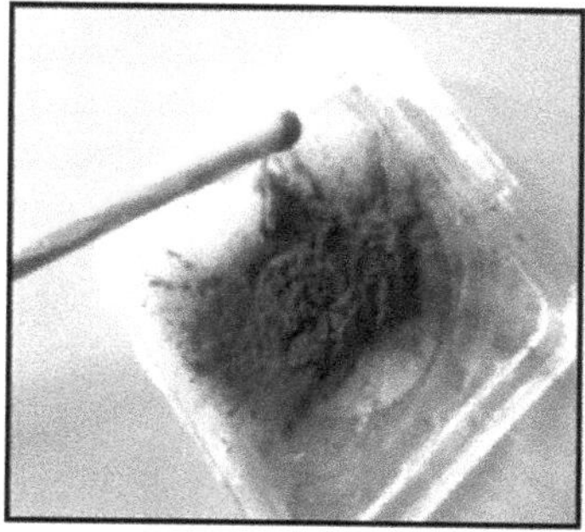

Fig. 1.1 The process of mixing the MRE on the left and the mixture of the MRE loads into the mold on the right

1.2.1 Sample Preparation

Firstly, the silicon oil and rubber were mixed using a mixer at a constant speed for 10 min for homogeneity. After that, iron particles were mixed in the composition and stirring was continued for 30 min. Then, the composition was poured into a 110 × 20 mm dimension mold for the curing process, as shown in Fig. 1.1. The mold was stored at ambient temperature for 24 h.

The fabrication process of MRE composites is similar to the fiberglass. The catalyst and resin were mixed using the composition ratio of 1:10 by its weight percentage for 30 s. Meanwhile, the tissue mat was sliced to obtain the dimension 110 × 20 mm, and the mixture was applied between the surface to form two layers of fiberglass. A roller was used on the tissue mat to remove air trapped in the fiberglass. Lastly, the fiberglass was cured at room temperature for 24 h. Figure 1.2 shows the fabricated fiberglass used as skins in the MRE sandwich structured plate.

1.2.2 Fabrication of the Sandwich Structural Plate

The fabricated MRE sandwich structural plate consists of two pieces of fiberglass and one piece of MRE composite with a dimension of 110 × 20 × 1 mm as shown

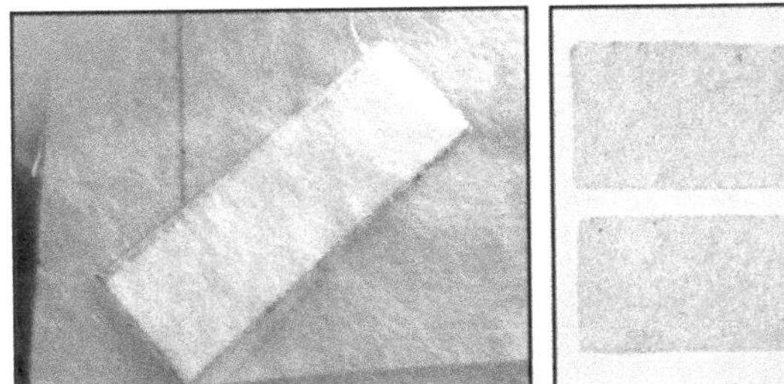

Fig. 1.2 Skin composites of fiberglass

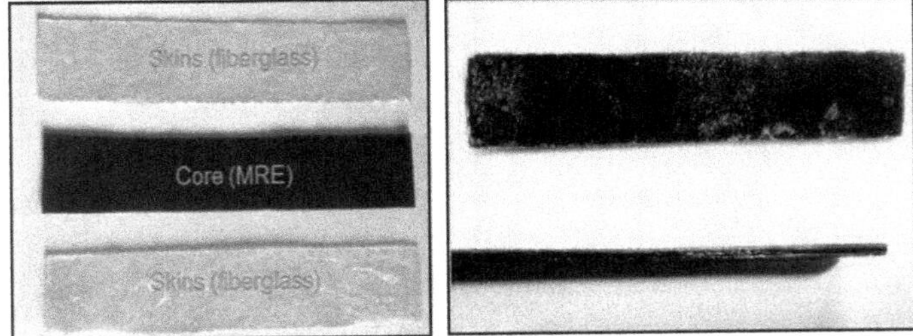

Fig. 1.3 Layers of the MRE composite on the left and MRE sandwich structural plate after attached together on the right

in Fig. 1.3. The skin layers and the core have been attached using a commercial adhesive.

1.2.3 Vibration Response Test Setup

The vibration response test setup was developed as shown in Fig. 1.4. The current from the DC power is supplied to the coil to generate the magnetic field to produce a forced excitation to the MRE sandwich plate. An accelerometer is attached to the top layer of the MRE sandwich plate, and it senses the vibration response. The vibration response is measured by DAQ (LMS SCADAS Mobile) and connected to the computer. The LMS Lab Test software is used to run the measurement of the vibration response.

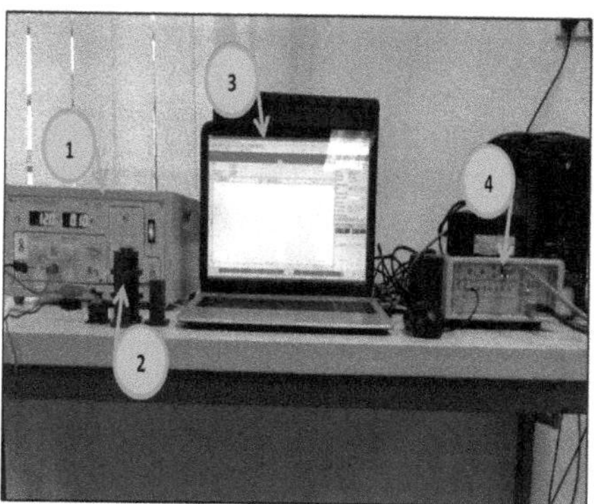

Fig. 1.4 Overall experimental setup on the left and the main vibration response test setup on the right

1.3 Result and Discussions

1.3.1 Magnetic Field Determination

Firstly, the value of the magnetic field was determined by adjusting the distance (2, 4, and 6 mm) between the magnet and the MRE sandwich plate using the finite element method magnetics (FEMM) software. The Neodymium N42 with a dimension of 10 × 2 × 50 mm was used as a magnet. Figure 1.5 shows the magnetic field contours and the value calculated was 0.6, 0.9, and 1.3 T for 2, 4, and 6 mm, respectively.

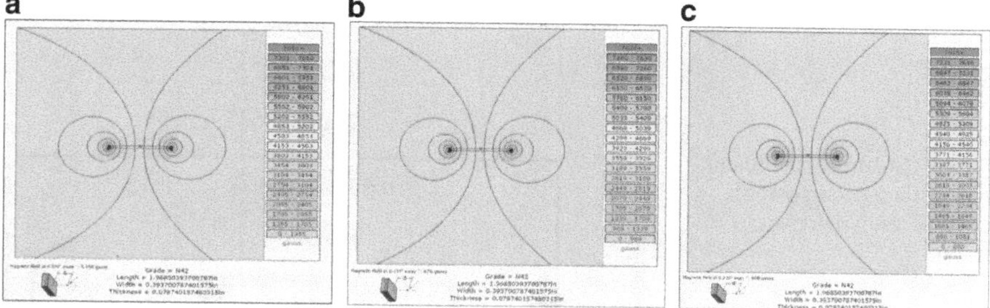

Fig. 1.5 Magnetic field contour for **a** 2 mm, **b** 4 mm and **c** 6 mm gap between Neodymium N42 and sandwich plate

1.3.2 *Effects of Iron Loading on the Vibration Responses*

Figure 1.6 shows the natural frequency of 30/70, 40/60, 50/50, 60/40, and 70/30% of MRE sandwich structural plates. The natural frequency was obtained in the range of 19.68–23.97 Hz for 0.6, 0.9, and 1.3 T of the applied magnetic field. In comparison, the system tuning frequency varies from 25.8 to 37.4 Hz under a magnetic field of 0.316 T (Selvaraj and Ramamoorthy 2020; Komatsuzaki et al. 2016).

The natural frequency was observed to increase with increasing iron particles in the MRE composition, and applied magnetic field. In agreement with previous studies, the natural frequency and the electrical resistance of the elastomer can be affected by the external magnetic field (Selvaraj and Ramamoorthy 2020; Komatsuzaki et al. 2016).

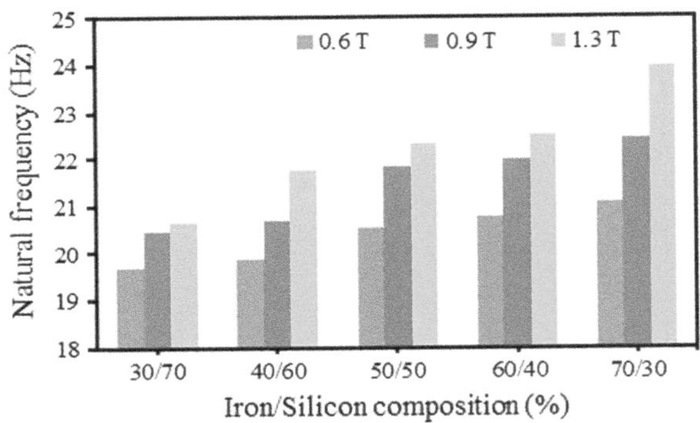

Fig. 1.6 Natural frequency of iron/silicon composition for MRE sandwich plate

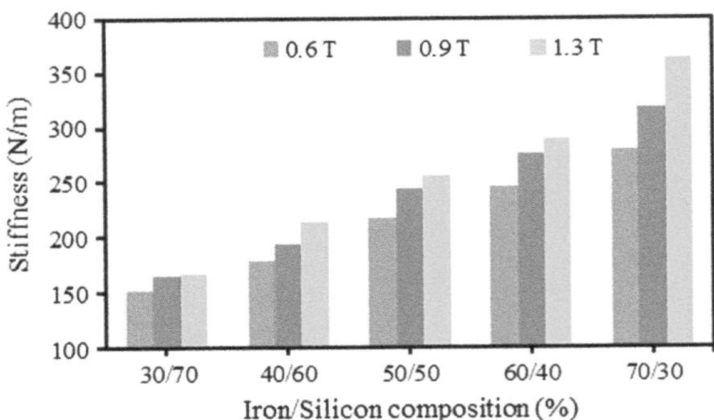

Fig. 1.7 Stiffness coefficient of iron/silicon composition for MRE sandwich plate

1.3.3 Effects of Iron Loading on the Stiffness Coefficient

Figure 1.7 shows the stiffness coefficient of 30/70, 40/60, 50/50, 60/40, and 70/30% of the MRE sandwich structural plates. The natural frequency was obtained in the range of 152.6–23.97 N/m for 0.6, 0.9, and 1.3 T of the applied magnetic field. The material stiffness increased with increasing iron particles in the MRE composition.

1.4 Conclusion

The MRE composites as core in the sandwich structural plate were successfully fabricated with a composition of iron particle and silicon rubber of 30/70, 40/60, 50/50, 60/40, and 70/30 by its weight percentages. The natural frequency and stiffness of the MRE sandwich structural plates were 19.68–23.97 Hz, and 152.6–23.97 N/m, respectively, for 0.6, 0.9, and 1.3 T of the applied magnetic field. In summary, the natural frequency and stiffness were highly affected by the composition of iron particles at high magnetic fields applied.

Acknowledgements The authors gratefully acknowledge the financial support for this research work by Yayasan Tengku Abdullah Scholarship (YTAS), Universiti Kuala Lumpur Malaysian Spanish Institute (UniKL MSI), and System Engineering and Energy Laboratory (SEELab).

References

Aguib, S., Nour, A., Zahloul, H., et al.: Dynamic behavior analysis of a magnetorheological elastomer sandwich plate. Int. J. Mech. Sci. **87**, 118–136 (2014)

Babu, V.R., Vasudevan, R.: Dynamic analysis of tapered laminated composite magnetorheological elastomer (MRE) sandwich plates. Smart Mater. Struct. **25**(3), 035006 (2016). https://doi.org/10.1088/0964-1726/25/3/035006

Chen, L., Gong, X.L., Li, W.H.: Microstructures and viscoelastic properties of anisotropic magnetorheological elastomers. S Smart Mater. Struct. **16**(6), 2645–2650 (2007)

Eshaghi, M., Sedaghati, R., Rakheja, S.: Dynamic characteristics and control of magnetorheological/electrorheological sandwich structures: a state-of-the-art review. J. Intell. Mater. Syst. Struct. **27**(15), 2003–2037 (2016)

Ismail, R., Ibrahim, A., Hamid, H.A.: A review of magnetorheological elastomers: characterization properties for seismic protection. In: Hassan, R., Yusoff, M., Ismail, Z., Amin, N., Fadzil, M. (eds.) InCIEC 2013. Springer, Singapore (2014)

Kallio, M.: The Elastic and Damping Properties of Magnetorheological Elastomers. Dissertation. VTT Technical Research Centre of Finland (2005). http://www.vtt.fi/inf/pdf/publications/2005/P565.pdf Accessed 25 May 2020

Kolekar, S., Venkatesh, K., Oh, J.-S., et al.: Vibration controllability of sandwich structures with smart materials of electrorheological fluids and magnetorheological materials: a review. J. Vib. Eng. Technol. **7**(4), 359–377 (2019)

Komatsuzaki, T., Inoue, T., Iwata, Y.: Experimental investigation of an adaptively tuned dynamic absorber incorporating magnetorheological elastomer with self-sensing property. Exp. Mech. **56**(5), 871–880 (2016)

Liu, T., Xu, Y.: Magnetorheological elastomers: Materials and applications. Smart Funct. Soft Mater. (2019). https://doi.org/10.5772/intechopen.85083

Rajhan, N.H., Hamid, H.A., Azmi, I., et al.: Material compositions of magnetorheological elastomers: a review. Appl. Mech. Mater. **695**, 255–259 (2015)

Selvaraj, R., Ramamoorthy, M.: Recent developments in semi-active control of magnetorheological materials-based sandwich structures: a review. J. Thermoplast. Compos. Mater. (2020). https://doi.org/10.1177/0892705720930749

Yeh, J.Y.: Vibration analysis of sandwich rectangular plates with magnetorheological elastomer damping treatment. Smart Mater. Struct. **22**(3), 035010 (2013). https://doi.org/10.1088/0964-1726/22/3/035010

Chapter 2
Nonlinear Control of a Magneto-Rheological Fluid Electrohydraulic Positioning System

Siti Lydia Rahim, Sulaiman Mohd Zulkifli, and Muhamad Husaini Abu Bakar

Abstract Herein we report the identification of magneto-rheological fluid (MRF) electrohydraulic system and the development of a fuzzy logic controller for the MRF directional valve. The available system is lacking a proper controller and highly nonlinear. The nonlinearities involve hysteresis effects due to magnetic properties of the fluid and the stiction phenomenon in the actuator. In order to implement the fuzzy logic controller, the input and output of the system is obtained by experiment and then identified using the Hammerstein–Weiner model. The identified model is 81.64% fit to the actual system and is used to develop and tune the fuzzy controller. Results show that with the developed fuzzy controller, the response time of the system has improved to 0.3 s, percentage overshoot, and the error is reduced considerably to 10%, respectively.

Keywords Electrohydraulic · Control valve · Magneto-rheological fluid · Fuzzy logic

2.1 Introduction

Magneto-rheological fluids (MRF) (Rabinow 1948; Weiss and Carlson 1994; Dyke et al. 1996) are part of smart materials and their main property is very useful in practical applications. The properties consist of viscosity modification in case of exposal to a magnetic field. Their rheological properties are controlled by applying

S. L. Rahim (✉) · S. M. Zulkifli
Manufacturing Section, Malaysian Spanish Institute, Universiti Kuala Lumpur, 09000 Kulim, Kedah, Malaysia
e-mail: sitilydia@unikl.edu.my

S. M. Zulkifli
e-mail: slayy0266@yahoo.com

S. L. Rahim · M. H. Abu Bakar
System Engineering and Energy Laboratory, Universiti Kuala Lumpur Malaysian Spanish Institute, Kulim Hi-Tech Park, 09000 Kulim, Kedah, Malaysia
e-mail: muhamadhusaini@unikl.edu.my

M. H. Abu Bakar et al. (eds.), *Progress in Engineering Technology III*,
Advanced Structured Materials 148, https://doi.org/10.1007/978-3-030-67750-3_2

an external magnetic field. A magnetic field generates groups which are aligned elongated particles along the magnetic field lines between the poles.

A magneto-rheological (MR) fluid is composed of micro-sized magnetic particles which are suspended in hydrocarbon oil. The rheological properties of an MR fluid can be fast and reversibly altered when an external magnetic field is applied. An MR fluid is a controllable fluid. MR fluid has received a great deal of attention over the past ten years, because they offer the promise in relation to a valve with no moving parts, low-cost directional control valves, and miniature size. The MR fluid can be interfaced between magnetic field and fluid power without the need for mechanical moving parts like spools in directional control valves (Salloom 2013). When the electric field strength or the magnetic field strength reaches a certain value, the suspension solidifies with high yield stress (Truong et al. 2009). The changing process is very quick and can easily be controlled by small amounts of energy on the order of several watts (Dyke et al. 1996).

The other part that is equally important in an electrohydraulic system is the directional control valve. There are many configurations of directional control valves. A directional control valve has a complex construction, such as the moving spool to control the direction of the actuator and the desired speed. A magneto-rheological (MR) fluid is one of the controllable fluids. Utilizing the MR fluid properties, a direct interface can be realized between the magnetic field and fluid power without the need for moving parts like a spool in directional control valves (Salloom 2013).

Currently, the MRF directional control valve is operated using a simple switch to open or close the channels in order to redirect the fluid flow so that the actuator moves up, down, or stop. However, in demand for accuracy and speed variation of hydraulic system the ON-OFF controller is obsolete. A new controller is required in order to continuously change the speed of the actuator and at the same time to preserve the accuracy of the system. Therefore, the fuzzy controller for the MRF electrohydraulic system was developed to study the performances of the control system. The performance needs to be observed are such as response time, percent overshoot and steady-state error.

The hysteresis effect due to magnetic properties in the fluid will lead to saturation in the input and output of the system. The stiction phenomenon also has a great potential in making the system more complex and difficult to control. Due to its inherently hysteretic and highly nonlinear dynamics, identification of the MRF system is significantly difficult (Truong et al. 2009; Askari and Davaie-Markazi 2008). To control the system and reduce the nonlinearities, fuzzy controller has proven to be better compared to a conventional PID controller (Rashid et al. 2008).

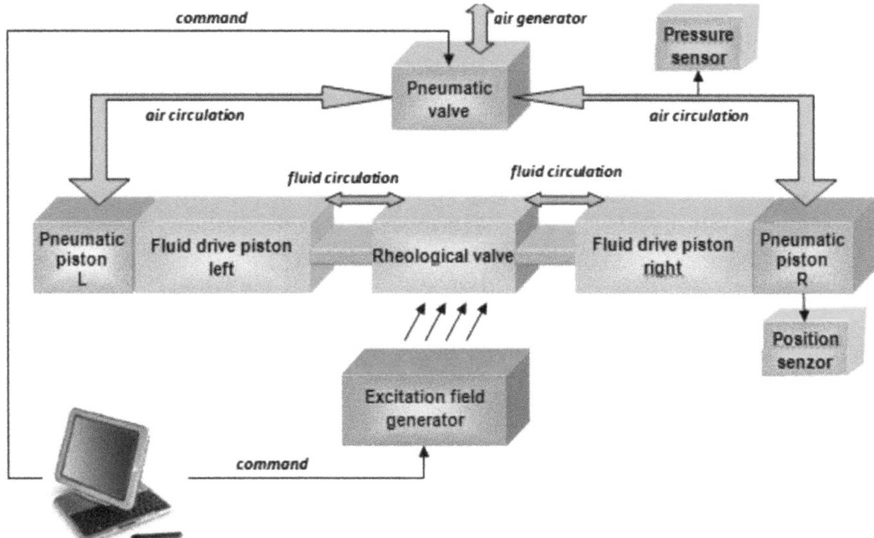

Fig. 2.1 General scheme of the platform

2.2 Literature Review

MRF has many applications and the directional control valve represents one of its most known applications. According to Vladu et al. (2012), because of the non-Newtonian flow of the MRF, its mathematical modeling is very difficult and it must be experimentally determined.

To determine the parameters of the MRF directional control valve experimentally, an experimental platform was developed as shown in Fig. 2.1. The presented platform is designed to determine the experimental parameters of the MRF directional control valve. As a concept, the platform is composed of a block which moves the rheological fluid, the block which generates the excitation field and the MRF valve.

2.3 Experiment Setup

For the experiment setup that is shown in Fig. 2.2, an Arduino Uno is used to control the MD10C current driver and the computer program MATLAB is used to convert the image position feedback into readable information and feed the control signal algorithm to the Arduino Uno. The MDC10C is used as current driver for the MRF valve solenoid based on signals from Arduino Uno. It is capable to supply an electrical current up to 13 A. Based on the position feedback and input given, this system can be identified and a suitable controller can be developed as shown in Fig. 2.3.

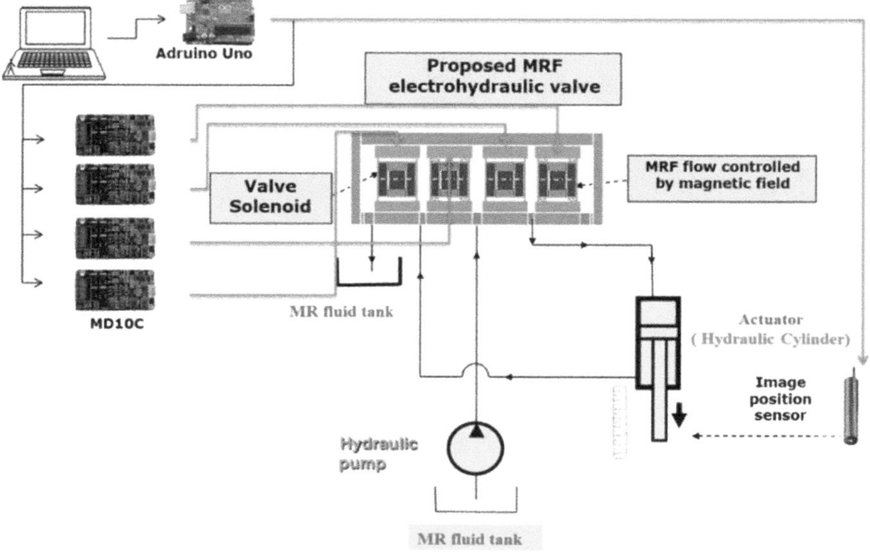

Fig. 2.2 Schematic of experiment setup

Fig. 2.3 Actual experimental setup

2.4 Result and Discussion

Running the experiment with sine input and random step input gives enough reading to identify the system for controller development. Figure 2.4 shows the sine output graph, while Fig. 2.5 shows the random step output graph from the system.

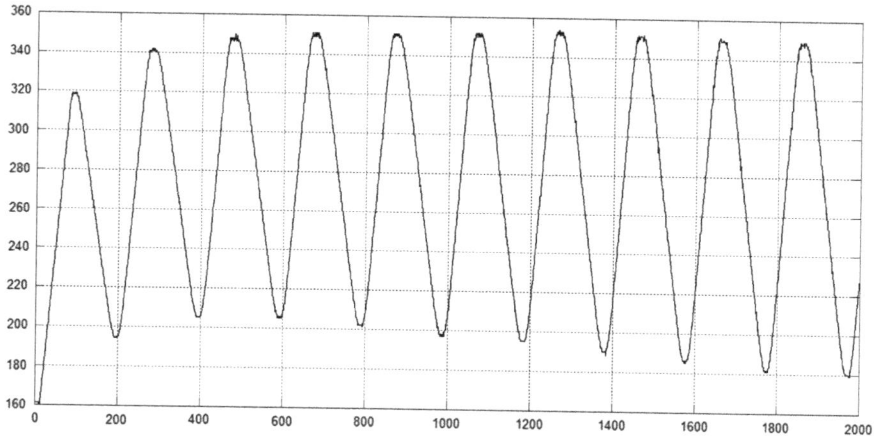

Fig. 2.4 Sine output from the system

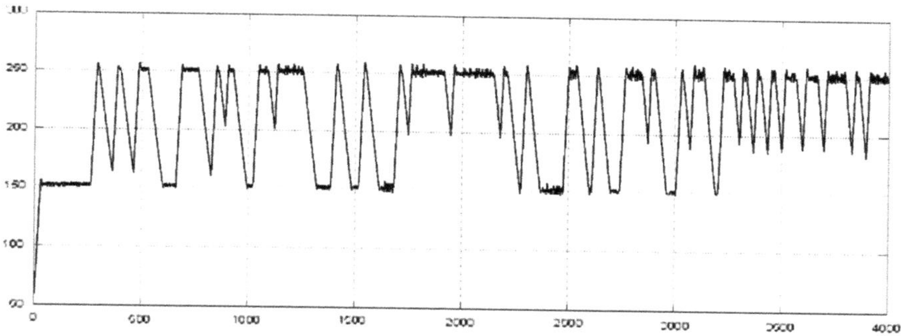

Fig. 2.5 Random step output from the system

Next, an identification process has been generated by using the input and output data for the system. Part of the data is selected as validation data and all the data is selected as working data. The model estimation is done using a system identification tool as shown in Fig. 2.6.

Figure 2.7 shows the comparison of estimated (a) space model and (b) Hammerstein–Wiener model with the actual model. The Hammerstein–Wiener model closer to the actual model compared to the space model with 81.64% and 65.27%, respectively.

A fuzzy controller was developed after the system identification. The comparison of the step response between before and after fuzzy implementation as shown in Fig. 2.8.

Table 2.1 Indicates the response time, percentage overshoot, and steady state errors for before and after the fuzzy implementation in the system. The fuzzy implementation in the system has enhanced the performances.

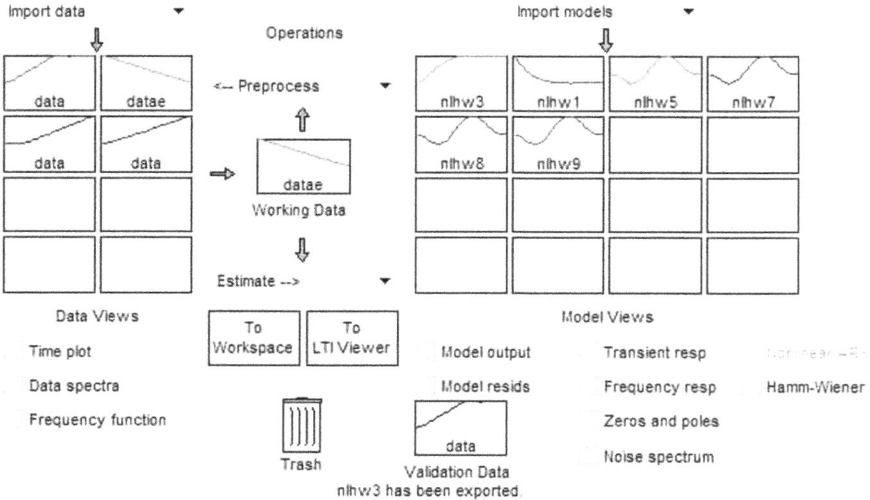

Fig. 2.6 System identification tool

2.5 Conclusion

As a conclusion, the MRF electrohydraulic system of the dynamic model can be identified using the MATLAB system identification tool using the experiment data. The Hammerstein–Wiener model is closer to the actual model compared to the space model with 81.64% and 65.27%, respectively. The response time, percentage overshoot, and steady state error of the developed control system was 0.18 s, 15% and 40%, respectively. The implementation of fuzzy in the system resulted in the fact that the response time, percentage overshoot, and steady state error improved by 0.3 s, 10% and 10%, respectively.

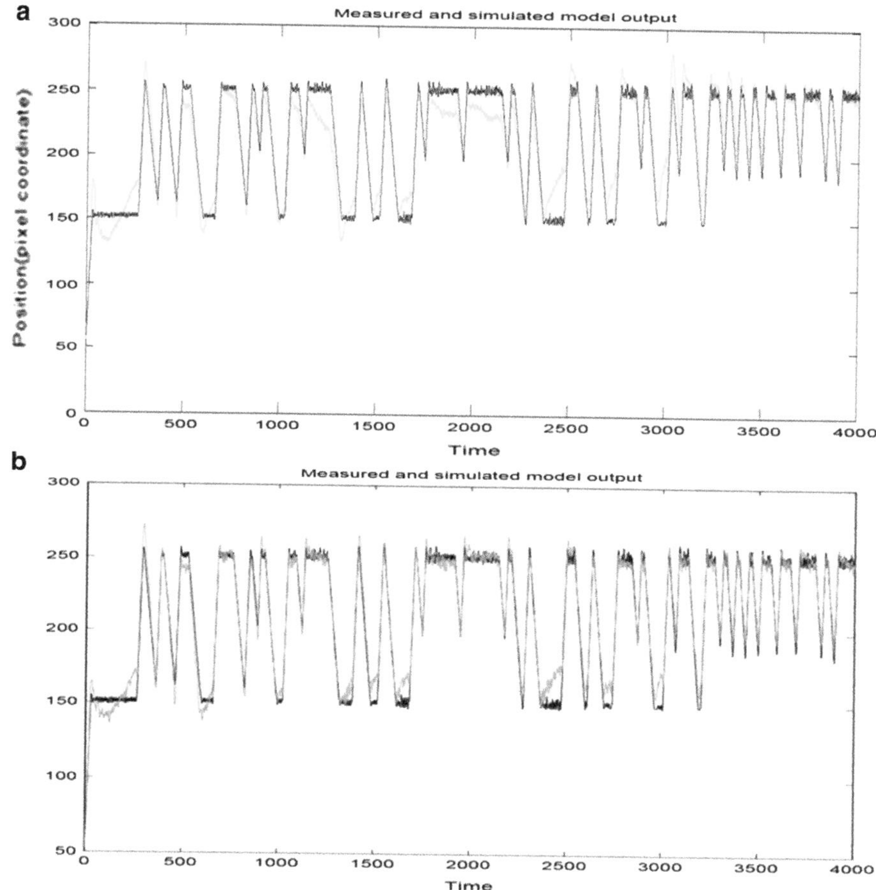

Fig. 2.7 **a** Estimated state space model (yellow line). **b** Estimated Hammerstein–Wiener (green line) and the actual model (black line)

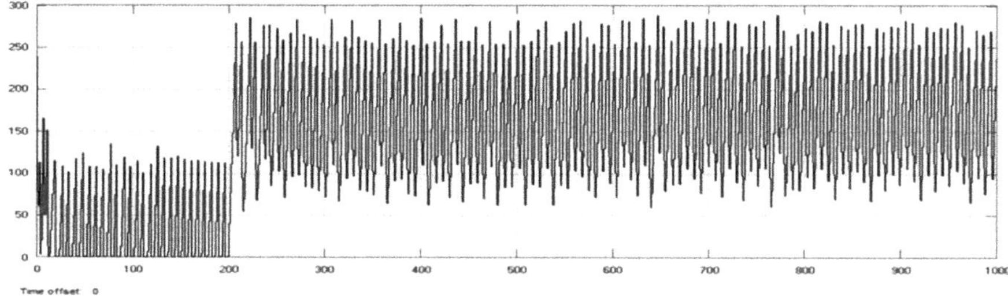

After Fuzzy Controller Implementation

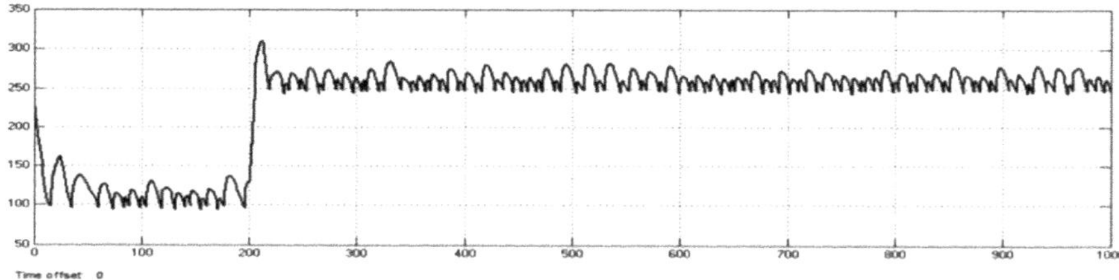

Fig. 2.8 Before and after fuzzy implementation for step response to the system

Table 2.1 System performance comparison

Controller	No fuzzy	Fuzzy
Response time (s)	0.18	0.3
Percentage overshoot (%)	15	10
Steady state error (%)	40	10

Acknowledgements The authors of this chapter acknowledge the Malaysian Spanish Institute, Universiti Kuala Lumpur (UniKL MSI) for funding this research that resulted in publication of the obtained results. Also supported by selected case study industry and anonymous reviewers to improve the quality of this research article are highly appreciated.

References

Askari, M., Davaie-Markazi, A.H.: Application of a new copmact optimized T-S fuzzy model to nonlinear system identification. In: 2008 5th International Symposium on Mechatronics and Its Applications, pp. 1–6 (2008)

Dyke, S.J., Spencer, B.F., Sain, M.K., Carlson, J.D.: Modeling and control of magnetorheological dampers for seismic response reduction. Smart Mater. Struct. **5**(5), 565–575 (1996)

Rabinow, J.: The magnetic fluid clutch. Trans. Am. Inst. Electr. Eng. **67**(2), 1308–1315 (1948)

Rashid, M.M., Rahim, N.A., Hussain, M.A., Mohamed, F., Rahman, M.A.: Development and testing of hybrid fuzzy logic controller for car suspension system using magneto-rheological damper. In: 2008 IEEE Industry Applications Society Annual Meeting, pp. 1–8 (2008)

Salloom, M.: Intelligent magneto-rheological fluid directional control valve. Int. J. Innov. Manag. Technol. **4**(4), 406–409 (2013)

Truong, D.Q., Ahn, K.K., Yoon, J.I., Thanh, T.Q.: Identification and verification of a MR damper using a nonlinear black box model. In: 2009 IEEE International Symposium on Computational Intelligence in Robotics and Automation—(CIRA), pp. 435–440 (2009)

Vladu, I., Strîmbeanu, D., Ivănescu, M., Vladu, I.C., Bîzdoaca, E.N.: Experimental results for magneto-rheological stop valve. In: 2012 16th International Conference on System Theory, Control and Computing (ICSTCC), pp. 1–6 (2012)

Weiss, K., Carlson, J.D.: A growing attraction to magnetic fluids. J. Mach. Des. **66**(15), 61–64 (1994)

Chapter 3
Development of a Magneto-Rheological Fluid Powertrain for Electric Vehicle Applications

Muhammad Faiz Shaefe, Muhammad Iqbal Mustakim Zainun, Mohd Nurhidayat Zahelem, and Muhamad Husaini Abu Bakar

Abstract The mechanical, rheological, and magnetic properties of magneto-rheological (MR) fluids were investigated for potential engineering applications. The typical modes of exploiting this technology were shown and discussed. An increasing number of industrial applications illustrate how the particular properties magneto-rheologic fluids may peculiar properties be used to provide optimal performance in torque transmitting devices. A torque transfer mechanism for electric vehicle transmissions that yields low losses, while still satisfying the conflicting requirements of compactness, quick response, high power density, and the most important is lightweight since the electric vehicle is running with a reduced total mass with no engine are explored in this contribution. Moreover, a systematic approach to MRF transmission design is proposed. The various design variants were chosen to serve two purposes, i.e., to demonstrate the foundation and to showcase approaches and solutions to specific problems that have a direct application in transmission design. The phenomenon of "spin loss" is well known in the automotive industry, which affects all transmissions. One energy sink identified in this regard lies in conventional wet clutches where it creates a drag on the transmission because of the oil churning around and between the rotating friction plates.

M. F. Shaefe · M. N. Zahelem
Mechanical Section, Malaysian Spanish Institute, Universiti Kuala Lumpur, 09000 Kulim, Kedah, Malaysia
e-mail: faizshaefe@gmail.com

M. N. Zahelem
e-mail: mnurhidayat@unikl.edu.my

M. I. M. Zainun (✉)
Manufacturing Section, Malaysian Spanish Institute, Universiti Kuala Lumpur, 09000 Kulim, Kedah, Malaysia
e-mail: iqbal.zainun@s.unikl.edu.my

M. I. M. Zainun · M. N. Zahelem · M. H. Abu Bakar
System Engineering and Energy Laboratory, Universiti Kuala Lumpur Malaysian Spanish Institute, Kulim Hi-Tech Park, 09000 Kulim, Kedah, Malaysia
e-mail: muhamadhusaini@unikl.edu.my

M. H. Abu Bakar et al. (eds.), *Progress in Engineering Technology III*,
Advanced Structured Materials 148, https://doi.org/10.1007/978-3-030-67750-3_3

Keywords Magneto-rheological fluid · Torque · Powertrain · Rheological · Transmission

3.1 Introduction

A magneto-rheological fluid (MRF) transmission is a device that transmits torque by the shear force of the MRF. The fluid is inserted between the rotating and fixed discs; hence a magnetic field is imposed on the fluid. In this paper, a complete test system for an MRF transmission was introduced. In this contribution also, an investigation was conducted to measure the transmitting torque. During the torque delivering process, the results obtained were different according to the current input to the MRF transmission system. A theoretical analysis for both, i.e. the MR transmission and the mechanical system is developed and is solved numerically using the Solidworks and FEMM software. The effect of the current input on the MRF transmission, magnetic flux density, and design parameters is taken into consideration.

The demand in the automotive industry for electric vehicles keeps rising over the years even though the engine was eliminated with batteries. The weight of the vehicle was considered high due to the presence of the powertrain system. The approach to solve the drawback is to develop a CAD model of the MRF gearbox system. Once the design was completed, the MRF powertrain was fabricated according to the CAD design. Lastly, a task was to evaluate the power transmission efficiency of the MRF powertrain. The approach of these objectives is illustrated in Fig. 3.1.

The first stage of the research, a CAD model of the magneto-rheological powertrain system was designed by using the software Solidworks. Initially, some designs were sketched which present the best and simple design. The next stage was to plan and set up a fully functioning experiment via appropriate tools to gain results. The experimental setup leads to the validation of this method so that it can be recognized.

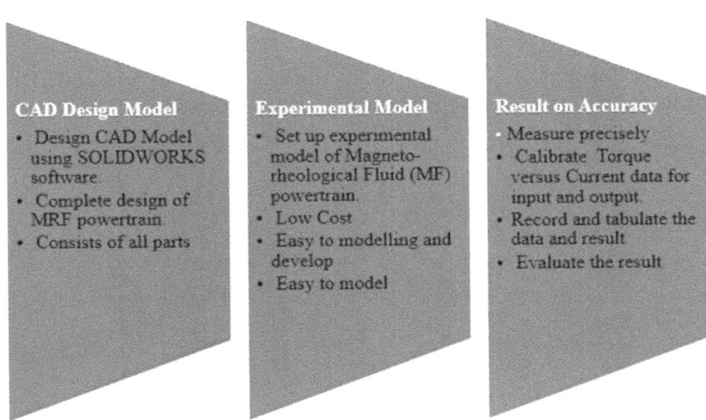

Fig. 3.1 Three stages of the approach

In the process of developing an MRF transmission for electric vehicle application, this contribution was developed. The CAD software Solidworks was used to produce the design of the experimental device. The aim of this project was to develop the MRF transmission system with low cost, lightweight, and high efficiency. Thus, in achieving the target, a design should be practical and small. The process starts with developing a CAD model in a proper way; SOLIDWORKS is a good solution for this first objective. Since this project utilizes the control of current to solidify and desolidify the MRF and to control the DC motor, the application of the MD10C is more important due to its accuracy and controllability. However, to control the system, an Arduino UNO plays an important role in its way. Usage of the lathe turning machine and a rapid prototyping machine is needed in the production of the MRF transmission system. The functionality and effectiveness of the system was verified with a strain gauge load cell and mathematical calculations.

3.2 Literature Review

The powertrain of an electric vehicle (EV) is a compound system with electrical subsystems, such as batteries, inverters, and electrical motors, as well as mechanical sub-systems, including transmissions, differential, and wheels. Since the electrical systems directly affect the vehicle driving performance and dynamics of an EV, integrated modelling considering both the mechanical and electrical systems is essential to assess the ultimate kinetic and dynamic characteristics of an EV in terms of input electrical quantities (Martín-González et al. 2013).

The series-parallel hybrid electric vehicle (HEV) has been well applied in the industry due to its significant advantages and the use the power split and mode switching transmission configuration (Green et al. 2011). The power split transmission is a planetary gear set that removes the need for a traditional stepped gearbox and transmission components. It acts as a continuously variable transmission (CVT) but with a fixed gear ratio. The limited driving range and low energy storage cause the EVs to lose their popularity among the users. The efficiency of EVs can be improved by applying transmission systems. An electric motor can operate with high efficiency for a long period by applying a multi-speed transmission (MST) and decrease in the energy consumption. A dual-clutch transmission (DCT) is acceptable to be used in the HEV as it is used in conventional vehicles. EVs also use the dual-motor multi-speed transmission, clutchless automated manual transmission (CLAMT), and two-speed planetary transmission as the multi-speed transmission. The configuration of the CLAMT is simple to control with high operating efficiency but during the gear change event a torque interruption is present.

The MRF technology is an old "newcomer" reaching the market very fast. MRFs respond fast and have an interface that is simple between the mechanical and electrical input power widely used in many applications (Kolekar et al. 2019; Cao et al. 2014; Attia et al. 2017). The rheological behavior of the MRF in case of the absence of the magnetic field is the same as the carrier fluid pattern except for the metal powder of the MRF which makes the liquid viscosity slightly higher (Attia et al. 2017; Hema Latha, K., Usha Sri, P., Seetharamaiah, N.: Design and manufacturing aspects of magneto-rheological fluid (MRF) clutch. Mater. Today Proc. 4(2, Part A) 1525). Several industries such as automotive use the MRF in different applications. The next generation of product design whereby accuracy, dynamic performance, and power density are the important features will use MRFs to achieve these goals. In addition, the different viscosities of the MRF should improve the functionality and cost needed to control the fluid motion. Direct shear mode (used in brakes and clutches) and valve mode (used in dampers) are two sectors of technology that have been researched widely in various application. Features such as high response, interface that is simple between the mechanical and electrical input power make the MRF to be needed in many applications (Green et al. 2011; Spaggiari, A.: Properties and applications of Magnetorheological fluids. Frat ed Integrità Strutt 7(23 SE), 48–61 2013).

MRFs are also applied in the vehicle suspension dampers. The MR damper has a built-in MR valve across which the MRF is forced. An MRF damper is a material filled with a magneto-rheological fluid, which is controlled by a magnetic field, usually using an electromagnet. This allows the damping characteristics of the shock absorber to be continuously controlled by varying the power of the electromagnet (Olabi and Grunwald 2007). This type of shock absorber has several applications, most notably in semi-active vehicle suspensions which may adapt to road conditions, as they are monitored through sensors in the vehicle, and in prosthetic limbs (Mangal, S.K., Sharma, V.: On state rheological characterization of MRF 122EG fluid using various techniques. Mater. Today Proc. 4(2, Part A), 637–644 2017).

3.3 Methodology Experimental SetUp

This section comprises the method of designing, fabrication, and evaluation of the magneto-rheological fluid transmission for electric vehicle applications. The methodological framework of the project is divided into three phases based on the objectives and illustrated in Fig. 3.2.

3.3.1 CAD Modelling

The MRF transmission body is the main part to make it a complete transmission system since it holds two shafts, and the inlet and outlet for filling up with the MRF. This part is the main part of this project, and its quality must be the best. Considering the magnetic field and the heat from the coil winding, the used material is extremely essential since it must support a heavy load. The chosen material has 74 MPa of flexural strength and 46 MPa tensile strength, other than that, it can resist oil leaks,

START

Development

Design and development:

- CAD design of MR fluid transmission system
- CAD design of experimental Setup
- Material and equipment used.

Evaluation of :

- Collecting data from FEMM and SOLIDWORKS.
- Calibrate the data of different voltage and current with torque transmittance

Consist of

Phase 1

Design MR fluid transmission system using CAD software → 1

Purpose / Goal

To design a CAD model of the MR fluid transmission system.

Phase 2

Set up an experimental model to simulate the actual operation system → 2

Purpose / Goal

To determine the functionality of the prototype which is driven by a 12 V high torque motor.

Phase 3

To evaluate the performance of the torque transmission → 3

Purpose / Goal

To evaluate the value of torque transmission by the system (by using strain-gauge load cell)

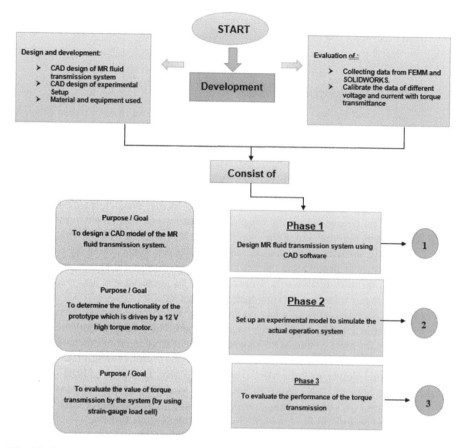

Fig. 3.2 Research flowchart

Fig. 3.3 Rapid prototyping machine product

this is because palm oil is used as a carrier oil in MRF. Figure 3.3 shows the part build from ABS material using a rapid prototyping machine.

3.3.2 *MRF Preparation Process*

An MRF is a solution that is non-colloidal. MRFs have micro-sized suspension and polarized magnetic particles. An MRF comprises of three fragments, namely, the polarized magnetic particles, the liquid carrier, and stabilizing agents (de Vicente et al. 2011). The particles are made up of high iron purity and have magnetic saturation. The diameter of the particle is between 0.1 and 500 μm. However, for this research, the preferred size is between 1 μm and 100 μm. In this project, an iron particle diameter of 10 μm is used since the magnetic saturation is crucial for a high strength MRF. This means that the size of the iron particles is important. A large iron particle size will rapidly settle the suspension. Iron particles are cheap and easy to obtain. Hydrocarbon oil was used as the carrier liquid. Hydrocarbon oil can be of mineral, synthetic, or a mixture of two oils. Generally, the oil must have a low viscosity so that the MRF is not transformed into energized grease.

There are two types of MRFs and an amount of 100 ml was mixed for each type as shown in Table 3.1. The ingredients were weighed. The table presents the amount of iron powder for MR122 and MR132 whereby the difference is at the weight of the hydrocarbon oil. This will manipulate the percentage difference in volume. Lithium grease was added slowly until the MRF was easier to mix. The resulted MRFs showed a fluid content between 20% and 32% volume of iron. The MRFs have the highest yield strength and the same magnetic properties same as the Rheonetic MRF 132-DL MRF produced commercially available by Lord Corporation (2005).

The suspension of the carbonyl iron powder in the host solution was achieved using additives, which inhibit settling, and accumulation. The appropriate amounts of oil, grease, and iron filling were measured. Next, the grease was added to the oil and mixed thoroughly. Mixing is most effectively accomplished with a rotary blender. The homogeneous mixture could rest for a few hours and then mixed. Finally, the iron powder was added to the oil and grease mixture. Adding then, about half from the total weight of the iron powder was added and mixed with the liquid by using a stirring stick. The balance of the iron powder was added and continuously stirred until the mixture appeared without lumps. At this point, one should continue to mix

Table 3.1 Ingredients of MR122 and MR132

Ingredients	MR 122	MR 132
Iron powder (grams)	150	150
Hydrocarbon oil (grams)	55	31
Lithium grease (grams)	5	5
Percentage volume (%)	20–22	30–32

using the rotary blender again. The mixture was remixed occasionally as the iron particle will slowly settle and leave a layer of clear oil on the top.

3.3.3 Strain Gauge Load Sensor Setup

The load cell was assembled to the HX711 module. The load cell consists of four wires, which must connect four pins from the HX711 module. The red wire was assembled to the E+, the black wire connects to the E−, the white wire connects to the A−, and the green wire connects to the A+. If there is a need to connect a second load cell, then the remaining pins should be utilized. The module needs to be connected to the Arduino Uno. The ground pin of both the module and Arduino Uno is connected. DT and SCK must connect to the digital pins of the Arduino. In this contribution, DT is connected to the digital pin number 4 and SCK is connected to the digital pin A5. The remaining pin VCC must be connect to the 5 V pin of the Arduino. As the LCM1602 module also requires a connection to the 5 V pin, a breadboard is used in-between to split the Arduino's 5 V signal. As the last step, the LCM1602 module's SDA and SCL pins must be connected to the corresponding SDA and SCL pins of the Arduino Uno. Moreover, the GND pin must be connected to one of the Arduino's GND pins and the VCC pin must be connected to the 5 V signal of the breadboard. Figure 3.4 shows the complete sketch of the MRF powertrain system.

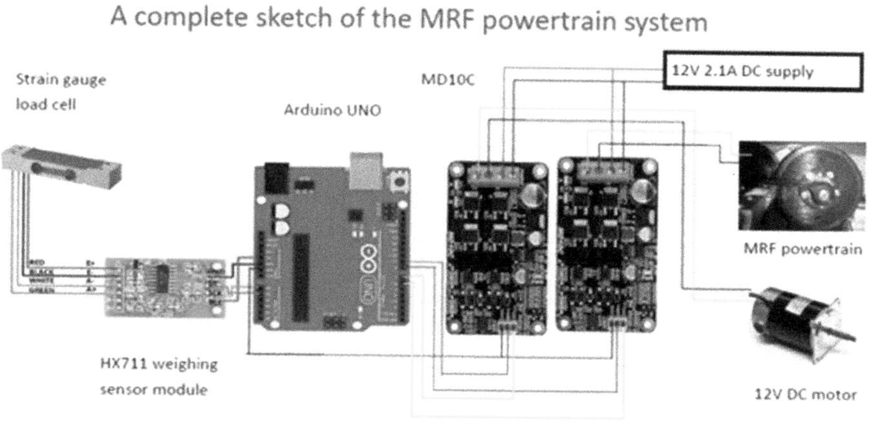

Fig. 3.4 Complete sketch of the MRF powertrain system

3.3.4 Finite Element Method Magnetics (FEMM) Analysis

A Finite element simulation is performed with the free simulation software FEMM, a software-tool which is based on the finite element method. In preventing the dimension error or out of range of the dimension tolerance, a CAD model was cut to the cross-section view (see Fig. 3.5) and exported in the DXF format to the FEMM. With exactly the same dimensions and gaps, the coil turning was set to 1200 turns, and the current applied was set to 1.15 Amp based on the value from the calibration curve of the PWM versus current.

Before proceeding to the real setup, the most crucial step is to identify the strength of the magnetic flux and the magnetic field flow because the MRF's functionality depends on the strength of the magnetic flux to become a solid state. Further, a different current ranging from 0.2 Amp to 6.0 Amp is supplied to the electromagnet to activate the MRF (Attia et al. 2017). Thus, the analysis of the MRF by using FEMM was performed with the lowest Amp value to prove the existence of the magnetic flux.

3.4 Results and Discussion

This part covers the result of the copper coil calibration, current versus torque transmittance, current versus power efficiency and distribution of magnetic flux density. The output was discussed according to the previous set objectives.

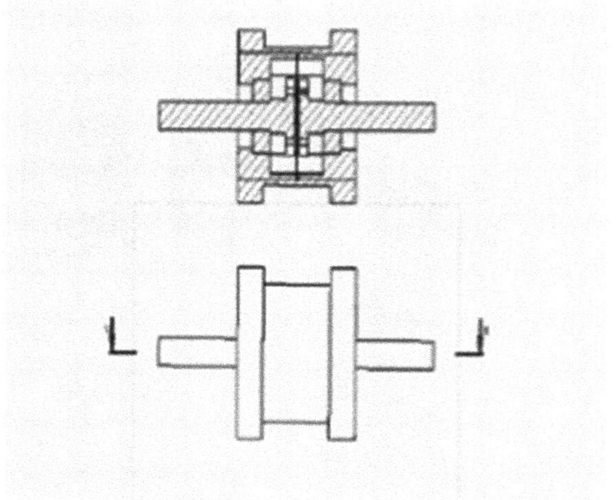

Fig. 3.5 Cross-section of the MRF transmission system

3.4.1 Copper Coil Calibration

There are 26 steps in increasing the PWM value starting from 10 until 255. The measure was rising from the low value of 10 PWM to improve the reading accuracy of the current compared to an initial reading value of 50 PWM. The higher accuracy means, the smaller the error of the readings. However, in the project setup, the MD10C was supplied with a 12 V 2.1 A DC power supply and controlled with an Arduino. There are two systems of the MRF transmission, named transmission A and transmission B and had been calibrated for the Current (mA) versus pulse width modulation (PWM). Figures 3.6 and 3.7 show the graphs of the calibration.

The graph in Fig. 3.6 for transmission A is slightly linear increasing from 10 PWM = 8.16 mA until 255 PWM = 1.15 A and the graph in Fig. 3.7 for transmission B is also slightly linear increasing from 10 PWM = 5.05 mA until 255 PWM = 1.15 A. In controlling the magnetic flux of the copper coil, the current value must be changed in the Arduino sketch with the MD10C controller, the unit for controlling current output from MD10C is in PWM.

Both MRF transmission systems give the same reading since it was coiled by SWG30 with 1200 turns. Figure 3.8 shows the reading comparison of the current in mA, versus PWM for both transmissions.

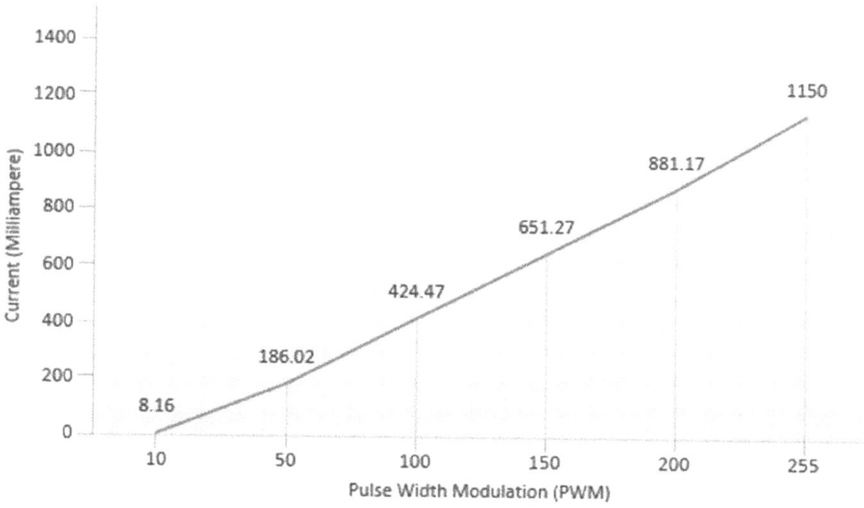

Fig. 3.6 Graph of the current (mA) versus pulse width modulation (PWM) calibration of the MRF transmission A

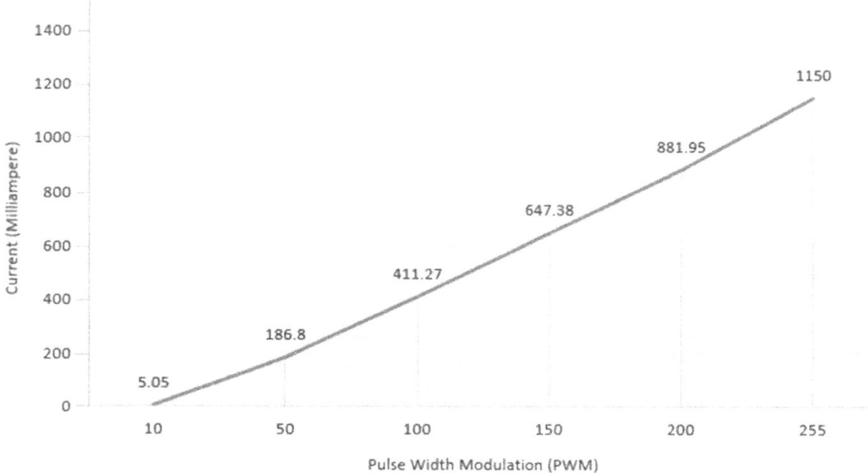

Fig. 3.7 Graph of the current (mA) versus pulse width modulation (PWM) calibration of the MRF transmission B

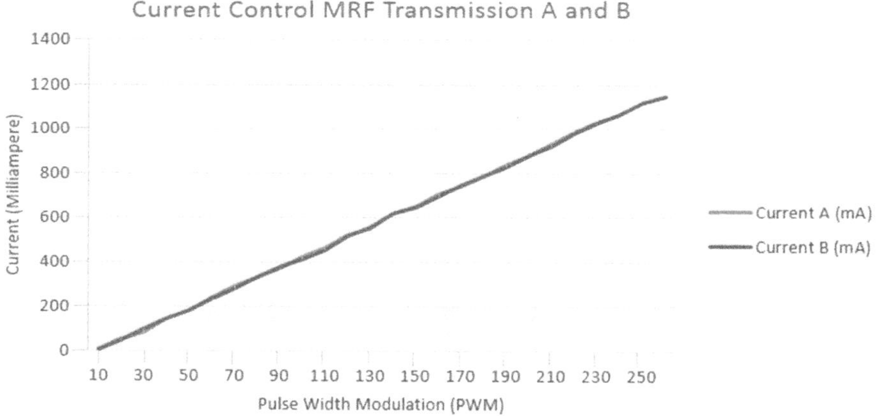

Fig. 3.8 Comparison of the current versus PWM calibration of MRF for transmission A and B

3.4.2 Current Versus Torque Transmittance

Following the standard characteristics of a conventional gearbox, when supplied with 10 PWM, it should be classified as low speed and a high torque mechanism. The input shaft's speed is higher than that for the output shaft, and 255 PWM is a direct drive, a condition where the speed and torque transfer from the input shaft and the output shaft is the same. Moreover, it is an electric vehicle powered by an electric

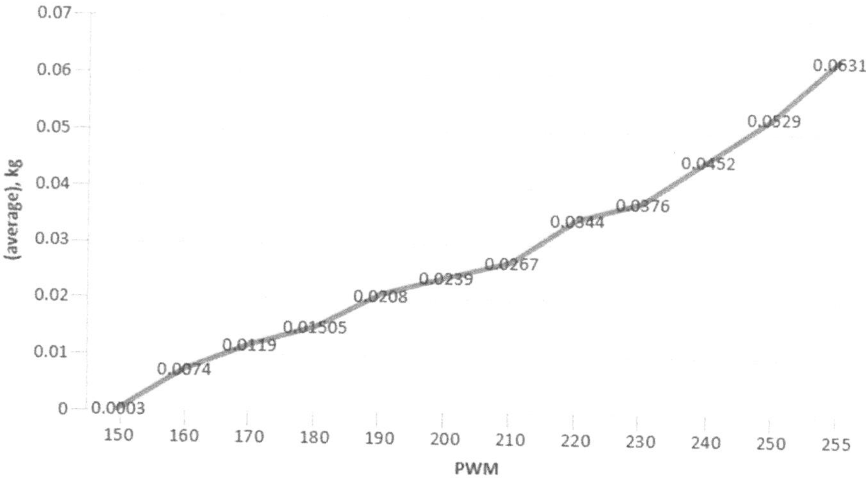

Fig. 3.9 Graph of the average weight exerted on the strain gauge load cell versus PWM

high torque motor. Thus, the speed of the output shaft can be varied by controlling the current supply to a variable input to output ratio.

This MRF transmission is possible to be implemented in a full gearbox transmission system since the first gear indicates as a high torque gear and the highest gear demonstrates as a direct drive or overdrive gear. But in the real project, the result shown in Fig. 3.9 was reflected in theory. Equation (3.1) is the force exerted on the strain gauge load cell and Eq. (3.2) shows the torque transmission.

$$F = ma \tag{3.1}$$

where, the mass exerted on the strain gauge load cell m is 0.0631 kg.

$$T = F \times R \tag{3.2}$$

The MRF transmission is possible to be implemented in a full system gearbox transmission since the first gear indicates as a high torque gear and the highest gear demonstrates as a direct drive or overdrive gear. Based on Fig. 3.10, the maximum torque transfer to the output shaft is lower than the expected value which is 0.0186 N m only. The calculation of the mass exerted on the strain gauge load cell and torque is shown in Eqs. (3.1) and (3.2).

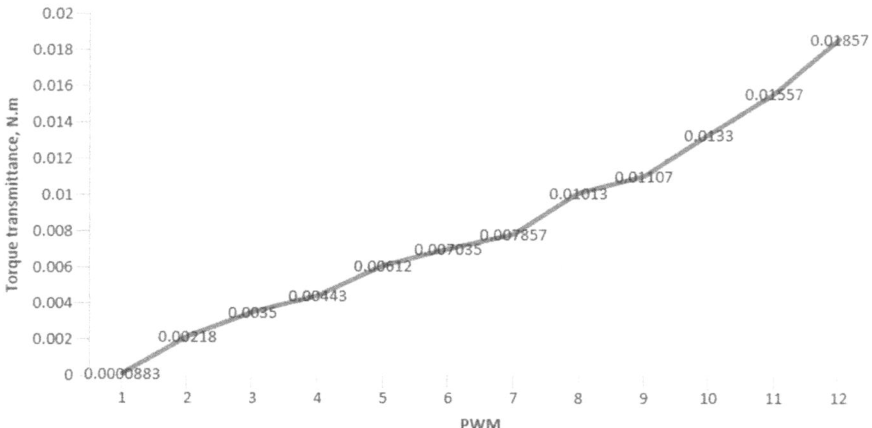

Fig. 3.10 Torque transmittance (N m) versus PWM

3.4.3 *Current Versus Power Transmission Efficiency*

The current applied to the coil varies the output torque of the output shaft, as discussed above. The power transmission efficiency verifies the quality of the power delivery from the input shaft to the output shaft. This point is mainly to focus on the MRF solidification. The input shaft is driven by a 5 N m torque of a DC motor, and the torque transmittance was qualified. The torque value obtained from Eq. (3.2) is substituted in Eq. (3.3) to compute efficiency.

$$\eta = T \times 100 \tag{3.3}$$

Based on the calculation from Eq. (3.3) above, the torque transfer efficiency is 0.3714%, and the balance of 99.6286% losses occurs at the oil seals, where the inside diameter of the oil seal is 9.5 mm (standard clearance of the 10 mm internal diameter oil seal). Because of that clearance, the rotation of the shaft becomes slightly heavy. The biggest problem in the MRF is about sealing the fluid from leaking and setting up the magnetic flux to freeze the liquid. Starting from sealing the liquid, there is a part used to achieve excellent sealing properties and maintains its functionality.

Figure 3.11 shows the five pieces of the insert needed to improve the sealing purpose and prevent the stuck of the shaft when the magnetic flux was applied. Other than that, the volume of the MRF also plays an important role in keeping the functionality of the MRF transmission system. The project was conducted with 10 mL fluid in each system and based on the result the volume used is not enough since the efficiency was not achieved according to the target. For the next project, the volume of the MRF used must be in the proper way, and any calculation may be applied.

Fig. 3.11 Five slots inside the body of the MRF transmission system

3.4.4 *Distribution of the Magnetic Flux Density*

The magnetic field distribution and vector plot obtained by FEMM at 1 A current load operation is shown in Fig. 3.12. The contour lines represent the magnetic field flow inside the shaft and the purple regions means the area has filled with the maximum magnetic flux density. The magnetic flux density is higher close to the low carbon steel shafts.

Magnetic flux density depends on the turning number and the diameter copper wire size (measured in American wire gauge, AWG or standard wire gauge, SWG). More turning results in a higher magnetic flux density. The coil was turned as 1200 turns with SWG30 of the copper wire and FEMM shows a better magnetic field distribution

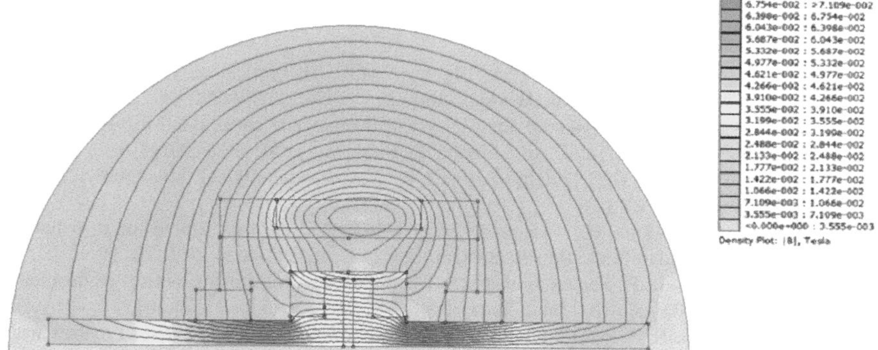

Fig. 3.12 Magnetic field distribution obtained by FEMM

and magnetic flux density for MRF to solidify. The magnetic flux density is highly concentrated in the region where the MRF was occupied so that the magnetic flow efficiency was good ($B = 6.754\text{E}{-}2$). The magnetism of the iron group of metals is a rare and remarkable property. It is not due to any inherent magnetic propensities of the atoms, but due to the structure of the metal. Other substances with similar structures also have similar magnetic properties. A magnetic field is created by electric currents; thus, it has its own direction. Apparently, the vector plot determines the fluxes flow direction where it supposedly tends to maximize at the pure iron.

3.5 Conclusions

The major conclusion is that the main outcome of the work has been completely achieved. By completing this project, the conclusion of the first objective, design a CAD model of an MRF powertrain, the second objective, fabrication of the MRF powertrain system and the third objective, which is the evaluation of the efficiency of the system, have been achieved. The world is full of potential MRF applications. For every system where it is desirable to control motion using a fluid with changing viscosity, a solution based on MRF technology may be an improvement in functionality and costs. Simplicity and more intelligence in the functionality are key features of the MRF technology.

Acknowledgements The authors of this article acknowledge the Universiti Kuala Lumpur Malaysian Spanish Institute (UniKL MSI) and System Engineering and Energy Laboratory (SEELab) for funding this research that resulted in publishing of this article. Also supported from selected case study industry and automotive reviewers to improve the quality of this research article are highly appreciated.

References

Attia, E.M., Elsodany, N.M., El-Gamal, H.A., Elgohary, M.A.: Theoretical and experimental study of magneto-rheological fluid disc brake. Alexandria Eng. J. **56**(2), 189–200 (2017)

Cao, Q., Li, L., Lai, Z., et al.: Dynamic analysis of electromagnetic sheet metal forming process using finite element method. Int. J. Adv. Manuf. Technol. **74**(1), 361–368 (2014)

de Vicente, J., Klingenberg, D.J., Hidalgo-Alvarez, R.: Magnetorheological fluids: a review. Soft Matter **7**(8), 3701–3710 (2011)

Green, R.C., Wang, L., Alam, M.: The impact of plug-in hybrid electric vehicles on distribution networks: a review and outlook. Renew. Sustain. Energy Rev. **15**(1), 544–553 (2011)

Hema Latha, K., Usha Sri, P., Seetharamaiah, N.: Design and manufacturing aspects of magneto-rheological fluid (MRF) clutch. Mater. Today Proc. **4**(2, Part A), 1525–1534 (2017)

Kolekar, S., Venkatesh, K., Oh, J.-S., et al.: Vibration controllability of sandwich structures with smart materials of electrorheological fluids and magnetorheologi-cal materials: a review. J. Vib. Eng. Technol. **7**(4), 359–377 (2019)

Mangal, S.K., Sharma, V.: On state rheological characterization of MRF 122EG fluid using various techniques. Mater. Today Proc. **4**(2, Part A), 637–644 (2017)

Martín-González, M., Caballero-Calero, O., Díaz-Chao, P.: Nanoengineering thermoelectrics for 21st century: energy harvesting and other trends in the field. Renew. Sustain. Energy Rev. **24**, 288–305 (2013)

Olabi, A.G., Grunwald, A.: Design and application of magneto-rheological fluid. Mater. Des. **28**(10), 2658–2664 (2007)

Spaggiari, A.: Properties and applications of Magnetorheological fluids. Frat ed Integrità Strutt **7**(23 SE), 48–61 (2013)

Chapter 4
GPU-Accelerated Vehicle Detection for Roads

Abd Munim Abd Halim, Ahmad Shafiq Fikri Ishak, Muhamad Husaini Abu Bakar, and Pranesh Krishnan

Abstract Nowadays, as people's demands and lifestyles change the need for advancing the type of technology we use is high. Everything we used has been innovated to a better standard. Intelligent and autonomous vehicles are promising solutions to improve road safety, traffic problems, and passenger comfort in advanced driving assistant systems (ADAS). Such applications require advanced computer vision algorithms that require high-speed computers. In some cases, it is still a significant challenge to keep smart vehicles on the road right up to their destination, especially when driving at high speed. The first main task is robust navigation, often based on system vision, to obtain RGB road images for further processing. Depending on the position, speed, and direction of the vehicle, the second task is the dynamic control. This document presents precise and efficient road borders and an intelligent and autonomous vehicle detection algorithm for painted lines. It combines the Hough Transform and Canny edges detector to initialize the algorithm at every time necessary. Lastly, you only look once (YOLO), a new detection approach, is presented. Classifiers for the detection of objects are repurposed prior to work. We instead frame object detection into separate spatial boundaries and associated class probabilities as a regression problem.

A. M. Abd Halim · M. H. Abu Bakar (✉)
System Engineering and Energy Laboratory, Universiti Kuala Lumpur Malaysian Spanish Institute, Kulim Hi-Tech Park, 09000 Kulim, Kedah, Malaysia
e-mail: muhamadhusaini@unikl.edu.my

A. M. Abd Halim
e-mail: amunim.halim@s.unikl.edu.my

A. S. F. Ishak
Electrical Electronic and Automation Section, Malaysian Spanish Institute Universiti Kuala Lumpur, Kulim Hi-Tech Park, 09000 Kulim, Kedah, Malaysia
e-mail: asafiq.ishak@s.unikl.edu.my

P. Krishnan
Intelligent Automotive Systems Research Cluster, Electrical Electronic and Automation Section, Malaysian Spanish Institute Universiti Kuala Lumpur, Kulim Hi-Tech Park, 09000 Kulim, Kedah, Malaysia
e-mail: pranesh@unikl.edu.my

M. H. Abu Bakar et al. (eds.), *Progress in Engineering Technology III*,
Advanced Structured Materials 148, https://doi.org/10.1007/978-3-030-67750-3_4

Keywords Lane detection · Pipeline · YOLO algorithm

4.1 Introduction

These days many cars in Malaysia are already equipped with advanced driving assistant systems (ADAS) which have a camera. By fully utilizing the camera that comes together with the vehicle, many useful applications can be developed to enhance the car features. Vehicle detection in dynamic situations can be helpful for various study development; this also will be contributing toward autonomous driving. By doing so, Malaysia's road and vehicle database also can be generated which will be useful for various studies. Several types of sensors such as visual sensor, radar and light detection and ranging (LIDAR) sensors are employed in the advanced driving assistance system for the development of autonomous vehicles (Xique et al. 2018). Recently, the optical sensor had significant potential as the visual sensor becomes inexpensive and lighter. These sensors are most preferred in driving assistance as they provide rich sensing information of the environment (Christiansen et al. 2017). Few cases include parking assistance (Fernández-Llorca et al. 2014), vehicle identification (Huang et al. 2017), traffic sign recognition (Luo et al. 2018), vehicle speed detection, and front collision warning (Lin et al. 2012), are available. Of late, due to the usage of the graphics processor unit (GPU), the speed of image processing has improved vastly (Stolz et al. 2010; Fassold2016; Pérez 2019). These hardware advancements make the image processing mission quick and proper for ADAS (Galko et al. 2014; Wang et al. 2019; McLean et al. 2020; Borrego-Carazo et al. 2020). The you only look once (YOLO) algorithm is a new object detection, classificatory are repurposed to detect objects (Redmon et al. 2016). A single neural network predicts directly from full images in the observation for bounding boxes and class probabilities. Because the entire detection pipeline is just one network, end-to-end detection performance can be optimized directly (Wu and Dai 2016). Drivers shift lanes after gauging the nearby traffic. They also check the rearview and side mirrors before deciding a turn. Nevertheless, through following the standard procedures, accidents occur due to the blind-sector of vehicles which is the foundation of danger.

The developed vehicle detection system contains an inter-vehicle distance assessment, which is made out of a few subsystems, including image pre-processing, data collection, detection of vanishing point, two-division of road sections, opposite vehicle recognition, and approximation of the inter-vehicle distance. Besides, the color data is not used in this work, so the input picture needs to be changed first into greyscale. At that point, downsampling, and elimination of the motionless region of casing are presented for expanding the computational productivity (Chiu and Lee 2010).

Faster R-CNN (Liu et al. 2016) is presently an established model for deep learning-based object detection. It roused numerous detection and segmentation models that came after it, including the two others that will be investigated in this work. Unfortunately, it cannot start to see Fast R-CNN without understanding its ancestors and

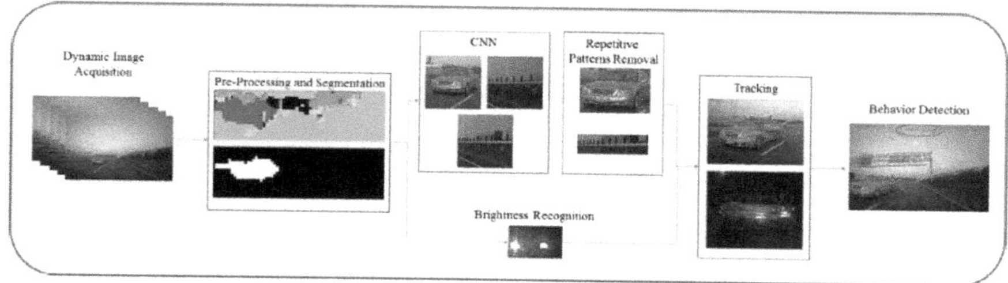

Fig. 4.1 Flowchart of the proposed system

R-CNN and Fast R-CNN. The authors state that the fast R-CNN technique has a few merits:

1. Advanced recognition class (mAP) than R-CNN, SPPnet.
2. Training is one-stage, by a multiple-task loss.
3. Training can appraise all network layers.
4. No disk storage is essential for feature storing.

Christiansen et al. (2017) showed a real-time system for safe path change operations using a monocular camera placed on the back of the vehicle. In this research, the probable surpassing vehicle is first sited and depends on movement prompts. The flowchart of the proposed system is shown in Fig. 4.1 and depicts the surpassing recognition method. It includes the following four parts:

1. To pre-process and segment the images.
2. Development of a convolutional neural network.
3. To remove repetitive pattern.
4. To track and detect the behavior.

Simultaneously, Fig. 4.2 shows a single coevolutionary network that forecasts several boxes and class probabilities for such boxes. The YOLO trains full pictures and improves the detection performance directly. This unified model has various advantages over traditional object detection methods (Redmon et al. 2016).

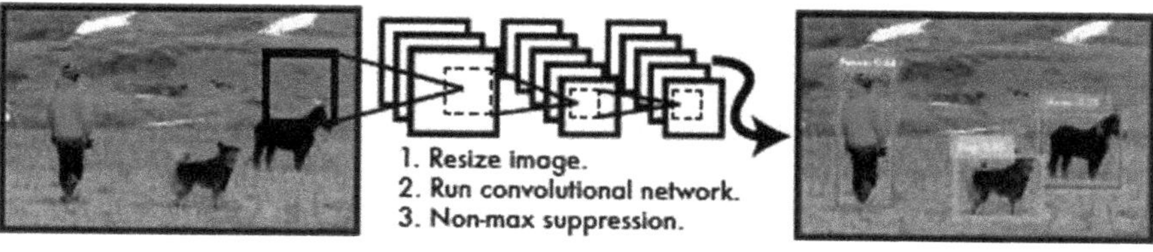

Fig. 4.2 The YOLO detection system

4.2 Methods

4.2.1 Vehicle Detection

Regression-based algorithms predict classes and bounding boxes for the whole picture instead of picking exciting parts of an image. The YOLO is used for the real-time recognition of objects which is a good example of regression-based algorithms. Every bounding box is described by means of four descriptors which are the center of a bounding doc (*Bx*, *By*), width (*Bw*), height (*BH*), and value c corresponding to vehicles in Fig. 4.3.

Figure 4.4 shows that it usually takes a grid of 19 × 19 to divide up an image instead of dividing it into cells. It is up to each cell to predict five bounding cases. Figure 4.5 shows that most cells and boxes will not have an object inside, which is why pc must be predicted. In the next step, one needs to remove boxes with low object

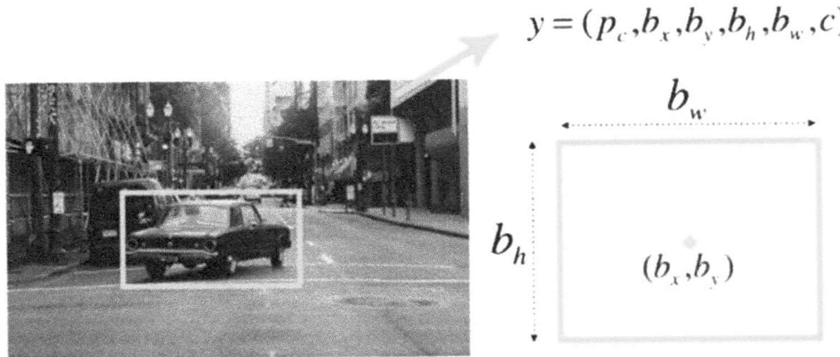

Fig. 4.3 The YOLO vehicle detection system

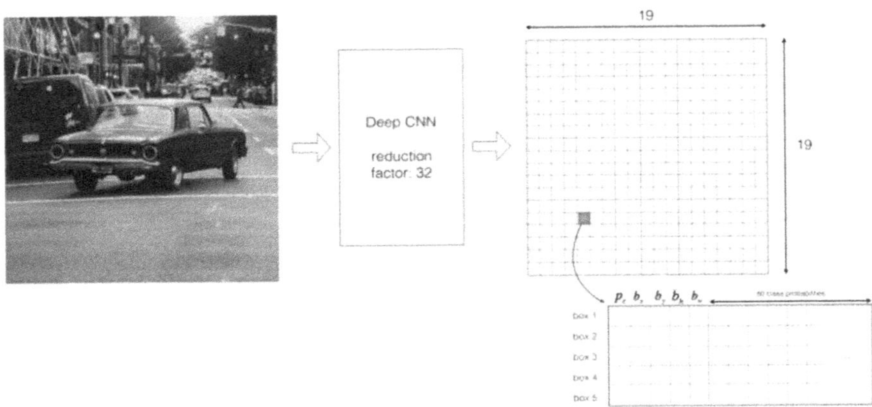

Fig. 4.4 Splitting the image into a cell

Fig. 4.5 Non-max suppression

likelihood and binding boxes with highest shared area and this process is known as the non-max suppression.

4.2.2 Lane Detection

Figure 4.6 shows the overall stages for lane detection pipeline. The following six steps are the pipeline that will lead to the detection of the lane:

1. Read and greyscale the image.
2. Apply Gaussian smoothing.
3. Implement Canny edge recognition.
4. Devise Hough transform on masked edge detected image.

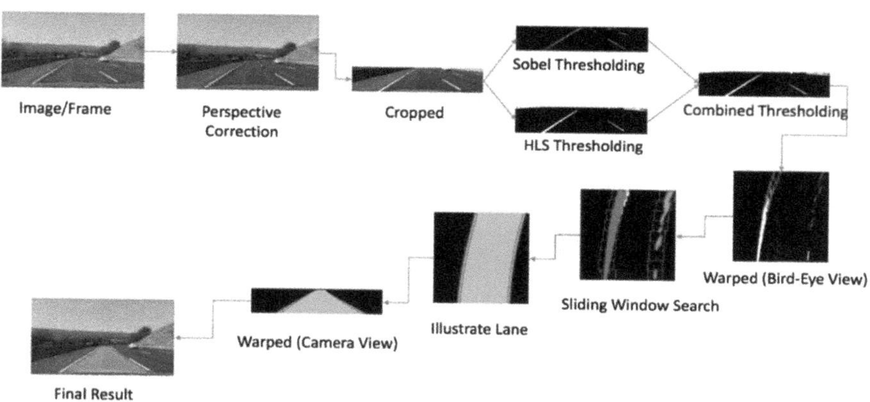

Fig. 4.6 Lane detection pipeline

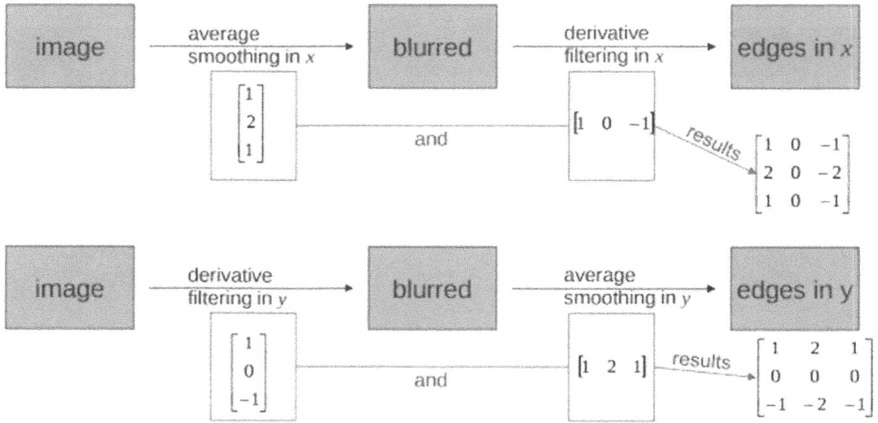

Fig. 4.7 Lane thresholding

5. Draw the line segments.
6. Combine the line image with the original.

Figure 4.7 shows the frameworks on the algorithm of lane thresholding for steps in lane detection pipeline.

4.3 Result and Discussion

4.3.1 Vehicle Detection During the Day

The algorithm for a day shows its workability at its maximum efficiency with the confidence above 80% for the vehicle nearest with the agent vehicle. The furthest vehicle resulting below 80% confidence showing that the pipeline is still working to detect the vehicle as shown in Fig. 4.8.

4.3.2 Multiple Class Vehicle Detection

The YOLO algorithm is also capable to class out the vehicle more than three courses in time as shown in Fig. 4.9 with the confidence percentage above 70%. This shows the pipeline workability during the peak hour in the road thus making the system pipeline acceptable to be used for roads.

Class	Confidence	Time / condition	result	Average
2	98.94%	day with direct sun-ray	6-out	Class 2 avg = 0.8664
2	94.17%	day with direct sun-ray	6-out	
2	94.41%	day with direct sun-ray	6-out	
2	87.29%	day with direct sun-ray	6-out	
2	97.63%	day with direct sun-ray	6-out	
2	69.24%	day with direct sun-ray	6-out	
2	75.33%	day with direct sun-ray	6-out	
2	76.09%	day with direct sun-ray	6-out	

Fig. 4.8 Vehicle detection during the day

Class	Confidence	Time / condition	result	Average
2	99.57%	day	12-out	Class 0 avg = 0.8432
7	56.52%	day	12-out	Class 2 avg = 0.9221
7	57.11%	day	12-out	Class 3 avg = 0.9228
0	92.62%	day	12-out	Class 7 avg = 0.6191
0	76.03%	day	12-out	
7	72.09%	day	12-out	
2	90.17%	day	12-out	
3	90.93%	day	12-out	
2	86.88%	day	12-out	
3	93.64%	day	12-out	

Fig. 4.9 Multiple class vehicle detection

4.3.3 Vehicle Detection During Night

The detection algorithm also was tested during the night to compute the algorithm efficiency, noted that the highways in Kuala Lumpur are mostly equipped with road lamps during night time. The algorithm becomes helpful to give a clearer input image for processing vehicle detection using YOLO. Figure 4.10 shows the result obtained from the night time. The algorithm was used to detect the vehicle from the nearest to the furthest distance possible. Figure 4.10a shows the result for the vehicle detected at the closest distance. While in Fig. 4.10b the car exposed during night at the furthest distance possible, i.e., ranging from 50 to 100 m, was detected with the average confidence above 80%.

4.3.4 Detection of Lane

The final lane detection is the projected lane which fills the lane with a color gradient with an end-to-end line separating from the other lane, as shown in Fig. 4.11. The vehicle offset from the center also is computed which will determine the car position in the lane in mimicking the lane departure warning system.

4.3.5 Vehicle and Lane Detection

Figure 4.12 shows the result from the combined pipeline of vehicle detection and lane detection tested during daylight. The detection rating is 17–19/fps while running on the CPU and GPU results in a higher detection rate of 40–45/fps.

4.4 Conclusion and Recommendation

This chapter presented a vehicle and lane detection system developed using the YOLO algorithm. The following are the outcomes of the experiment:

- The vehicles on Malaysian roads, especially on highways can be recognized and categorized as car, motorcycle, and truck.
- The YOLO algorithm uses a single neural network that results in quicker detection and is suitable for real-time application.
- The GPU enhances the capabilities of the pipeline for detecting the maximum number of vehicles in time.
- This experimental method proved its suitability for Malaysian roads.

Class	Confidence	Time / condition	result	Average
2	92.60%	Night with road light	20-out	Class 2 avg = 91.52
2	84.45%	Night with road light	20-out	
2	97.51%	Night with road light	20-out	

Class	Confidence	Time / condition	result	Average
2	71.98%	Night with road light	17-out	Class 2 avg = 0.8191
2	80.67%	Night with road light	17-out	
2	51.78%	Night with road light	17-out	
2	89.62%	Night with road light	17-out	
2	93.06%	Night with road light	17-out	
2	78.47%	Night with road light	17-out	
2	97.46%	Night with road light	17-out	
2	92.23%	Night with road light	17-out	

Fig. 4.10 Vehicle detection during the night with road lights

Fig. 4.11 Lane detection during the day

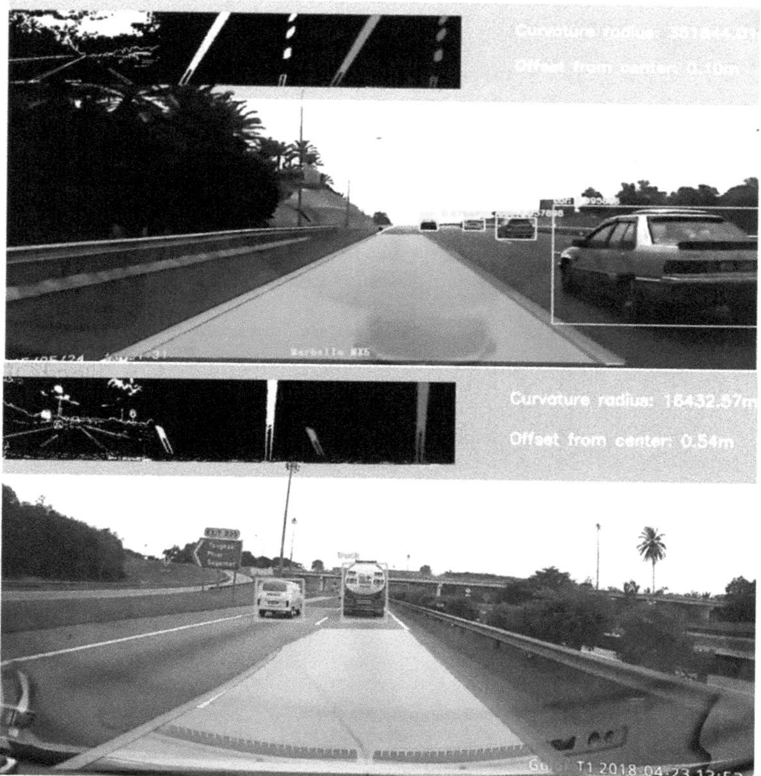

Fig. 4.12 Vehicle and lane detection

Acknowledgements The authors gratefully acknowledge the financial support for this research work by Yayasan Tengku Abdullah Scholarship (YTAS), Universiti Kuala Lumpur Malaysian Spanish Institute (UniKL MSI), and System Engineering and Energy Laboratory (SEELab).

References

Borrego-Carazo, J., Castells-Rufas, D., Biempica, E., Carrabina, J.: Resource-constrained machine learning for ADAS: a systematic review. IEEE Access **8**, 40573–40598 (2020)

Chiu, H.-W., Lee, C.-H.: Prediction of machining accuracy and surface quality for CNC machine tools using data driven approach. Adv. Eng. Softw. **114**:246–257 (2010)

Christiansen, R.H., Hsu, J., Gonzalez, M., Wood, S.L.: Monocular vehicle distance sensor using HOG and Kalman tracking. In: 2017 51st Asilomar Conference on Signals, Systems, and Computers, pp. 178–182 (2017)

Fassold, H.: Computer vision on the GPU—tools, algorithms and frameworks. In: 2016 IEEE 20th Jubilee International Conference on Intelligent Engineering Systems (INES), pp. 245–250 (2016)

Fernández-Llorca, D., García-Daza, I., Martínez-Hellín, A., et al.: Parking assistance system for leaving perpendicular parking lots: experiments in daytime/nighttime conditions. IEEE Intell. Transp. Syst. Mag. **6**(2), 57–68 (2014)

Galko, C., Rossi, R., Savatier, X.: Vehicle-hardware-in-the-loop system for ADAS prototyping and validation. In: 2014 International Conference on Embedded Computer Systems: Architectures, Modeling, and Simulation (SAMOS XIV), pp. 329–334 (2014)

Huang, D.-Y., Chen, C.-H., Chen, T.-Y., et al.: Vehicle detection and inter-vehicle distance estimation using single-lens video camera on urban/suburb roads. J. Vis. Commun. Image Represent. **46**, 250–259 (2017)

Lin, H., Chen, L., Lin, Y., Yu, M.: Lane departure and front collision warning using a single camera. In: 2012 International Symposium on Intelligent Signal Processing and Communications Systems, pp. 64–69 (2012)

Liu, X., Liu, W., Mei, T., Ma, H.: A deep learning-based approach to progressive vehicle re-identification for urban surveillance. In: Leibe B., Matas J., Sebe N., Welling M. (eds) Computer Vision—ECCV 2016. ECCV 2016. Lecture Notes in Computer Science, vol. 9906. Springer, Cham (2016). https://doi.org/10.1007/978-3-319-46475-6_53

Luo, H., Yang, Y., Tong, B., Wu, F., Fan, B.: Traffic sign recognition using a multi-task convolutional neural network. IEEE Trans. Intell. Transp. Syst. **19**(4), 1100–1111 (2018)

McLean, S.D., Craciunas, S.S., Alexander Juul Hansen, E., Pop, P.: Mapping and scheduling automotive applications on ADAS platforms using metaheuristics. In: 2020 25th IEEE International Conference on Emerging Technologies and Factory Automation (ETFA). Vol. 1, pp. 329–336 (2020)

Pérez, S., Pérez, J., Arroba, P., et al.: Predictive GPU-based ADAS management in energy-conscious smart cities. In: 2019 IEEE International Smart Cities Conference (ISC2), pp. 349–354 (2019)

Redmon, J., Divvala, S., Girshick, R., Farhadi, A.: You only look once: unified, real-time object detection. In: 2016 IEEE Conference on Computer Vision and Pattern Recognition (CVPR), pp. 779–788 (2016)

Stolz, L., Endt, H., Vaaraniemi, M., Zehe, D., Stechele, W.: Energy consumption of Graphic Processing Units with respect to automotive use-cases. In: 2010 International Conference on Energy Aware Computing, pp. 1–4 (2010)

Wang, X., Huang, K., Knoll, A.: Performance optimisation of parallelized ADAS applications in FPGA-GPU heterogeneous systems: a case study with lane detection. IEEE Trans. Intell. Veh **4**(4), 519–531 (2019)

Wu, X., Dai, W.: Research on machining allowance distribution optimization based on processing defect risk. Procedia CIRP **56**, 508–511 (2016)

Xique, I.J., Buller, W., Fard, Z.B., et al.: Evaluating complementary strengths and weaknesses of ADAS sensors. In: 2018 IEEE 88th Vehicular Technology Conference (VTC-Fall), pp. 1–5 (2018)

Chapter 5
Anomaly Vehicle Detection Using Deep Neural Network

Abd Munim Abd Halim, Mohd Hamizan Yaacob, Muhamad Husaini Abu Bakar, and Pranesh Krishnan

Abstract This project is focusing more on educational purposes. The number of vehicles keeps increasing on the road since almost all people have a personal vehicle. However, tally with the increment of the vehicles on the road, the case of accidents on the road also increases and this makes the road not safer for being used even with extra care. Accidents are caused by people that are careless and do not follow the traffic law on the road. This project is proposed to detect and classify the anomaly vehicle since the anomaly vehicle leads to vehicle accidents on the road. The primary objective is to develop a video processing algorithm to detect and classify vehicles. The objective also extends to construct an anomaly vehicle detection algorithm and propose an alert system for anomaly vehicle detection using a deep neural network. The vehicles are detected using algorithms developed using the python programming language, along with few open-source libraries, namely, OpenCV and the Visual Studio code. A convolutional neural network is used for image processing to reframe and resize the image thus train the image for detection. The bounding box was drawn by using the YOLO object detection method. The data is visualized using the Orange3 software found in the Anaconda python distribution. The Orange3 software is also used for machine learning and data analytics which is used to identify the anomaly vehicle.

A. M. Abd Halim · M. H. Abu Bakar (✉)
System Engineering and Energy Laboratory, Universiti Kuala Lumpur Malaysian Spanish Institute, Kulim Hi-Tech Park, 09000 Kulim, Kedah, Malaysia
e-mail: muhamadhusaini@unikl.edu.my

A. M. Abd Halim
e-mail: amunim.halim@s.unikl.edu.my

M. H. Yaacob
Electrical Electronic and Automation Section, Malaysian Spanish Institute Universiti Kuala Lumpur, Kulim Hi-Tech Park, 09000 Kulim, Kedah, Malaysia
e-mail: mhamizan.yaacob@s.unikl.edu.my

P. Krishnan
Intelligent Automotive Systems Research Cluster, Electrical Electronic and Automation Section, Malaysian Spanish Institute Universiti Kuala Lumpur, Kulim Hi-Tech Park, 09000 Kulim, Kedah, Malaysia
e-mail: pranesh@unikl.edu.my

M. H. Abu Bakar et al. (eds.), *Progress in Engineering Technology III*,
Advanced Structured Materials 148, https://doi.org/10.1007/978-3-030-67750-3_5

Keywords Anomaly detection · Deep neural network · Vehicle detection

5.1 Introduction

Video processing using a deep neural network (DNN) has been used widely to enhance the effectiveness of the system based on the video or camera vision. The DNN has been acknowledged because of the performances and has been applied to various fields such as computer vision (CV), natural language processing (NLP), speech recognition (SR) and audio recognition (AR), machine translation (MT), social network filtering (SNF), board game programs, drug design, and bioinformatics. For the analysis of anomaly vehicles, DNN also achieved excellent performance. Anomaly means something beyond the standard, regular, or expected. For example, the vehicle not moving in usual ways like to keep moving left and right, stopping in the middle of the road, or making U-turn on the straight road.

The anomaly vehicle is one of the reasons for an increased number of road accidents. Due to the increased increment of transportation on the road, it is hard for the authorities like Jabatan Pengakutan Jalan Malaysia (JPJ) and police traffic to do sufficient monitoring and to reduce the rate of accidents to happen. So by applying anomaly vehicle detection on the traffic closed-circuit television (CCTV), all the monitoring will be more efficient because the authorities can be warned and accidents or injuries can be treated earlier thus decreasing the rate of dying people.

Majlis Perbandaran Pulau Pinang (MPPP) stated that there are already about 200 CCTV in Pulau Pinang and it is planned to install about 3000 more CCTV in the future. They proposed the anomaly vehicle detection using the DNN to increase the safety on the road in Pulau Pinang state and to develop Pulau Pinang to become the first smart city in Malaysia.

5.1.1 Image Processing

Image processing has been used in many fields since it has been discovered that it can be used for a bigger purpose. It can be used in machine vision, color processing, UV imaging, video processing, audio processing, and others. This pushes image processing algorithms in the direction of parallel programming according to Saahityan (2017). Based on a paper by Rudofer et al. (2018), big companies like Amazon and Netflix make extensive use of this paradigm that lead them to build a robust system for complexity which has never been possible to imagine ten years ago. As image processing has been used in medical application, digital imaging has altered the way the dental office obtains and stores according to Yoon et al. (2018). It has created exciting chances to generate, enhance, and analyze radiograph.

5.1.2 Video Analysis

Based on Tsakanikas et al. (2018), video analysis has been used for many things such as a surveillance system. In the last decade, video surveillance has spun from simple display systems and video recording to a semi-intelligent autonomous system which is skilled in performing complex actions. Now, a video surveillance system can use some of the intricate video and image algorithms such as classification, decision-making, pattern recognition, and image enhancement. A modern surveillance system comprises a video and image acquisition device, storage unit, components, and data processing analysis modules which are crucial for the system's workflow.

The graphics processing units (GPUs) also have been used with video processing and analysis since they are more powerful than the central processing units (CPUs). It can use a massive number of cores that are optimized for fast arithmetic massive calculations required by an algorithm. The design flow allows greater numerical precision due to the native dependence on the floating-point arithmetic and draws significantly more power based on Blair et al. (2016).

5.1.3 Machine Learning

Huang et al. (2004) developed a new idea for network learning that inspires all over the world to facilitate educational innovation. Many nations pay attention to computer technology and expect to successfully facilitate education improvements. It is famous for the application of computer and internet teachings tools and compared to conventional teaching, it needs a few alterations. Besides, the research and development of proper learning models must extremely reflect the mutual communication among the users and the computers, the instructor and the learners, and the interaction among the learners. Implementing the associated research problems to the above procedure, high-quality research outcomes can be anticipated.

Deep transfer learning is one of the machine learning algorithms. The paper by Zhang (2011) states that, when learning an intricate task, people often need huge data to train as to achieve exceptional presentations. However, training samples are hard to recover. Human learning experiences the same difficulty, but people use past knowledge and understandings to handle the problem. Transfer learning is a learning pattern that intends to trail the same principle for constructing improved learning systems. In transfer learning, associated learning tasks and data are discovered to help learn a dissimilar goal task. The earlier is regularly known as secondary tasks or data and the latter is called the target task. Nevertheless, there have been many kinds of research attempts in this direction; the extent to which knowledge transfer can be applied is quite limited. Numerous current transfer learning methods assume that the secondary and the target tasks are alike. For instance, in the instance-based transfer pattern, the instances in the auxiliary data are directly used as training instances for the target learning.

Machine learning also has been used for anomaly detection. Anomaly detection denotes the tricky part of detecting patterns in data that do not conform to predictable behavior, where the outlines are usually stated as anomalies and outliers (Collins et al. 2018). In prognostics and health management (PHM), the status of anomaly detection is because anomalies in data translate to important information about the product's health status. In general, most anomaly detection methods construct a profile of standard instances, then identify anomalies that do not conform to the standard profile. For illustration, the useful data (also called training data or baseline data) were obtained from the cycling of a ball grid array (BGA) and used to model the BGA's health status. By capturing the discrepancies between the model's estimates and in-situ observations, the authors detected the BGA's anomalous behavior. For anomaly detection, methods can be categorized into distance-based, clustering-based, classification-based, and statistical anomaly detection methods. Distance-based methods use the nature of anomalies located far from the data collected from healthy nominally products. Clustering-based methods assume that regular observations will belong to the same cluster(s) (Jiang et al. 2006).

5.2 Methods

5.2.1 Image Processing Using Convolutional Neural Network

A Convolutional Neural Network (CNN) is used to process images in the video, as shown in Fig. 5.1. A ConvNet/CNN) is a Deep Learning algorithm which takes images as input, appoints significance (learnable weight and predispositions) to

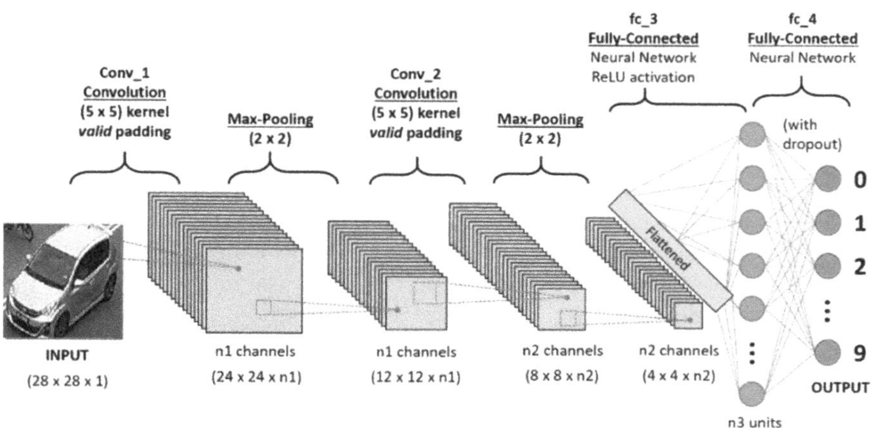

Fig. 5.1 A sequence of CNN to classify the type of vehicle

different perspectives/questions in the picture and to have the option to differentiate one from the other. The pre-preparing required in a ConvNet is much lower when contrasted with other order calculations. While in crude techniques channels are hand-built, with enough preparing, ConvNets can get familiar with these channels/qualities.

The design of a ConvNet is analogous from that of the availability example of neurons in the human brain and was enlivened by the association of the visual cortex. Singular neurons react to improvements just in a limited district of the visual field known as the receptive field. An accumulation of such fields covers to cover the whole visual region.

A ConvNet can effectively catch the spatial and temporal dependencies in a picture through the utilization of essential channels. The engineering plays out a superior fitting to the picture dataset because of the decrease in the number of parameters included and the re-usability of loads. Further, the network can be trained to comprehend the refinement of the image better. A video or image is taken as a matrix of pixel values so that the image is flattened as shown in Fig. 5.2 and feed it to a Multi-Level Perceptron for classification purpose. In cases of an original binary image, the method may demonstrate a standard precision score while performing the prediction of classes. However, it would have next to zero precision with regards to an intricate image having pixel dependencies all through.

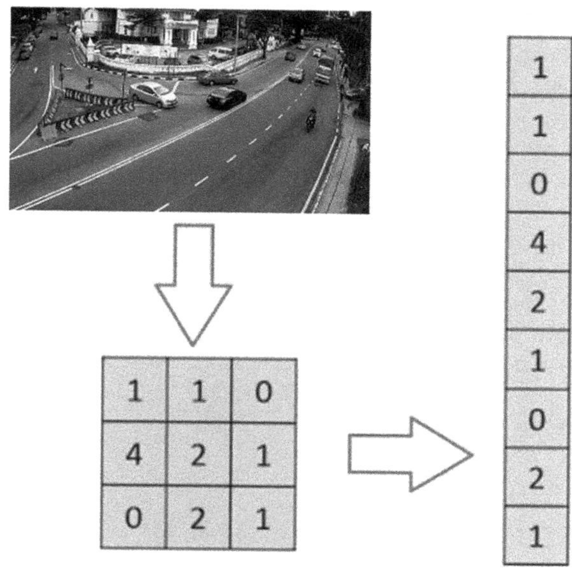

Fig. 5.2 Flattening the image using the example of a 3×3 image in a 9×1 vector

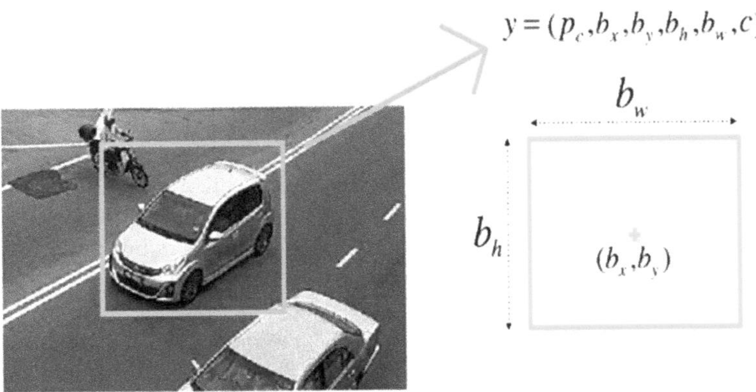

Fig. 5.3 Reference to calculate box coordinates

5.2.2 *Building the Box for Object Detection*

The initial step to utilize the YOLO is that it is expected to see how it encodes its output. The input picture is isolated into a $S \times S$ grid of cells. For each object that is available in the picture, one matrix cell is said to be "responsible" for anticipating it. That is where the center point of the image falls.

Every grid cell predicts *B* bounding boxes just as *C* class probabilities. The bounding box prediction has five segments: (*x*, *y*, *w*, *h*, certainty). The (*x*, *y*) coordinates represent the center point of the case, concerning the matrix cell area (recall that, if the center point of the grid does not fall inside the framework cell, at that point is not in charge of it). These directions are standardized to fall somewhere in the range of *0* and *1*. The (*w*, *h*) box measurements are additionally standardized to [0, 1], relative to the picture size as shown in Fig. 5.3.

It is required to forecast the class possibilities, *Pr*(*Class*(*i*)|*Object*). This possibility is trained on the grid cell comprising one object. In practice, it means that if no object is present in the grid cell, the loss function does not penalize for a wrong class estimation as discussed in the later section. The network only foresees one set of class possibilities per cell, irrespective of the number of boxes *B*. That makes $S \times S \times C$ class chances in total. Adding the class predictions to the output vector, we get an $S \times S \times (B * 5 + C)$ tensor as output as shown in Fig. 5.4.

5.2.3 *Anomaly Detection Model*

The detected object or vehicle is shown in the bounding box in the video frame with the label of classes. By using the bounding box from the detection, we can use it as a reference and know the position of the object or vehicle by printing the data as coordinates. The coordinates are stored in the CSV file. By using the Orange3

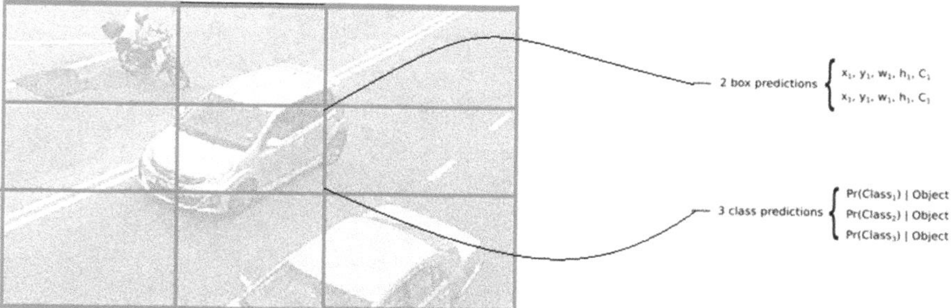

Fig. 5.4 For each grid cell makes B bouncing box estimations and C class predictions ($S = 3$, $B = 2$ and $C = 3$ in this example)

software, the data is visualized like Fig. 5.5. By getting the data, the anomaly can be identified. The height and width are taken from the data to differentiate either is an anomaly or normal. Equation (5.1) shows the difference between a standard and anomaly vehicle. As can be seen, the height and weight of the standard and anomaly vehicle are different, refer to Fig. 5.6.

$$\frac{h}{w} = \text{anomaly\_ vehicle} \tag{5.1}$$

where "h" is the height, and "w" is the weight of the anomaly vehicle.

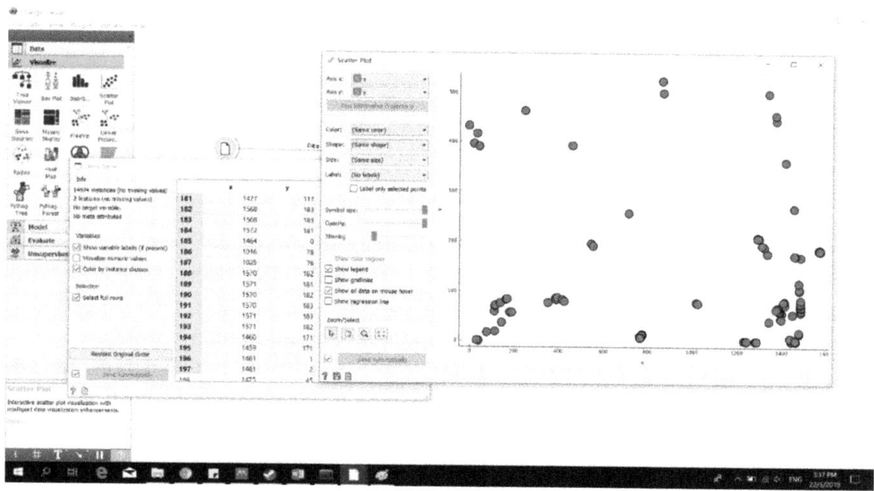

Fig. 5.5 Plotted data in Orange3 software

Fig. 5.6 Normal and anomaly vehicle path prediction

5.3 Result and Discussion

The result of the accuracy of the object detection model has been analyzed. The differences between frames per second using GPU and no GPU. The machine learning software has been used to analyze the vehicle normal and abnormal behavior by data mining. From the data mining, the data shows the frequency of the vehicle on the image and accuracy value on the object detection in the image. The following discussion is derived from the result obtained.

5.3.1 Accuracy of the Detection at Nighttime and Different Weather

A few videos have been used to test the accuracy of the object detection for this project in day time and nighttime conditions. The video has been processed using the same method but applied in different places and time. The object detection is assumed to have high accuracy and has been implemented in many projects.

From the data, i.e., the density versus accuracy, it turns out that both figures are inconsistent. However, the detection still detects the object even with just low-confidence value of accuracy. The data in Fig. 5.7 is at night times, shows that the density of the accuracy 0.5–0.9 and the density around 1–3 is not so frequent, but the data from 0.9 to 1.0 is frequent. The data in Fig. 5.8 from rainy days shows that the accuracy is from 0.5 to 0.9 and that the density around 0 to around 3 is high, while the others lower the noise. The noise in this environment is the rainwater on the CCTV lens it makes the image captured blur. There is also a strong wind that makes

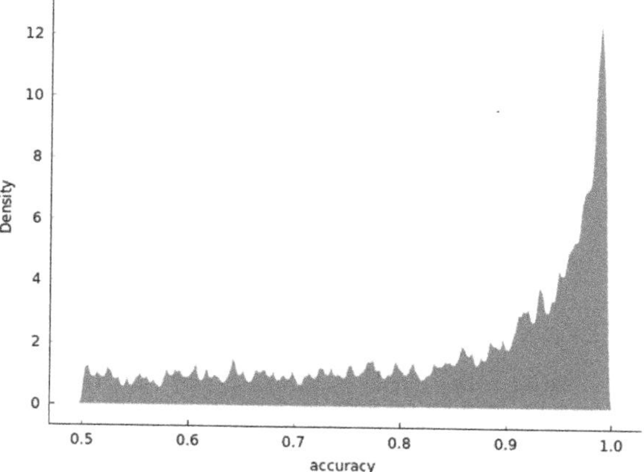

Fig. 5.7 Graph of data from night video

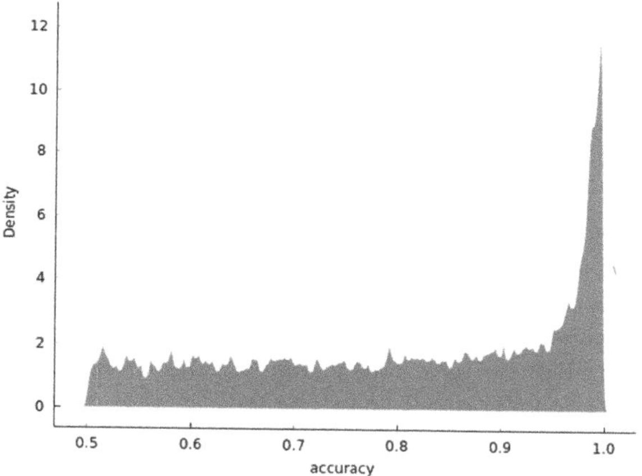

Fig. 5.8 Graph of data during heavy rain and with heavy noise

the CCTV not static. The result showed that the accuracy of the detection object at night is higher than on rainy days, the detection for this project can be used at night since it can detect well in low light conditions while the accuracy of this detection is pretty low under the influence of heavy noise.

5.3.2 CPU Versus no GPU

The laptop used for this project is an MSI GP 60. The specification for the laptop is as follows: the CPU is a 4th generation Intel® Core™ i7 Processor, the chipset of Intel HM86, memory DDR3L, up to 1600 MHz, 2 Slots, max 16 GB, the graphics card is an NVIDIA GeForce GT740M and 2G VRAM DDR3. In Fig. 5.9, the average

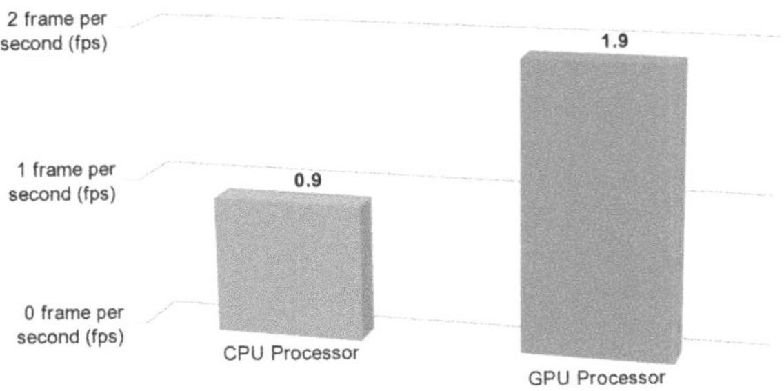

Fig. 5.9 CPU and GPU comparison

fps without using GPU is the result of the frame per second when this project is run. The fps without using the GPU is low, only at 0.9 fps by using the MSI GP60 specification. While the average fps using the GPU is 1.9 frames per second. The results show that frames per second increase about 1 frame per second by using the GT 740 M GPU. The frame per second does not increase a lot because this laptop GPU is not powerful enough to do the processing at high speed.

5.4 Conclusion and Recommendation

Object detection and anomaly detection has been considered and established by many researchers to detect, for example, anomaly road, human behavior, and monitoring systems. Besides, vehicle detection also approaches the areas of machine vision and machine learning. The main objective of this research is to advance a video processing algorithm for vehicle detection and classification, to construct an anomaly vehicle detection algorithm, and to propose an alert system for anomaly vehicle detection using a deep neural network. As been discussed in the methodology, this project was divided into three phases, which apply detection and classification, data analytics, and anomaly detection. Thus, the results were presented and then discussed and briefly explained. The results have been proven to be practicable and competent to be used for the anomaly vehicle monitoring system.

Acknowledgements Authors gratefully acknowledge the financial support for this research work by Yayasan Tengku Abdullah Scholarship (YTAS), Universiti Kuala Lumpur Malaysian Spanish Institute (UniKL MSI), and System Engineering and Energy Laboratory (SEELab).

References

Blair, C.G., Robertson, N.M.: Video anomaly detection in real time on a power-aware heterogeneous platform. IEEE Trans. Circuits Syst. Video Technol. **26**(11), 2109–2122 (2016)

Collins, J., Howe, K., Nachman, B.: Anomaly detection for resonant new physics with machine learning. Phys. Rev. Lett. **121**(24), 241803 (2018). https://doi.org/10.1103/PhysRevLett.121.241803

Huang, C.-J., Liu, M.-C., Chu, S.-S., Cheng, C.-L.: Application of machine learning techniques to Web-based intelligent learning diagnosis system. In: Fourth International Conference on Hybrid Intelligent Systems (HIS'04), pp. 242–247 (2004)

Jiang, S., Song, X., Wang, H., Han, J.-J., Li, Q.-H.: A clustering-based method for unsupervised intrusion detections. Pattern Recognit. Lett. **27**(7), 802–810 (2006)

Rudorfer, M., Krüger, J.: Industrial image processing applications as orchestration of web services. Proc. CIRP **76**, 144–148 (2018)

Saahithyan, V., Suthakar, S.: Performance analysis of basic image processing algorithms on GPU. In: 2017 International Conference on Inventive Systems and Control (ICISC), pp. 1–6 (2017)

Tsakanikas, V., Dagiuklas, T.: Video surveillance systems-current status and future trends. Comput. Electr. Eng. **70**, 736–753 (2018)

Yoon, D.C., Mol, A., Benn, D.K., Benavides, E.: Digital radiographic image processing and analysis. Dent. Clin. North Am. **62**(3), 341–359 (2018)

Zhang, J.: Deep transfer learning via restricted boltzmann machine for document classification. In: 2011 10th International Conference on Machine Learning and Applications and Workshops, vol 1, pp. 323–326 (2011)

Chapter 6
GPU Accelerated Speech Recognition

Mohamad Rosyidi Ahmad, Ahmad Fakhri Arif Mat Zaid, Muhamad Husaini Abu Bakar, Mohd Fauzi Alias, and Pranesh Krishnan

Abstract Speech recognition technology is one of the quickly developing advanced technologies of engineering. It has various applications in different zones and offers potential points of interest. There is a correlation to pre-requirement for speech recognition and machine learning which is that grammar classification and a method for extraction phonemes from utterances are required. For Speech architecture, three models are utilized in speech recognition to do the preparing like a phonetic dictionary, acoustic model, language model. There are four stages of the recognition process: analysis, feature extraction, modeling and matching. Deficiency factors were brought forward which is the dataset threshold and feature extracting method. The higher dataset produces a lower word error rate (WER) which gives more accuracy in the recognition process.

Keywords Acoustic model · CMU sphinx · Grapheme · Hidden markov model · Language model · Mel-frequency cepstral coefficients · Phonetic dictionary

---

M. R. Ahmad (✉) · A. F. A. M. Zaid · M. F. Alias
Electrical Electronic and Automation Section, Malaysian Spanish Institute Universiti Kuala Lumpur, Kulim Hi-Tech Park, 09000 Kulim, Kedah, Malaysia
e-mail: mrosyidi@unikl.edu.my

A. F. A. M. Zaid
e-mail: afakhri.zaid@s.unikl.edu.my

M. F. Alias
e-mail: fauzialias@unikl.edu.my

M. H. A. Bakar
System Engineering and Energy Laboratory, Universiti Kuala Lumpur Malaysian Spanish Institute, Kulim Hi-Tech Park, 09000 Kulim, Kedah, Malaysia
e-mail: muhamadhusaini@unikl.edu.my

P. Krishnan
Malaysian Spanish Institute Universiti Kuala Lumpur, Intelligent Automotive Systems Research Cluster, Electrical Electronic and Automation Section, Kulim Hi-Tech Park, 09000, Kulim, Kedah, Malaysia
e-mail: pranesh@unikl.edu.my

M. H. Abu Bakar et al. (eds.), *Progress in Engineering Technology III*,
Advanced Structured Materials 148, https://doi.org/10.1007/978-3-030-67750-3_6

6.1 Introduction

Speech recognition is the process of capturing utterances by telephone or microphone and converting them into a digitally stored set of words. The quality of a speech recognition system is assessed according to two factors, i.e. its accuracy which is a word error rate (WER) in converting spoken words to digital data and speed on how well the software can keep up with a human speaker. Typically, speech is processed by software have limitation. The usual conversation contains more than 80% of the information. Most of the voice-recognition programs are in English, and there have been limited studies conducted on other languages. Thus, recognition for accents by multiracial speakers for the Malay language is limited. In other words, Malay speech recognition will benefit Malaysians of various ethnic groups who speak the language. The dictionary also created with only under 50 words for the early research stage will be insufficient for a robust speech recognition system but will do the job for this stage.

The methods for extracting voice or phonemes will use a sampling method of Fast Fourier Transform (FFT) gained from the microphone. For each sampling, there will be graphemes trained with the Hidden Markov Model (HMM).

Speech recognition or automatic speech recognition (ASR) is a machine or program's ability to identify and convert words and phrases into a machine-readable format. The automatic word recognition system (ASR) can be defined as the ability of a machine to emulate the spoken utterance of a speaker into a text string or other applications. Speech recognition works using acoustic and language modeling by algorithms. Acoustic modeling represents the relationship between linguistic units of speech and audio signals; language modeling matches sounds with word sequences to help distinguish between words that sound similar. The grapheme to phoneme or G2P alludes to the conversions of grapheme to phoneme. The G2P is the way toward utilizing principles to produce a word pronunciation to make a pronunciation dictionary. The rules are formed by an automated statistical analysis of the dictionary of pronunciation. G2P translates a sequence of words into a series of phonemes. Three models are utilized in speech recognition to do the preparing:

- Phonetic dictionary: contains mapping words to phones.
- Acoustic model: holds acoustic properties to each sub phonetic. There are context-independent models containing features.
- A language model is employed to limit the word search. It defines that a word may follow antecedently recognized words and helps to considerably prohibit the matching process by baring words that do not seem to be probable.

The Hidden Markov Model is a finite set of states, each associated with a (usually multidimensional) distribution of probability. State transitions are governed by a set of possibilities known as transition probabilities. According to the associated probability distribution, a result of observation may be generated in a particular state. It is only the result, not the visible state to an external observer, and therefore rules are "hidden" to the exterior.

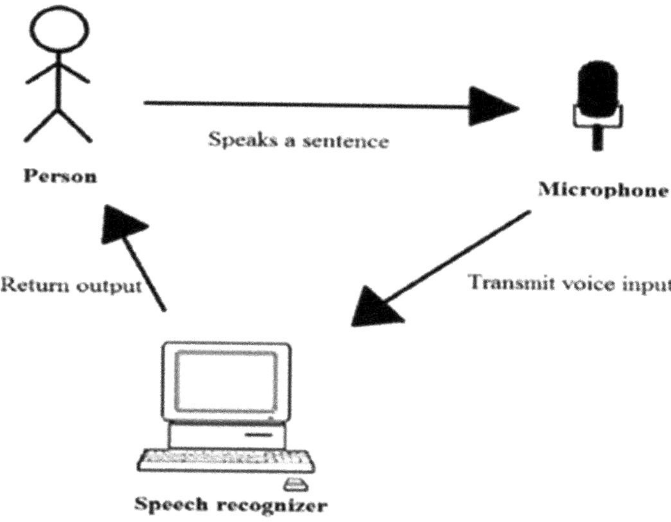

Fig. 6.1 System overview for the speech recognition system

6.2 Methods

This topic discusses the method used and the requirement for a Malay speech recognition system by building a phonetic dictionary, language model and acoustic model for pre-processing and then to assemble and validate the system.

System Overview

Figure 6.1 outlines the proposed framework for perceiving the spoken phrases from an individual as input and deciphers the content as output. The primary user of this device is the individual. The individual will interact with the speech recognizer by way of speaking through the microphone. The individual can talk to the speech recognizer to interpret the uttered words into words. This procedure is known as speech-to-text.

6.3 Speech Recognition Techniques

A speech recognition system's main objective is to be capable of listening, understanding and then acting on the spoken information. There are four main stages in a speech recognition system that are further classified, as shown in Fig. 6.2.

1. Analysis: Analysis is the first stage. When the speaker speaks, various types of information are included in the speech that helps identify a speaker. Due to the vocal tract, the source of excitation and the behavior feature, the data is

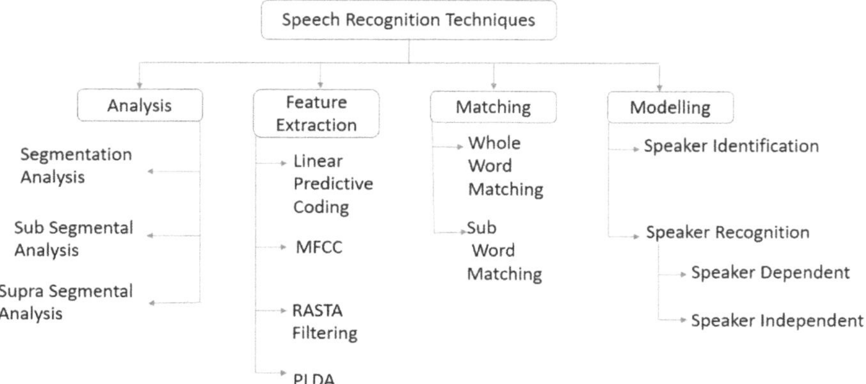

Fig. 6.2 Speech recognition technique

different. Segmentation analysis is one of the methods used to testing extort speaker information which is done using frame size and shift between 10 and 30 ms (ms).

2. Feature Extraction Technique: Extraction of features is a major part of the speech recognition system. It is regarded as the system's core. The work of this is to extract from the input speech (signal) those features that help the system identify the speaker. Extraction of the feature compresses the magnitude of the input signal (vector) without harming the power of the voice signal. The extraction technique used is Mel-frequency cepstrum coefficient (MFCCs). MFFCs is based on the known variations of critical bandwidths of the human ear at frequencies below 1000 Hz. The MFCC processor's primary purpose is to copy human ear behavior. The following steps are used to derive the MFCC.
3. Modeling techniques: Modeling techniques are aimed at producing speaker models by using the extracted features (feature vector). The modeling techniques are further categorized into the recognition and identification of speakers, as shown in Fig. 6.2. Recognition of speakers can be classified as reliant speakers and independent speakers. Speaker identification is a process where the system can identify who the speaker is based on the voice signal information extracted. Pattern recognition modeling techniques are used. It involves two steps: Pattern comparison and pattern training. It is further classified into template-based and stochastic approaches. This approach makes use of robust mathematical formulas and develops speech pattern representations.

6.4 Implementation and Analysis

Several tools are required for developing the prototype for the speech recognition system. The system is developed under a Linux operating system, i.e. Ubuntu. It is quite recommended to use a Linux operating system due to the fact that the CMUSphinx toolkit is not fully compatible with different platforms such as Windows and Mac. Visual Studio Code is used to write the programming codes for the project. It is chosen because it is faster to compile and run the software instead of having to open the terminal every time modifications are made to the program. Audacity is used to record and edit sounds because it is a free, open-source and cross-platform software program.

The first step is by downloading and installing the essential toolkit from the CMUSphinx download page. There are four components of the CMUSphinx toolkit that were used for this project:

- Sphinxbase
- Sphinxtrain
- Pocketsphinx
- Sphinx4

Sphinxbase is a support library required by pocket sphinx and sphinxtrain. Sphinxtrain is an acoustic model training tool which was used for the constructing of the acoustic model. Pocketsphinx is a recognizer library written in the C language. Sphinx4 is an adjustable and modifiable recognizer written in the Java language. After these four components have been installed, three models are required in speech recognition to do the matching:

- Phonetic dictionary
- Language model
- Acoustic model

6.5 Creating the Phonetic Dictionary

The phonetic dictionary consists of a mapping from words to phonemes. A phoneme is one of the units of sound that distinguish one word from every other in a language. In this project, a dictionary of 6 words is created that includes words from the Malay language in a.txt file. There need to be solely one word per line. Then, for the phonemes, the letters of each phrase are separated with the aid of space between the letters by Malay regulations of designation (Figs. 6.3 and 6.4).

Creating the Statistical Language Model

Language models are used to constrain search in a decoder using limiting the range of viable words that need to be considered at any point in the search. The outcome is quicker execution and higher accuracy. There are two sorts of language models. The first type is the grammar language model. Grammar is appropriate for command and control, and when you have very little speech data. The second type of language model is a statistical language model. For this project, the statistical language model is used as it is appropriate for free-form input, such as spontaneous speech. A statistical language model is a probability distribution over sequences of words.

Fig. 6.3 Malay phonetic dictionary

```
APA AE P AH
AWAK    AO AH K
BUKA    B Y UW K AH
HELLO   HH AH L OW
KASIH   K AE S IY
LAMPU   L AE M P UW
NAMA    N AE M AH
SAYA    S EY AH
TERIMA  T EH R AH M AH
TOLONG  T AA L AH NG
TUTUP   T Y UW T AH P
```

Fig. 6.4 Transcription text

```
<S> SAYA BUKA LAMPU </S>
<S> AWAK BUKA LAMPU </S>
<S> DIA BUKA LAMPU </S>
<S> KITA BUKA LAMPU </S>
```

To start building the language model, CMUclmtk needed to be downloaded and installed. CMUclmtk stands for Carnegie Mellon University Cambridge Language Modeling Toolkit. The ARPA model training proceeds with the CMUclmtk. The process for creating a language model is as follows:

- Prepare a reference text that will be used to generate the language model. The language model toolkit expects its input to be in the structure of normalized text files, with utterances delimited by using <s> and </s> tags. The reference text file is named as "other_lm.txt". Figure 6.4 is an example of the reference text.
- Generate the vocabulary file by using typing this command in the terminal. A vocabulary file is a listing of all the phrases in the file: text2wfreq <other_lm.txt|wfreq2vocab> other_lm.vocab.
- Generate the ARPA format language model with the commands: text2idngram -vocab malay_lm.vocab -idngram other_lm.idngram < other _lm.txt idngram2lm -vocab_type 0 -idngram other_lm.idngram—vocab other_lm.vocab –arpa other_lm.lm.
- Generate the CMU dump form: sphinx_lm_convert -I other_lm.lm -o other_lm.lm.DMP.

Creating the Acoustic Model

A new folder is created and called "other" to start developing a new acoustic model. Then two folders in the other folder were created and called "etc." and "wav." In the "etc." folder, there are eight files:

- other.dic (Phonetic dictionary)
- other.phone (List of phonemes)

Table 6.1 Audio format

Channel	Frequency	Audio format
Mono channel	16 kHz	WAV 16 bit

- other.lm (Language model)
- other.filler (List of fillers)
- other_train.fileids (List of audio file names)
- other_train.transcription (List of transcriptions of audio files)
- other_test.fileids (List of audio file names)
- other_test.transcription (List of transcriptions of audio files)

The audio files ought to be in the format mentioned in Table 6.1.

After that, the terminal was opened and the directory was changed to the malay_am folder and the following command was typed: sphinxtrain -t another setup as shown in Fig. 6.5. Then, saved the sphinx_train.cfg and typed in this command windows: sphinxtrain run (Figs. 6.6 and 6.7).

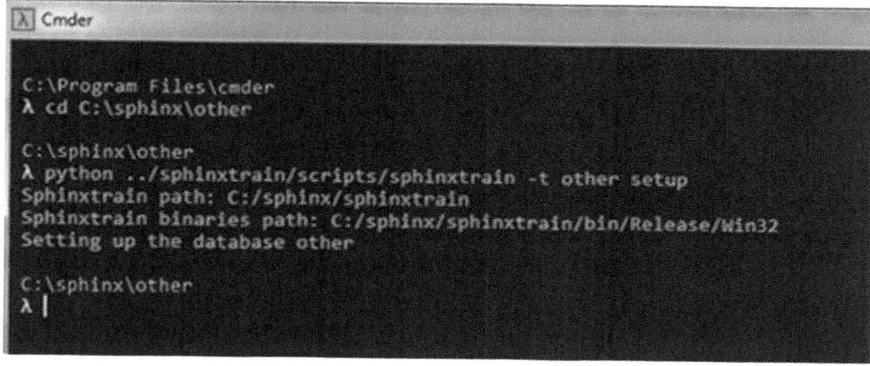

Fig. 6.5 Sphinxtrain setup

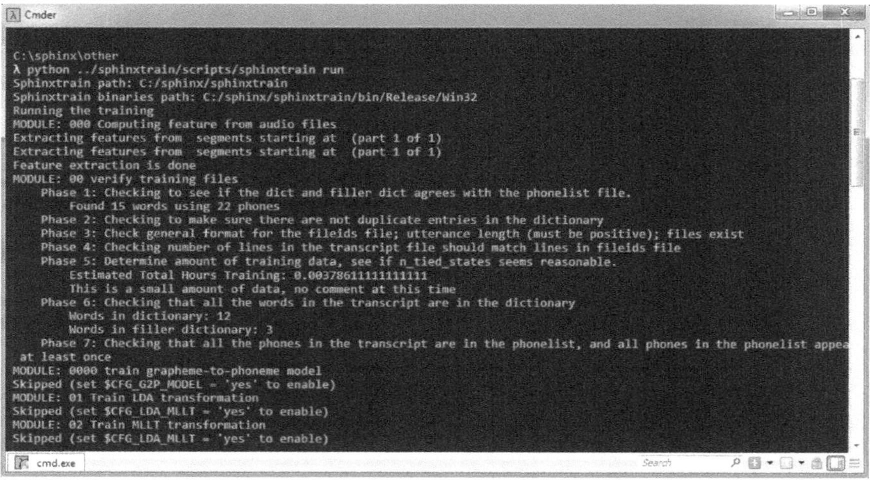

Fig. 6.6 Training process

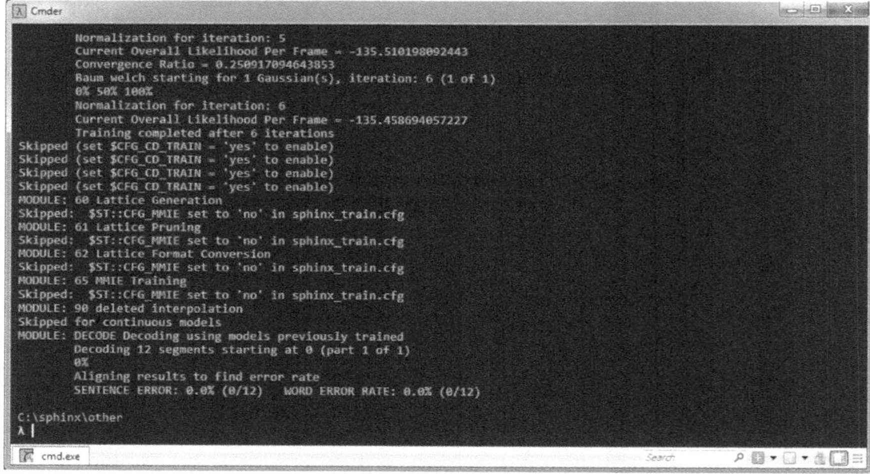

Fig. 6.7 Training result

Program Testing

The software is opened by using the command prompt, after selecting the bin directory, respectively. The code file pocketsphinx is compiled and then run. The result is shown in Fig. 6.8 and Fig. 6.9.

Accuracy Test

For the testing phase, there is a requirement for the number of the dataset which will calculate the word error rate (WER). The value of data is stepped by 200 until 1200.

```
Cmder
INFO: cmn_live.c(105): Update to   <  4.92 -3.09  3.03  1.33  8.44  8.65 10.22  1.03 -0.51 -2.60 -2.18 -1.93  0.67 >
INFO: cmn_live.c(120): Update from <  4.92 -3.09  3.03  1.33  8.44  8.65 10.22  1.03 -0.51 -2.60 -2.18 -1.93  0.67 >
INFO: cmn_live.c(138): Update to   <  4.75 -3.05  2.92  1.23  8.00  8.35 10.02  1.31 -0.32 -2.47 -2.15 -2.02  0.67 >
INFO: ngram_search_fwdtree.c(1550):      343 words recognized (2/fr)
INFO: ngram_search_fwdtree.c(1552):     3123 senones evaluated (19/fr)
INFO: ngram_search_fwdtree.c(1556):     1672 channels searched (10/fr), 715 1st, 388 last
INFO: ngram_search_fwdtree.c(1559):      388 words for which last channels evaluated (2/fr)
INFO: ngram_search_fwdtree.c(1561):       51 candidate words for entering last phone (0/fr)
INFO: ngram_search_fwdtree.c(1564): fwdtree 0.02 CPU 0.009 xRT
INFO: ngram_search_fwdtree.c(1567): fwdtree 1.57 wall 0.951 xRT
INFO: ngram_search_fwdflat.c(302): Utterance vocabulary contains 4 words
INFO: ngram_search_fwdflat.c(948):      118 words recognized (1/fr)
INFO: ngram_search_fwdflat.c(950):     2019 senones evaluated (12/fr)
INFO: ngram_search_fwdflat.c(952):      995 channels searched (6/fr)
INFO: ngram_search_fwdflat.c(954):      454 words searched (2/fr)
INFO: ngram_search_fwdflat.c(957):      126 word transitions (0/fr)
INFO: ngram_search_fwdflat.c(960): fwdflat 0.00 CPU 0.000 xRT
INFO: ngram_search_fwdflat.c(963): fwdflat 0.01 wall 0.006 xRT
INFO: ngram_search.c(1250): lattice start node <s>.0 end node </s>.148
INFO: ngram_search.c(1276): Eliminated 0 nodes before end node
INFO: ngram_search.c(1381): Lattice has 20 nodes, 9 links
INFO: ps_lattice.c(1376): Bestpath score: -7410
INFO: ps_lattice.c(1380): Normalizer P(O) = alpha(</s>:148:163) = -413089
INFO: ps_lattice.c(1437): Joint P(O,S) = -418762 P(S|O) = -5673
INFO: ngram_search.c(872): bestpath 0.00 CPU 0.000 xRT
INFO: ngram_search.c(875): bestpath 0.05 wall 0.027 xRT
APA
INFO: continuous.c(275): Ready....
INFO: continuous.c(261): Listening...
pocketsphinx_continuous.exe    Search
```

Fig. 6.8 Training of the speech recognition model (a)

```
Cmder
INFO: cmn_live.c(88): Update from <  5.02  1.62  1.35  3.69  6.75 12.37  8.76  3.64  3.90  3.53  0.09 -5.42 -0.57 >
INFO: cmn_live.c(105): Update to   <  5.03  2.31  1.92  4.14  7.34 12.55  8.58  3.51  4.55  3.79 -0.07 -5.34 -0.77 >
INFO: cmn_live.c(120): Update from <  5.03  2.31  1.92  4.14  7.34 12.55  8.58  3.51  4.55  3.79 -0.07 -5.34 -0.77 >
INFO: cmn_live.c(138): Update to   <  4.96  2.58  1.85  4.24  7.03 12.47  8.55  2.89  5.24  4.64  0.24 -4.42 -0.13 >
INFO: ngram_search_fwdtree.c(1550):      132 words recognized (1/fr)
INFO: ngram_search_fwdtree.c(1552):     1707 senones evaluated (10/fr)
INFO: ngram_search_fwdtree.c(1556):      702 channels searched (4/fr), 331 1st, 200 last
INFO: ngram_search_fwdtree.c(1559):      200 words for which last channels evaluated (1/fr)
INFO: ngram_search_fwdtree.c(1561):        8 candidate words for entering last phone (0/fr)
INFO: ngram_search_fwdtree.c(1564): fwdtree 0.05 CPU 0.027 xRT
INFO: ngram_search_fwdtree.c(1567): fwdtree 1.58 wall 0.912 xRT
INFO: ngram_search_fwdflat.c(302): Utterance vocabulary contains 2 words
INFO: ngram_search_fwdflat.c(948):      126 words recognized (1/fr)
INFO: ngram_search_fwdflat.c(950):     2046 senones evaluated (12/fr)
INFO: ngram_search_fwdflat.c(952):      904 channels searched (5/fr)
INFO: ngram_search_fwdflat.c(954):      343 words searched (1/fr)
INFO: ngram_search_fwdflat.c(957):       55 word transitions (0/fr)
INFO: ngram_search_fwdflat.c(960): fwdflat 0.02 CPU 0.009 xRT
INFO: ngram_search_fwdflat.c(963): fwdflat 0.01 wall 0.007 xRT
INFO: ngram_search.c(1250): lattice start node <s>.0 end node </s>.157
INFO: ngram_search.c(1276): Eliminated 0 nodes before end node
INFO: ngram_search.c(1381): Lattice has 10 nodes, 6 links
INFO: ps_lattice.c(1376): Bestpath score: -7588
INFO: ps_lattice.c(1380): Normalizer P(O) = alpha(</s>:157:171) = -425926
INFO: ps_lattice.c(1437): Joint P(O,S) = -428592 P(S|O) = -2666
INFO: ngram_search.c(872): bestpath 0.02 CPU 0.009 xRT
INFO: ngram_search.c(875): bestpath 0.05 wall 0.030 xRT
TUTUP
INFO: continuous.c(275): Ready....
INFO: continuous
pocketsphinx_continuous.exe    Search
```

Fig. 6.9 Training of the speech recognition model (b)

From Fig. 6.10, an accuracy seems its best at the most dataset used. However, during the mid-region of the dataset, the WER produced a higher drop compared to the first region which is using fewer dataset as shown in Fig. 6.11.

Comparison with Previous Works

There are some of the examples taken from the developer community that works on the previous method and number of the dataset in Fig. 6.12.

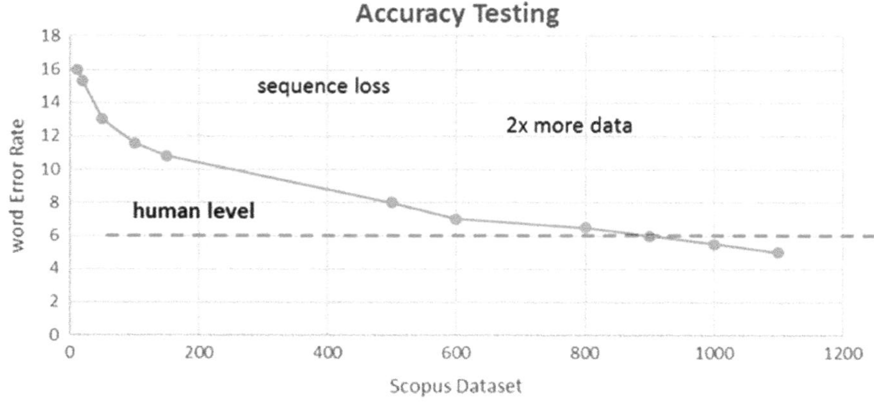

Fig. 6.10 Accuracy testing on different dataset threshold

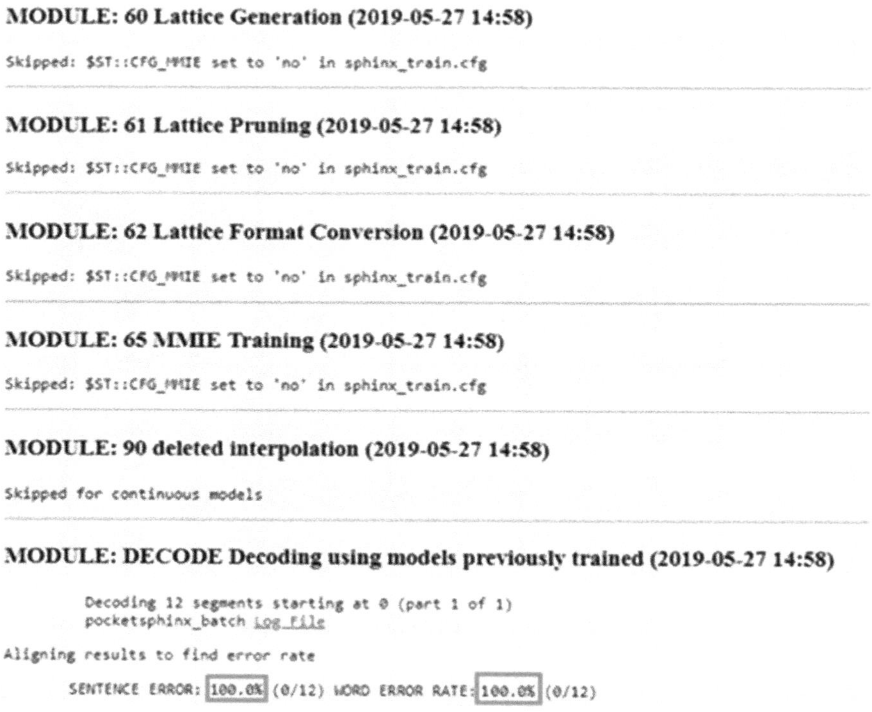

MODULE: 60 Lattice Generation (2019-05-27 14:58)

Skipped: $ST::CFG_MMIE set to 'no' in sphinx_train.cfg

MODULE: 61 Lattice Pruning (2019-05-27 14:58)

Skipped: $ST::CFG_MMIE set to 'no' in sphinx_train.cfg

MODULE: 62 Lattice Format Conversion (2019-05-27 14:58)

Skipped: $ST::CFG_MMIE set to 'no' in sphinx_train.cfg

MODULE: 65 MMIE Training (2019-05-27 14:58)

Skipped: $ST::CFG_MMIE set to 'no' in sphinx_train.cfg

MODULE: 90 deleted interpolation (2019-05-27 14:58)

Skipped for continuous models

MODULE: DECODE Decoding using models previously trained (2019-05-27 14:58)

Decoding 12 segments starting at 0 (part 1 of 1)
pocketsphinx_batch Log File

Aligning results to find error rate

SENTENCE ERROR: 100.0% (0/12) WORD ERROR RATE: 100.0% (0/12)

Fig. 6.11 Accuracy testing on small densities

```
p  I  T      t  s  b  u  r  g  H       (MMXG-CEN5-MMXG-B)
p  R  EIGHTY t  s  b  u  r  g  EIGHT   (MMXG-CEN5-MMXG-B)
Words: 10 Correct: 7 Errors: 3 Percent correct = 70.00% Error = 30.00% Accuracy = 70.(
Insertions: 0 Deletions: 0 Substitutions: 3
october twenty four nineteen seventy  (MMXG-CEN8-MMXG-B)
october twenty four nineteen seventy  (MMXG-CEN8-MMXG-B)
Words: 5 Correct: 5 Errors: 0 Percent correct = 100.00% Error = 0.00% Accuracy = 100.(
Insertions: 0 Deletions: 0 Substitutions: 0
TOTAL Words: 773 Correct: 587 Errors: 234
TOTAL Percent correct = 75.94% Error = 30.27% Accuracy = 69.73%
TOTAL Insertions: 48 Deletions: 15 Substitutions: 171
```

Fig. 6.12 Accuracy testing on large densities

6.6 Conclusion and Recommendation

This speech recognition thesis/project work began with a brief introduction of the technology and its applications in various sectors. The report's project part was based on the speech recognition software development. We discussed multiple tools at the later stage to bring this idea into practical work. It was finally tested and reviewed after the software development. A few deficiency factors were brought forward which are the dataset threshold and the feature extracting method. The software's advantages were described after the testing work, and suggestions for further enhancement and enhancement were discussed.

References

Fook, C.Y., Hariharan, M., Yaacob, S., Ah, A.: Malay speech recognition in normal and noise condition. In: Proceedings—2012 IEEE 8th International Colloquium on Signal Processing and Its Applications, CSPA 2012, pp. 409–412 (2012)

Ghahramani, Z.: An introduction to hidden markov models and Bayesian networks. Int. J. Pattern Recognit. Artif. Intell. **15**(01), 9–42 (2001)

https://medium.com/syncedreview/language-model-a-survey-of-the-state-of-the-art-technology-64d1a2e5a466. Accessed 12 Dec 2017

https://www.happyscribe.co/blog/history-voice-recognition. Accessed 30 Mar 2018

https://searchcrm.techtarget.com/definition/speech-recognition. Accessed 30 Mar 2018

Chapter 7
Aluminum Nanoparticles Preparation via Plasma Arc Discharge

Shahruzaman Sulaiman, Amir Ainul Afif Mahmud, Muhamad Husaini Abu Bakar, and Ahamad Zaki Mohamed Noor

Abstract The aluminum material has good material properties especially if the material is fabricated in nanoparticle size. This nanoparticle size suits well in electrolyte solutions in electric vehicle (EV) batteries. The problem with current EV batteries used in Tesla vehicles is high cost. Thus, there is a need for a cheaper process and material for production process of electrolyte in EV, which is the substitute by Al material. The next problem is in the nano synthesis method which is using common machines such as pulverized ball milling. This produces the nanoparticles in a larger size, i.e. larger than 10^9 m. The particles also tend to aggregate into larger particle clusters and lower the rates of dispersion. Synthesized nanofluids produce large particle diameters and not well-distributed suspension rates. Furthermore, the amount of nanoparticles produced is very limited. Hence, the objective of this research is to prepare an aluminum oxide (Al_2O_3) nanofluid via the plasma arc discharge nanofluid synthesis system (PADNSS) using Al material. The system of PADNSS was developed and the aluminum electrode was submerged in deionized water. The result obtained was three spots having different compositions, EDS spectrum view, and emission peak. From the three spots, shows that Spot 3 was selected due to emission peak and composition of aluminum supersedes compared to the other spots. Hence, this spot will be used for further development of the EV battery.

S. Sulaiman (✉) · A. A. A. Mahmud · M. H. A. Bakar · A. Z. M. Noor
Manufacturing Section, Universiti Kuala Lumpur Malaysian Spanish Institute, 09000 Kulim, Kedah, Malaysia
e-mail: shahruzaman@unikl.edu.my

A. A. A. Mahmud
e-mail: amirafif990@gmail.com

M. H. A. Bakar
e-mail: muhamadhusaini@unikl.edu.my

A. Z. M. Noor
e-mail: ahamadzaki@unikl.edu.my

S. Sulaiman · M. H. A. Bakar
System Engineering and Energy Laboratory, Universiti Kuala Lumpur Malaysian Spanish Institute, Kulim Hi-Tech Park, 09000 Kulim, Kedah, Malaysia

M. H. Abu Bakar et al. (eds.), *Progress in Engineering Technology III*,
Advanced Structured Materials 148, https://doi.org/10.1007/978-3-030-67750-3_7

Keywords Electric vehicle battery · Plasma arc discharge nanofluid synthesis system · Aluminum oxide nanofluid · Aluminum composition · Emission peak

7.1 Introduction

For the past decades, metal nanoparticles development attracts numerous researchers around the world because of their novel physical, electrical, thermal, catalytic properties, and interdisciplinary applications (Sharma et al. 2019). This paper presents a research on the analysis of liquid nanopowder aluminum oxide nanoparticle (Al_2O_3) by using the plasma arc discharge nanofluid synthesis system (PADNSS).

The study conducted was to obtain a high concentration of an aluminum nanofluid purity suspension in water. The particle's surface metallurgy of samples was studied to determine its size distribution. The nanopowder solution of Al_2O_3 was observed for its metallurgy characteristics and its compositions. The nanofluid of Al_2O_3 is used as an electrolyte for EV also known as tesla car. It was invented to improve the rechargeable rate. The problem with current EV batteries used in Tesla vehicle is high cost. Thus, there is a need for a cheaper process and material for the production process of the electrolyte in EV which is substituted by the Al material.

The next problem is in the nano synthesis method which is using common machines such as pulverized ball milling. This produces the nanoparticles in a larger size, i.e. larger than 10^9 m. The particles also tend to aggregate into larger particle clusters and lower the rates of dispersion. Synthesized nanofluids produce large particle diameters and not well-distributed suspension rates. Furthermore, the amount of nanoparticles produced is very limited, the evidence can be seen from the lower concentration of nanoparticles in the aqueous state. In comparison with the dry synthesis method, the wet synthesis method results in much smaller particles being synthesized, as well as nanofluids with better rates of dispersion. The objectives of this paper are to prepare an Al_2O_3 nanofluid via the PADNSS using Al material. Then to analyze metallurgy of Al_2O_3 nanoparticles using a scanning electron microscope (SEM). Lastly, to characterize the composition of aluminum nanoparticles using energy dispersive x-ray spectroscopy (EDS).

7.1.1 Plasma Arc Discharge Nanofluid Synthesis System

The plasma arc discharge nanofluid synthesis system (PADNSS) is a significant process improvement from the original system. The original system for the power distribution and power supply setup of the servo control system was improved. In addition, several important modifications were implemented.

Plasma synthesis in the conditions of the arc discharge with aluminum electrodes is one of the methods for obtaining metal nanoparticles of aluminum powders. For this purpose, the same material for anodes and cathodes in a rod form is used. Apart from metal carbide powders, carbide microcrystals, and carbon microfibers installed with carbide molecules can be formed in the discharge chamber in this case. The possible formation of unnecessary particles should also be taken into account because many metals are effective catalysts due to their formation.

The plasma arc discharge nanoparticle synthesis (PADNSS) also known as the spark erosion method, has been extensively researched and utilized. The arc discharge method is widely used in industry as a method for material synthesis. The principle behind the arc discharge method is based on the use of a spark in a narrow gap to create a short arc discharge (Hontañón et al. 2013). The arc discharge method may be used to synthesize from any raw material as long as the material in question is electrically conductive, regardless of the material's hardness. The method is perhaps one of the most commonly and broadly applied electric-based manufacturing methods. The PADNSS can produce aluminum nanoparticles reaching average particle size diameter ranging from 20 to 30 nm.

The PADNSS uses two electrodes, in the form of a metallic rod positioned in a dielectric solution. Under a sufficiently strong electric field, an arc will be created in the gap between the closely placed electrodes via the "Pincer" effect. With sufficiently high energy, the metallic rod will become extremely hot, producing a region of plasma between those metallic rods (Cong et al. 2018). The plasma region is composed of electrons, positive ions, and uncharged atoms. The temperature of the plasma produced may reach upward 150–200 °C (Stoiljković et al. 2000; Ananthapadmanabhan et al. 2004).

7.2 Methodology

To achieve the objectives of this paper, planning of the overall process must be in conjunction with the collection of data from explored research and research studies. There are several number of steps that were already planned before conducting the research.

Figure 7.1 shows the diagram that represents the workflow of a process for nanofluid synthetization production using certain methods. Various shapes and arrows are used in order to illustrate the process. The main purpose of a flowchart is to analyze different processes. The aims of the flowchart are for analyzing data, make decisions following by step-by-step actions until the entire task is finished. The beginning stage of the research is to select a suitable material as workpieces for the PADNSS. The material that was chosen was Al 6061. The 2nd process is determining the arc discharge parameters used for effective production. The parameters which were chosen are based on voltage, pulse duration, and spark generation when both electrodes touch each other due to the different polarity which is the anode(+) and cathode(−). The 3rd process is doing the machining process using the PADNSS. The last process is the result analysis about the appearance and metallurgy of the specimen. Data calibration takes place when the obtained result does not satisfy the objectives. Hence, the parameter needed to be manipulated. The 4th process is

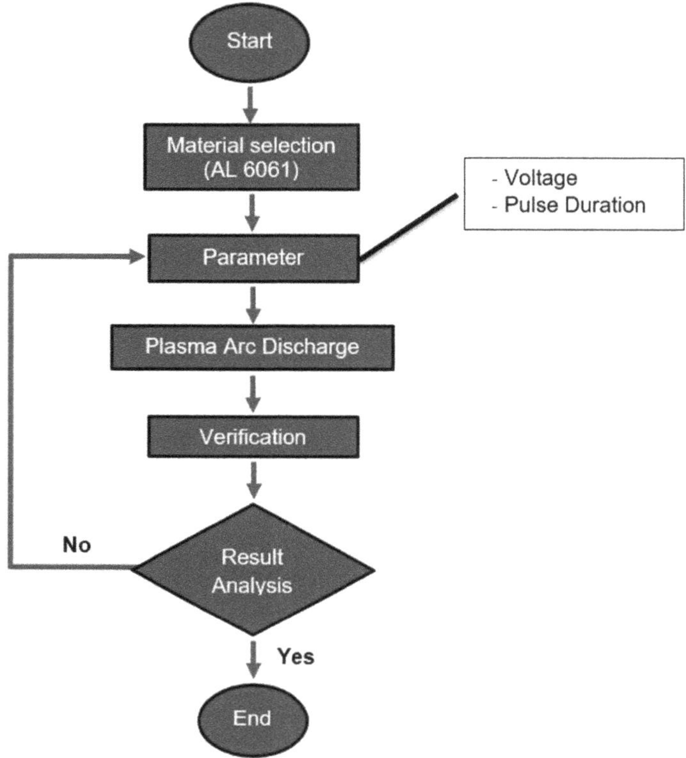

Fig. 7.1 The workflow of a process for nanofluid synthetization

to test the obtained Al_2O_3 powder by using the SEM to determine micrographs of specimens and EDS analysis.

The type of aluminum used was 6061-T6 as shown in Fig. 7.2. The diameter is 1 mm and the length is 200 mm. There are 4 bending points, the angles are 90°. The bending of the electrodes was done by using a vise and mallet in the workshop. The dimension was already measured for each bending points. After that, the electrodes were placed inside the beaker and checked whether the electrodes can fit inside the beaker or not. Bending process was continued until the required design was achieved.

The PADNSS machine is shown in Fig. 7.3a. The complete system of PADNSS includes an integrated circuit and Arduino software, which are shown in Fig. 7.3b. The principle of the PADNSS machine was similar to the EDM concept.

The research was conducted using two aluminum electrodes (Al-6061), each with a diameter of 1 mm. The first step was the preparation of two aluminum rods (Ø1 mm) by bending and accurately measuring the shape to fit inside the beaker. Both act as arc discharging conductivity mechanisms which are positive and negative poles. The other ends of the electrodes were submerged in a medium consisting of 40 ml of deionized water at room temperature as shown in Fig. 7.4. The machines

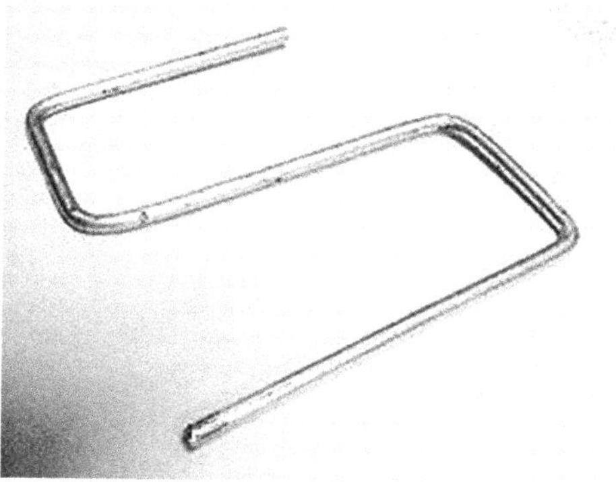

Fig. 7.2 Al 6061 rod

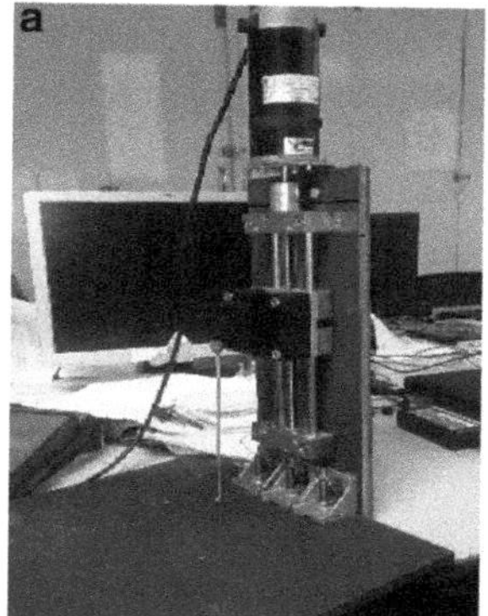

Fig. 7.3 **a** PADNSS machine, **b** complete system of PADNSS

used to directly analyze the structure and morphology of the synthesized aluminum nanoparticles are SEM and EDX analysis. The first system setup is set the DC power supply for 18 V. The voltage value was influenced by the synthesis process to get better concentration. Then, the jumper cables from the power supply terminal, which are positive and negative, were connected to both electrodes.

The top Al electrodes were pushed until they created a frictional force at the bottom of the Al electrode. Thus, creating rubbing friction against each other. Both Al electrodes made contact at each other surface by the applied little force, hence arc or spark erosion occurred. The principle behind the arc discharge method is based on the spark effect at a narrow gap to create a short arc discharge. The Al rods were etched in the aqueous water solution. The Al vapor condensed in the medium, thus creating a stable aqueous suspension. Spark erosion occurs due to friction of the anode and cathode at both electrodes as seen in Fig. 7.5. Hence, well-separated nano-size Al clusters in pure water seem to be thermodynamically stable for a longer time.

Fig. 7.4 Submerged electrode into deionized water

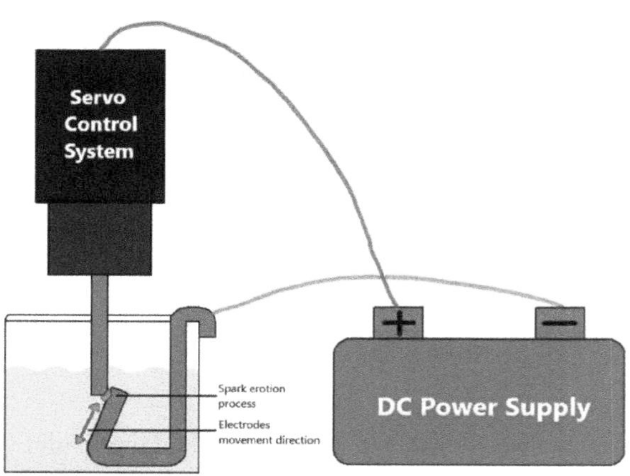

Fig. 7.5 Arch discharge glow and spark erosion

The focal point of local ablation while performing the arc discharge process will vaporize and result in melting. Hence the metallic rod surface undergoes ionization (Ananthapadmanabhan et al. 2004). The ions discharge in water will aggregate, creating metallic nanoparticles which stabilize as a suspension in the liquid dielectric (Chang et al. 2001). The arc discharge process ends when the pulse is discontinued, after which the process enters the rest state. When the pulse is discontinued, the local pressure and temperature will drop, resulting in a return to the original state of isolation between the metallic rod and the workpiece. The local area of arc discharge will be rapidly cooled by the surrounding plasma. After the system returns to the rest state, the arc discharge process may be repeated as necessary. Together, the arc discharge and rest state may constitute a cycle (Yue et al. 2009). This cycle of arc discharge and rest state may be repeated several times in a short duration.

The arc discharge state happens when the gap between the electrodes is narrowed to a distance of between 5 and 50 μm, or if the voltage is increased to raise the strength of the electric field, the isolating effect of the dielectric is greatly reduced, which results in the aggregation of electrons at the tip of the cathode electrode. Once a few electrons break the dielectric barrier and begin to bombard the anode from the aggregation point on the cathode, a discharge current will be created. At this time, the distance between the cathode and anode may be properly described as the arc discharge gap. The voltage and pulse current of the discharge are set at a stable value which are 18 V and 12 A. After the dielectric breaks down, the voltage drops down to 40 V. The peak current is held for the duration of around 50 s.

The next step after finishing all the process is to convert from the liquid to the solid state (powder) due to the limitations of the SEM machine which cannot test liquid specimens. In SEM analysis, the test specimens are dried at 60 °C until they become a crystalline powder which means that all the liquid already dried out. The specimens are directly placed on the surface of an aluminum sheet as the surface naturally undergoes oxidation, EDS analysis will reveal the presence of oxygen from the natural oxidation occurred. The best specimen is based on its composition (wt%) and concentration of liquid.

7.3 Result and Discussion

The analysis is split into three sections. The first section discusses the results of the diameter size distribution of colloidal aluminum nanoparticles. The second section discusses the results of spectral analysis of the resultant colloidal aluminum nanoparticles. The spectral analysis involves the use of an EDS machine to analyze and compare the quality of aluminum nanoparticles produced in an aqueous deionized medium. The last section analyzes the stability and quality of the resultant colloidal aluminum nanofluids by investigating the reasons behind the higher colloidal stability of Al_2O_3.

The specimen must make sure that the electron beam strikes only the aluminum particles in the sample. The colloidal aluminum has been analyzed for the composition of aluminum nanoparticles. EDS analysis is used to analyze the differences in the composite makeup between the smaller and larger particle clusters. This will explain the reason behind the production of bubbles during the synthesis of aluminum nanoparticles in a dielectric composed of pure water.

The analysis result of this research is to investigate the composition (wt%) and the physical appearance or metallurgy or also called SEM micrograph of the aluminum nanofluids synthesized by the PADNSS.

7.3.1 *Micrograph of SEM Images and Cliff-Lorimer Ratio Method*

This section will discuss the stability of the resultant nanofluids by using Cliff-Lorimer Ratio Method. The strong penetrative capability of the SEM is used to analyze the pure water colloidal aluminum samples. The results show 3 spots of micrograph SEM images of aluminum nanoparticles at a resolution or magnification lens at 10,000× operated at 10 kV by the Bruker Quantax EDX.

$$\frac{I_A}{I_B} = k_{AB}^{-1} \frac{n_A}{n_B} \tag{7.1}$$

where

- I_A The integrated EDS peak intensity of the element A.
- I_B The integrated EDS peak intensity of the element B.
- n_A The atomic density (in percentage) of the element A.
- n_B The atomic density (in percentage) of the element B.
- k_{AB} is Cliff-Lorimer sensitivity factor between elements A and B.

At spot 1, the EDS spectrum shown in Fig. 7.6 has emissions that peak at 1.4 keV and 0.26 keV, respectively. These peaks show that both aluminum and oxygen have the highest value of substance composition. We used the Cliff-Lorimer thin ratio section to count the composite ratio of both elements. The results are $n^2{}_A = 2.70$, $n^2{}_B = 1.429$, $k_{AB(Al)} = 25.2$, $k_{AB(O)} = 23.5$, $I_{A(Al)} = 1.4$ keV, and $I_{B(O)} = 0.26$ keV. So, the 1st spot contains 35.02% elements of Al, and 52.52% elements of O. Transformed to equivalent atomic calculation (at.%), there are 23.43% elements of Al, 59.24% elements of O after transforming into equivalent atomic calculation.

At spot 2, the EDS spectrum having emissions that peak at 0.48 keV and 1.4 keV, respectively. These peaks show that both aluminum and oxygen have the highest value of substance composition. Using Cliff-Lorimer thin ratio section to counting the composite ratio of both elements. The results are $n_A = 2.70$, $n_B = 1.429$, $k_{AB(Al)} = 24.5$, $k_{AB(O)} = 30.8$, $I_{A(Al)} = 1.4$ keV, and $I_{B(O)} = 0.48$ keV. So, the 2nd spot contain, 35.07% elements of Al, and 56.71% elements of O. There are 23.72% elements of Al and 64.66% elements of O after transformed into equivalent atomic calculation.

At spot 3, the EDS spectrum having emissions that peak at 0.45 keV and 1.5 keV, respectively. These peaks show that both aluminum and oxygen have the highest value of substance composition. Using Cliff-Lorimer thin ratio section to counting the composite ratio of both elements. The results are $n_A = 2.70$, $n_B = 1.429$, $k_{AB(Al)} = 21.6$, $k_{AB(O)} = 25.9$, $I_{A(Al)} = 1.5$ keV, and $I_{B(O)} = 0.45$ keV. So, the 3rd spot contains 36.20% elements of Al, and 56.48% elements of O. There are 24.70% elements of Al and 65.00% elements of O after transformed to equivalent atomic calculation.

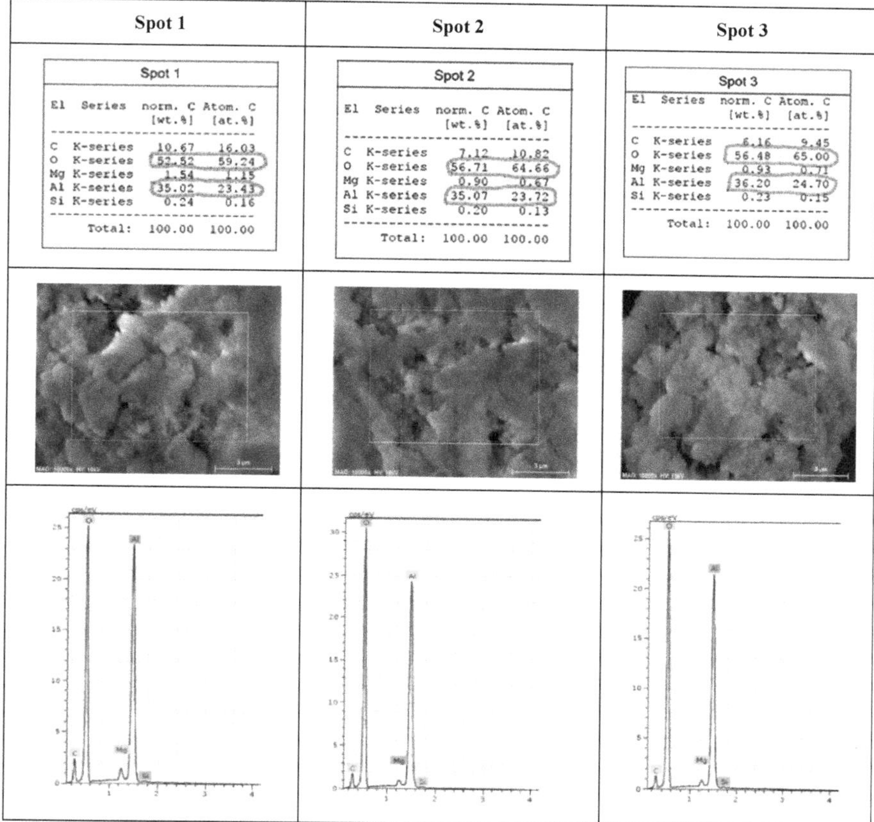

Spot 1

El	Series	norm. C [wt.%]	Atom. C [at.%]
C	K-series	10.67	16.03
O	K-series	52.52	59.24
Mg	K-series	1.54	1.15
Al	K-series	35.02	23.43
Si	K-series	0.24	0.16
	Total:	100.00	100.00

Spot 2

El	Series	norm. C [wt.%]	Atom. C [at.%]
C	K-series	7.12	10.82
O	K-series	56.71	64.66
Mg	K-series	0.90	0.67
Al	K-series	35.07	23.72
Si	K-series	0.20	0.13
	Total:	100.00	100.00

Spot 3

El	Series	norm. C [wt.%]	Atom. C [at.%]
C	K-series	6.16	9.45
O	K-series	56.48	65.00
Mg	K-series	0.93	0.71
Al	K-series	36.20	24.70
Si	K-series	0.23	0.15
	Total:	100.00	100.00

Fig. 7.6 Results of composition, SEM, and EDS for three spots

7.4 Discussion

The EDS spectrum indicates the ionization energy and the ordinate indicates the counts. The higher the counts of a particular element, the higher will be its presence at that point or area of interest. The EDS also displays the amount of each element in number of counts or in weight percentage. Depending on the equipment used, the readings of elements may not be accurate enough for quantitative analysis. But it should be able to indicate the existence of certain materials in the area. In the graph of the EDS spectrum, the X-axis indicates the energy (keV), while the Y-axis indicates the x-ray intensity counts (cps/eV). The highest energy peaks correspond to the Al and O elements in the sample.

From the results above, the graph shows that there is more composition (wt%) of oxygen, compared to composition (wt%) of aluminum. The oxygen contained inside the H_2O, aggregates into H and O. The H oxidation effect occurred and becomes gas, so there is no composition (wt%) of H. Other elements contained in the sample are copper, magnesium, and silicon which represent only a small value of (wt%).

In fact, the composite was found to be the same for all tested specimens from the pure water dielectric, regardless of particle cluster size. Clearly, the final end product synthesized from aluminum electrodes submerged in pure water is Al_2O_3. EDS analysis confirms the assumptions made and explains why large quantities of bubbles were produced during the synthesis of aluminum when pure water was used as the dielectric medium. The mark was produced by an electron beam when analyzing the particle sample during EDS analysis.

This result may explain why aluminum nanoparticles can remain as a suspension in pure water for an extended period of time at room temperature. When this sample is left at rest for a week, the nanoparticles settle out of the medium and the color of the colloid changes from grey to clear. In this study, PADNSS has the possibility to produce more aluminum nanoparticles and more hydrogen gas at the same time by using a larger aluminum rod. The parameters of the arc discharge also have the potential to improve the system. Larger aluminum rods have a more superficial measure for discharge, but the two aluminum rods also have more possibility to weld with each other by the extremely high-temperature plasma. In this reason, the frictional force between the two electrodes must be higher than this case.

The result also was discussed according to respective spots. The PADNSS is able to produce Al_2O_3 nanoparticles with an average size of 22 nm. The advantage of PADNSS in Al_2O_3 nanoparticles is that the aluminum rods were synthesized in pure water. Thus, the dispersion of nanoparticles in the dielectric medium is able to avoid aggregation into large particles. While PADNSS uses pure water as a dielectric medium to synthesize aluminum nanoparticles, the newly formed aluminum nanoparticles were oxidized by water to produce Al_2O_3 nanoparticles.

7.5 Conclusion

This research used the arc discharge technique to fabricate an Al_2O_3 nanoparticle suspension in deionized water. During the research, it was found that when pure aluminum nanoparticles react with water it may become Al_2O_3 nanoparticles and release hydrogen. Traditionally, preparation of electrolyzed water is costly. However, using nano aluminum and react with water produces cheaper cost in producing electrolyzed water.

Hydrogen fuel in the current application must consider the storage and transportation difficulties, whether it is in the hydrogen fuel cell power generation or hydrogen internal combustion engine field. Since hydrogen gas is corrosive and flammable, a good storage technology must be used that is ready for any risks that may cause damage to the equipment. High-pressure cylinders are the most common way for hydrogen fuel storage. Nanotechnology research has found that carbon nanotubes can store hydrogen as hydrogen may be adsorbed in the nanocarbon and become a solid state hydrogen storage device. However, the cost of carbon nanotubes are still very expensive, so this technology depends on the cost of carbon nanotubes.

Furthermore, this study found that the nano aluminum can be regarded as a solid fuel use. This method can become more feasibility with a large amount of recycled aluminum form businesses and industry, which can provide cheap aluminum. Also aluminum recycling could be done at solar photovoltaic farms or wind generator sites. Therefore, Al_2O_3 nanoparticles can remain as a suspension in pure water for an extended period of time at room temperature. Analysis reveals that aqueous-based dielectrics result in nanofluids composed of Al_2O_3 nanoparticles.

Acknowledgement The authors would like to thank the Ministry of Higher Education (MOHE) of Malaysia through Fundamental Research Grant Scheme (FRGS/1/2018/TK03/UNIKL/02/2) for providing the financial support and System Engineering and Energy Laboratory (SEELab) for the guidance of this project.

References

Ananthapadmanabhan, P.V., Thiyagarajan, T.K., Sreekumar, K.P., Venkatramani, N.: Formation of nano-sized alumina by in-flight oxidation of aluminium powder in a thermal plasma reactor. Scr. Mater. **50**(1), 143–147 (2004)

Chang, P.-L., Yen, F.-S., Cheng, K.-C., Wen, H.-L.: Examinations on the critical and primary crystallite sizes during θ- to α-phase transformation of ultrafine alumina powders. Nano Lett. **1**(5), 253–261 (2001)

Cong, H., Li, Q., Du, S., Lu, Y., Li, J.: Space plasma distribution effect of short-circuit arc on generation of secondary arc. Energies **11**(4), 828 (2018). https://doi.org/10.3390/en11040828

Hontañón, E., Palomares, J.M., Stein, M., et al.: The transition from spark to arc discharge and its implications with respect to nanoparticle production. J. Nanoparticle Res. **15**(9), 1957 (2013). https://doi.org/10.1007/s11051-013-1957-y

Sharma, G., Kumar, A., Sharma, S., et al.: Novel development of nanoparticles to bimetallic nanoparticles and their composites: a review. J. King Saud. Univ. Sci. **31**(2), 257–269 (2019)

Stoiljković, M.M., Holclajtner-Antunović, I.: Using the total voltage drop of a stabilized DC arc to assess plasma parameters. Plasmas Ions **3**(1), 83–88 (2000)

Yue, Y., Hao, Y., Feng, Q., Zhang, J., Ma, X., Ni, J.: Study of GaN MOS-HEMT using ultrathin Al_2O_3 dielectric grown by atomic layer deposition. Sci. China Ser. E Technol. Sci. **52**(9), 2762–2766 (2009)

Chapter 8
Three-Dimensional Image Reconstruction for Automated Defect Detection at Artificial Metallic Surface Specimens

Nor Liyana Maskuri, Muhamad Husaini Abu Bakar, and Ahmad Kamal Ismail

Abstract Non-destructive testing (NDT) is a testing and analysing method that is used nowadays for evaluation properties, material, system, or structure for differences characteristic of surface defects without causing any damages on the analysed parts. The flaws on product surfaces caused by some fault in the production lane will downgrade the product quality. The metallic surface defect is now becoming a trend in the world of research and development. That is because the human resource workforce is still applied nowadays to verify the anomalies and may lead to some human error. This paper is focusing on three-dimensional (3-D) image reconstruction for automated defect detection on artificial metallic surface specimens. The development of an algorithm is to reconstruct the model by using visual-based aid. The 3-D images were reconstructed from several infrared image thermograms captured during the experimental and analysis session by a specific developed algorithm. For better result analysis, the sample was drawn in a 2-D CAD program and then they have been analysed. The major finding from the implication and this significant study is increasing the quality of the product, decreasing the consumed time, human limitation, and error. The technology of this visual-based 3-D reconstruction system can overcome the current expensive method in production lanes. This system application will cover a fast quality checking process on a metallic component surface compared to conventional manual inspection. It can be concluded that the 3-D image reconstruction will fasten the automated quality checking process with better defect classification.

N. L. Maskuri (✉) · M. H. Abu Bakar · A. K. Ismail
System Engineering and Energy Laboratory, Universiti Kuala Lumpur-Malaysian Spanish Institute, Kulim Hi-Tech Park, 09000 Kulim, Kedah, Malaysia
e-mail: nliyana.maskuri@s.unikl.edu.my

M. H. Abu Bakar
e-mail: muhamadhusaini@unikl.edu.my

A. K. Ismail
e-mail: ahmadkamal@unikl.edu.my

M. H. Abu Bakar et al. (eds.), *Progress in Engineering Technology III*,
Advanced Structured Materials 148, https://doi.org/10.1007/978-3-030-67750-3_8

Keywords Non-destructive test · Infrared image · 3-D image reconstruction · Defect detection · Image classification

8.1 Introduction

Human technology has witnessed colossal development during several decades from the development of smart technology, a reliable and fast industrial inspection that is the main challenge in manufacturing scenarios (Zhang et al. 2018; Addepalli et al. 2019; Gholizadeh 2016). Every piece of equipment has been researched and developed at full speed in these technology ages. The process of checking the design and the quality product is one of the essential developing technologies.

Infrared thermography performs necessary steps in metallic-based inspection in the manufacturing process. The technology has high efficiency in process accuracy and flexibility with a cost-effective production (Sirikham et al. 2020). Hence, the proper control of specimen parameters is important to deliver an excellent quality product. There are varieties of procedures in active infrared thermography. The transient thermography, square pulse thermography, and lock-in thermography are the three most commonly used methods nowadays (Teza and Pesci 2019; Theodorakeas and Koui 2018). The square pulse thermography concept can be explained by short pulse, short inspection time, short duration time for heating up the specimen using a high heated source (Mansouri and Tavallali 2019). Then, the result is recorded from the thermal response (surface) during the cooling process. The information contains defect characterization with different size and depths. This method will give a fast analysis, but it requires a high peak power for a more in-depth analysis, and there are limitations on surfaces of non-uniform emissivity (Pena and Rapún 2019). Detection of surface flaws depends on raw sequences of the thermal image quality. This variation in surface emissivity affects the thermogram, and there is a need for reconstruction and filtering algorithms to improve the detection of flaws (Marani et al. (2019); Goel et al. 2017).

This research goal was aimed at utilizing the thermal images and a machine vision-based method to improve the quality of an inspection system. That is because this current technology is still facing challenges where it needs to deliver maximum output with the high-quality declaration, especially in metal-based products. As an example, a drilling process may cause some defect on the drilling results such as uneven hole formation, sharp cutting line surface cracks, and others may contribute to the anomalies on the product. The visual inspection is one of the most and standard quality control methods. In the metallic inspection case, the technique involved in sampling test modules for the visual appearance after the completed manufacturing process and the taken samples have to be judged whether the batch is accepted or not. Some of the drawbacks of the manual inspection process are that they are conducted and performed only by qualified quality inspectors. This process involves the challenging control. Furthermore, it is difficult to determine the defect class, and the process is full of distractions and time-consuming. This method is often inadequate with several limitations such as tiredness, lack of competency skills, and diagnostic capability issues in detecting defect variations.

8.2 Methodology

8.2.1 Proposed Method

In this section, at the existing quality control approach, there are too many disadvantages in detecting the metallic defects during the manufacturing process. By human factors, traditional manual inspection is often insufficient and is difficult to control. Moreover, defect in metallic plates are characterized by their tiny size and the detection process has major difficulties to detect them. Furthermore, statistical sampling does not reflect the indicate the milling process real-time condition of reliability to produce smooth metallic surface profiles. Times lapses are significant with analytical sample testing, i.e. the number of damages would already be enormous by the time discrepancies are found, and many products/pieces would be wasted.

Automated metallic surface defect detection using image 3-D reconstruction is the main research focus which includes the development of an algorithm to reconstruct an object from a 2-D image to a 3-D image model. The project focused on algorithm development for 3-D reconstruction using a metallic based plate as an example to detect the surface anomalies, stretching the detailed evaluation on the 3-D reconstructed image, and investigating the benefit of this suggested method compared to the current conventional method.

This system was implemented as a machine vision system and applied to analyse a metallic surface as the next possible approach for monitoring surface defects. The optical image processing method was integrated with the naked human eyes justification, and it was compared to the manual inspection. This high detection capabilities of powerful tools can detect product anomalies, thus increasing the quality checking reliability in the industry.

8.2.2 Experimental Setup

In this section, the experimental setup was fully developed to give the best data acquisition and presentation. Next, the portable thermal infrared camera was used to provide the image sample and able to identify abnormalities four times faster than the in-camera resolution. The thermal camera was selected because it produces thermogram images that have plenty of information compared to another digital image. The 50 mm $\times$ 70 mm defected aluminium alloy T6061 plate was the main item as a product sample. The milling process was chosen to produce various parameters of artificial holes. The images of the samples were taken by placing the specimen on

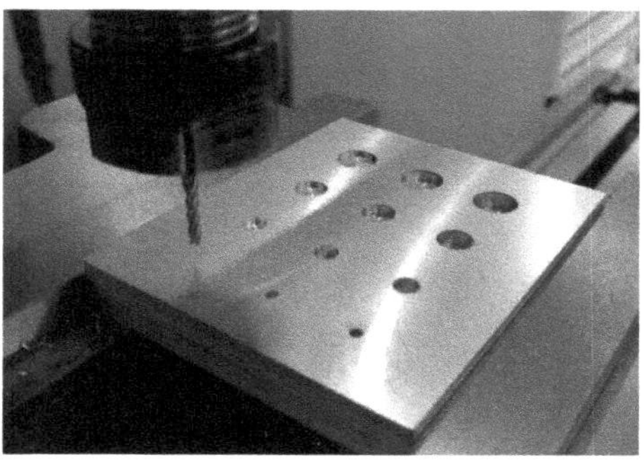

Fig. 8.1 Milling process of metallic specimen with different artificial defects (thickness and size)

a the black box provided to avoid the light that may interrupt the data and produce an additional error. Next, the heating process takes place, and the thermogram image was captured by the specific condition. The thermogram images taken must to fulfil the condition where all of it is condition that all of it is at the same distance, pixel contrast, and amount of light. This specific requirement is needed to provide consistent data for the research. Figure 8.1 is a metallic specimen with different artificial defects of differing depth and size produced by the milling process.

A general plan for the experimentations performed during the project phase shown in Fig. 8.2. The tests were completed using heated specimens each having up to 90 °C for 15 min period of time. The heating procedure was conducted in a microwave oven for the continual and well-preserved temperature. The thermogram images were captured by the infrared camera U5855A TrueIR Thermal Imager. Next, the data acquisition and processing stage were performed using a computer and software provided. The sensitivity of the thermal camera is 0.07 °C and the resolution gain is 320 × 240 in infrared pixels with eight frames per second. Each of the frames is cropped to 280 × 220 resolution as selected specimen region to reduce the time taken

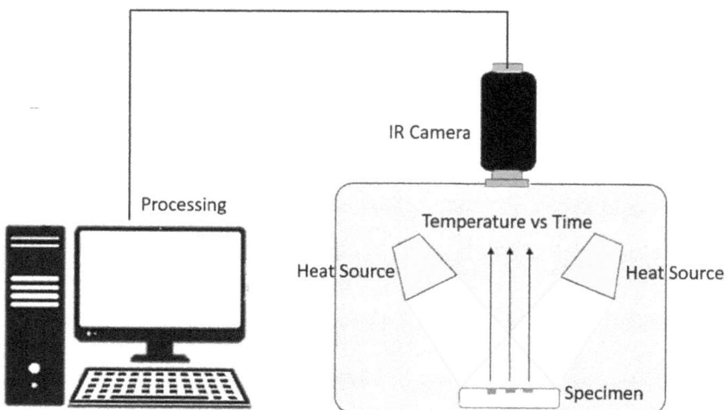

Fig. 8.2 General structure of the system for active pulse thermography

to process the full frame sequences. This region of interest (ROI) is the technique to define the object border consideration.

8.2.3 Simulation Development Using MATLAB

The process of algorithm development is done by using the MATLAB environment. MATLAB is a high-performance mathematical computing language. Computation, visualization, and programming is the combination of an environment where the problem and solutions are presented in conventional mathematical notation. MATLAB is able to run massive simulations in parallel on multicore desktop, computer clusters, or in a cloud without writing lots of code. MATLAB is an interactive framework, the primary data component of which is an array that does not require dimensioning. This allows users to solve many technical computing difficulties, mainly for those with vector and matrix formulations.

8.2.4 Image Processing

The thermographic camera device forms a video and an image from the infrared radiation that is known as the thermogram. For the first step, the camera captured an image based on the considered parameters. The thermogram is in RGB format and extracted to anothers image representation. Pre-processing of the captured image, such as RGB conversion and image enhancement processes, is carried out. The group of images is divided for each class as a reference. The image was cropped by a specific pixel declaration and divided into classes for further image classification of the defects. For a clearer defect presentation, the 2-D defect images were developed and analysed by using a CAD software for better visualization.

8.3 Results and Discussion

8.3.1 Image Presentation

The goal of this project phase was to concentrate on the image processing algorithm, which was developed as a MATLAB software code. Images taken should be simple enough to be processed, and the errors that may occur should be eliminated. The images were represented in 2-D image CAD drawing software to reveal more specific details. The selected 2-D image is cropped to the same size region of interest (ROI) by using MATLAB. Next, the ROI image was grouped into the selected class. Figure 8.3 is an example of a ROI image that was cropped by the same matrix. Some selected

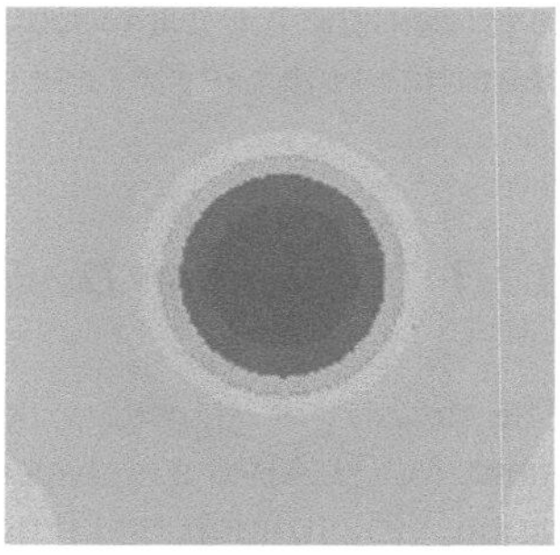

Fig. 8.3 2-D defect image cropped in ROI image

methods of pre-processing of the data are required to compensate for the thermal effect.

8.3.2 3-D Image Reconstruction

In this section, the use of the MATLAB environment has been conducted. All samples of selected images that were carried out from the sample were registered in their separate classes for sampling training and testing purposes. Figure 8.4 shows the method to facilitate the quality assessment according to their respective classes easily. The reference image is introduced in the system as an image that does not have any defects on its surface. The image subtraction method was applied to the selected image with the reference image to get the specific data. Next, image 1 to image 4 are variations of the image introduced in the system and represent their own flaws.

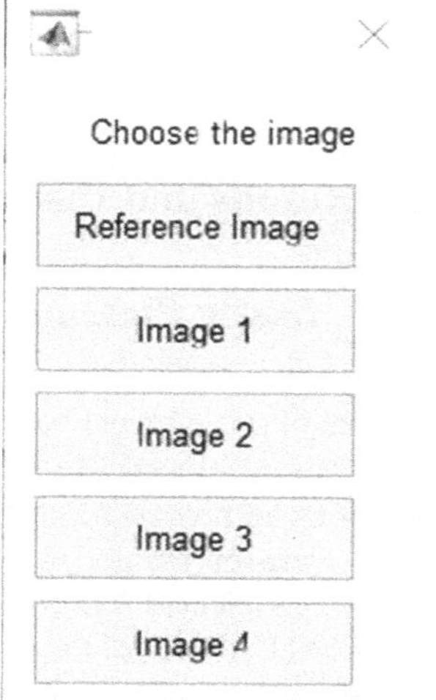

Fig. 8.4 The pop-up message sign to select the image

Figure 8.5 shows the pop-up sign result message that appears when the selected image has not any defect. As a result, through the manufacturing process, this system sign will create better quality control over flawed product escape. This is one of the fast and accurate methods in the quality check system application. The pop-up sign result may have a variety of result appearance following the image type feeded in the system and will decide the following result.

As shown in Fig. 8.6, the image taken is the result of the generated code. The figure is divided into two columns and four rows with the division as such that the reference image and the 3-D reconstruction image, the compared image and 3-D reconstruction image, the image generated after two images being subtracted and the image that is supposed to happen if the image quality check is passed. In this case, the reference 2-D image selected is an image that does not contain any sign of defect, so it will be declared as passing the quality check.

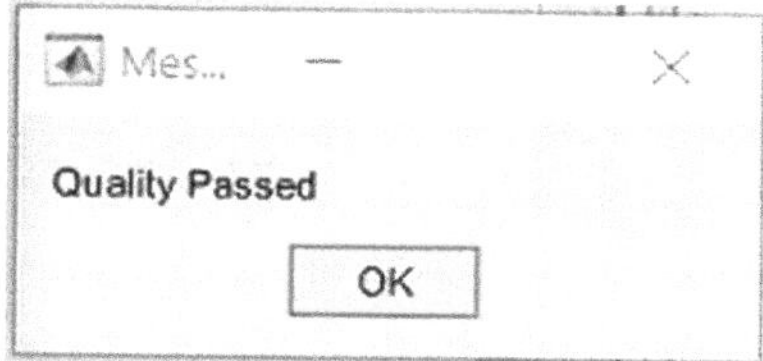

Fig. 8.5 The pop-up message sign appears as quality check passed the tested

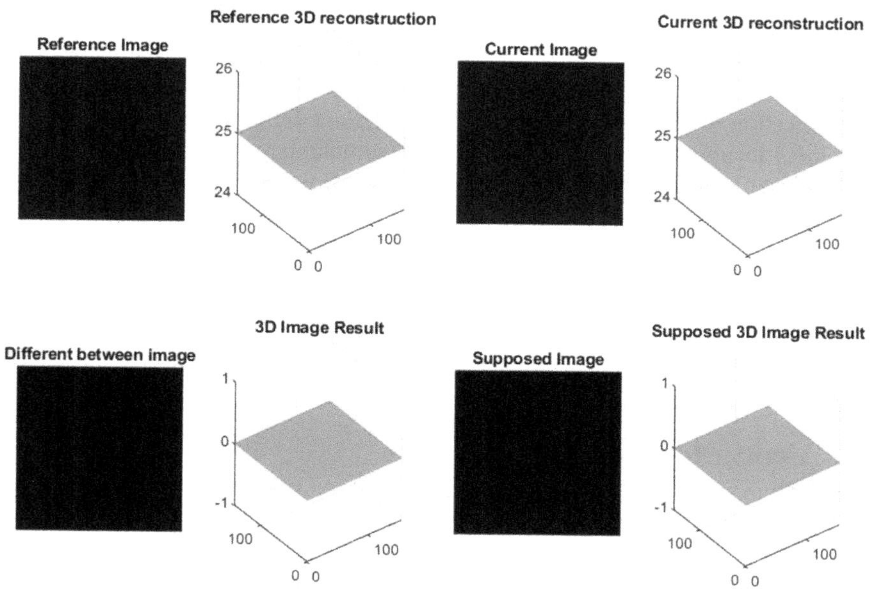

Fig. 8.6 The result of a good product sample image

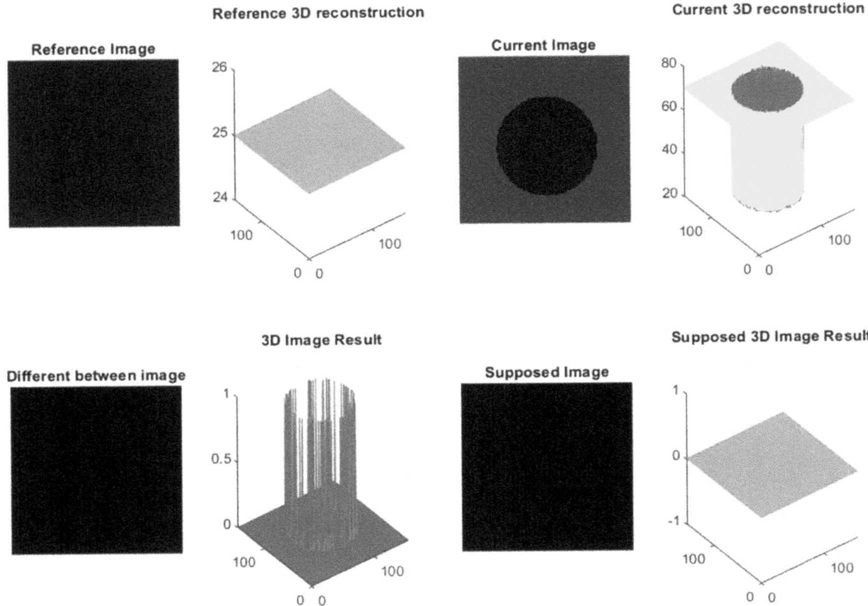

Fig. 8.7 The sample image taken was subtracted from the reference image

Figure 8.7 shows that an image subtraction has to take place if a sign of defect is present in the chosen image. Image subtraction or pixel subtraction is a process whereby the digital numeric value of one pixel or the whole image is subtracted from another image. This is primarily done for detecting changes between two images. This detection of changes can be used to tell if something in the image moved or something inappropriate occurred. The image subtraction results are also shown in Fig. 8.7. The 3-D reconstruction takes place to express the flaws detected in the image taken.

Figure 8.8 shows the result of generating the *XYZ* coordinates. Since the generated table has plenty of rows and column, the list coordinates are very extensive data. The 3-D image reconstruction also can manually select the pixel subtraction appearance as *XYZ* coordinates for better defect depth information by a simple command.

8.4 Conclusions

Quality control and inspection process play an important role in product quality assurance nowadays. Maintaining the high-quality of metallic surfaces with the correct classification sign of any flaws is a critical research area. The difficult process of 3-D image reconstruction of defect detection meets the qualification of ongoing research

Fig. 8.8 Result generated in *XYZ* coordinates

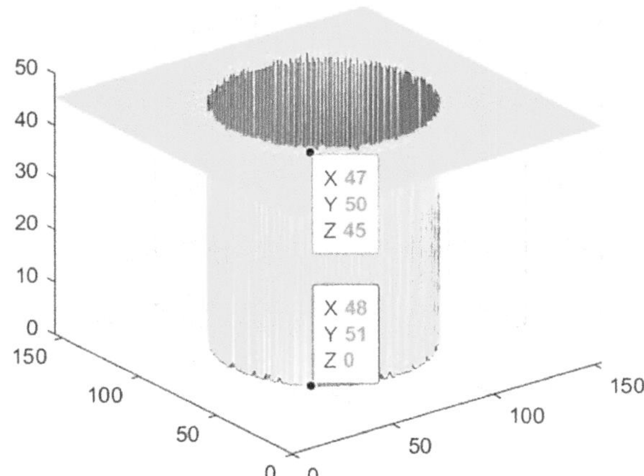

towards the current IR 4.0 era. This research was carried out for surface defect investigation of a metallic component by machine vision imaging techniques implementation. This process will be integrated in the continuous flow of the manufacturing quality checking process. The improvement of the 3-D reconstruction algorithm leads to a better characterization, defined and labelled with the correct defect classification by implementing machine vision imaging techniques. It can be concluded that the application of the developed system covers the 3-D image reconstruction, and defect classification was successfully developed. This creates a better quality control process and reduces the drawback on conventional manual inspection.

Acknowledgements All of the experiment and analysis were conducted at System Engineering and Energy Laboratory (SEELab), Universiti Kuala Lumpur-Malaysian Spanish Institute and received support from Yayasan Tengku Abdullah Scholarship (YTAS).

References

Addepalli, S., Zhao, Y., Roy, R., et al.: Non-destructive evaluation of localised heat damage occurring in carbon composites using thermography and thermal diffusivity measurement. Measurement **131**, 706–713 (2019)

Gholizadeh, S.: A review of non-destructive testing methods of composite materials. Proc. Struct. Integr. **1**, 50–57 (2016)

Goel, V., Singhal, S., Jain, T., Kole, S.: Specific color detection in images using RGB modelling in MATLAB. Int. J. Comput. Appl. **161**(8), 38–42 (2017)

Mansouri, S., Tavallali, S.M.: Heat transfer approximate modeling, parameter estimation and thermography of thermal pulsing in electrofusion joints of gas pipelines. Infrared Phys. Technol. **98**, 354–363 (2019)

Marani, R., Palumbo, D., Galietti, U., Stella, E., D'Orazio, T.: Enhancing defects characterization in pulsed thermography by noise reduction. NDT E Int. **102**, 226–233 (2019)

Pena, M., Rapún, M.L.: Detecting damage in thin plates by processing infrared thermographic data with topological derivatives. Adv. Math. Phys. (2019). https://doi.org/10.1155/2019/5494795

Sirikham, A., Zhao, Y., Liu, H., Xu, Y., Williams, S., Mehnen, J.: Three-dimensional subsurface defect shape reconstruction and visualisation by pulsed thermography. Infrared Phys. Technol. **104**, 103151 (2020)

Teza, G., Pesci, A.: Evaluation of the temperature pattern of a complex body from thermal imaging and 3D information: a method and its MATLAB implementation. Infrared Phys. Technol. **96**, 228–237 (2019)

Theodorakeas, P., Koui, M.: Depth retrieval procedures in pulsed thermography: remarks in time and frequency domain analyses. Appl. Sci. **8**(3), 409 (2018)

Zhang, H., Sfarra, S., Sarasini, F., Santulli, C., Fernandes, H., Avdelidis, N.P., Ibarra-Castanedo, C., Maldague, X.P.V.: Thermographic non-destructive evaluation for natural fiber-reinforced composite laminates. Appl. Sci. **8**(2), 240 (2018)

Chapter 9
Additive Manufacturing Processes, Challenges and Applications: A Review

Farooq I. Azam, Ahmad Majdi Abdul Rani, Muhammad Al'Hapis Abdul Razak, Sadaqat Ali, and Abdul'azeez Abdu Aliyu

Abstract Technology has become a major part of everyday life for all of us, from our smartphones, headsets, wearable technology, tablets to implantable lifesaving medical devices and the premium features in the cars we drive. Each of these cutting-edge products is manufactured using advanced manufacturing technologies to meet with the customer's demand for customized products and services. Additive manufacturing (AM) (also known as 3D printing) is one of these advanced manufacturing techniques that has been revolutionizing the manufacturing industry rapidly in the last few decades. The development of this manufacturing process has removed manufacturing constraints significantly and widened freedom of design. The AM technique manufactures complex geometries with exceptional dimensional accuracy, including assemblies. This chapter presents an extensive review on the available additive manufacturing technologies, challenges involved and their various applications.

Keywords Additive manufacturing · 3D printing · Rapid prototyping · Rapid manufacturing · Layer manufacturing

---

F. I. Azam (✉) · A. M. A. Rani
Mechanical Engineering Department, Universiti Teknologi PETRONAS, Bandar Seri Iskandar, 32610 Perak, Malaysia
e-mail: farooq.i_g03648@utp.edu.my

A. M. A. Rani
e-mail: majdi@utp.edu.my

M. A. A. Razak (✉)
SEELAB, Manufacturing Section, Universiti Kuala Lumpur Malaysian Spanish Institute, Kulim Hi-Tech Park, 09000 Kedah, Malaysia
e-mail: alhapis@unikl.edu.my

S. Ali
Mechanical Engineering Department, School of Mechanical and Manufacturing Engineering (SMME), National University of Sciences and Technology (NUST), Islamabad, Pakistan
e-mail: sadaqat.ali@smme.nust.edu.pk

A. A. Aliyu
Mechanical Engineering Department, Bayero University, Gwarzo Road, Kano, Nigeria
e-mail: aaaliyu.mec@buk.edu.ng

M. H. Abu Bakar et al. (eds.), *Progress in Engineering Technology III*,
Advanced Structured Materials 148, https://doi.org/10.1007/978-3-030-67750-3_9

9.1 Introduction

Additive manufacturing covers an extensive range of processes and techniques with a wide array of capabilities to manufacture the final product. However, what all the different processes and technologies have in common is the way production is carried out, that is one layer at a time to build up the final product. In additive manufacturing, the first step is to generate a 3D solid CAD model of the object. Then this CAD model is sliced into layers to create digital information based on the geometry of individual layer. This digital information is then used to control the tool path which combines the material layer by layer into the final product as shown in Fig. 9.1 (Zhai et al. 2014). AM is not just used for creating prototypes, with improvement in process, hardware and materials it has become possible to manufacture finished products. Several AM processes have been introduced in the market and the ones which were considered the most relevant in the past and have the greatest potential for different industries in the future are described below.

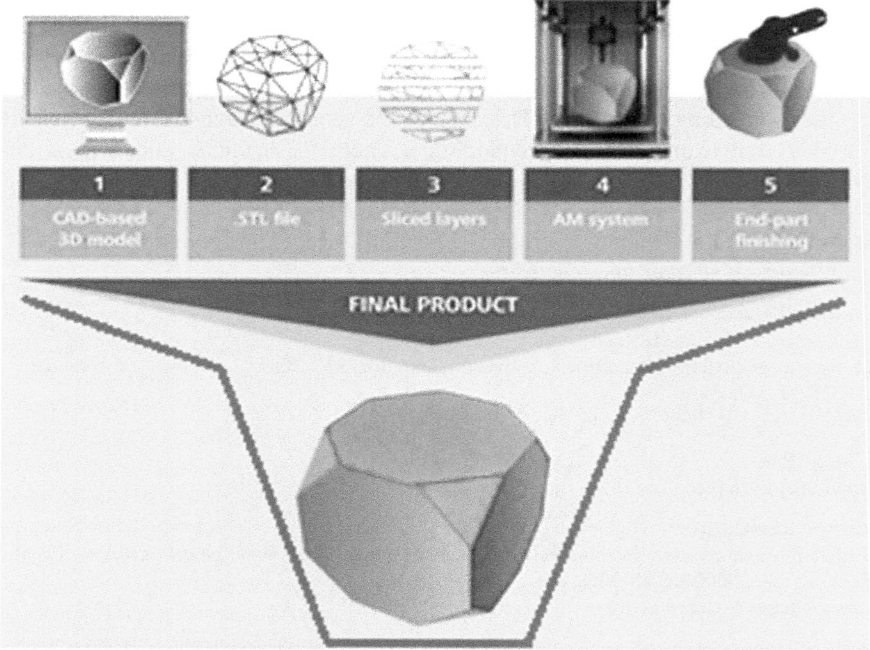

Fig. 9.1 Additive manufacturing process flow

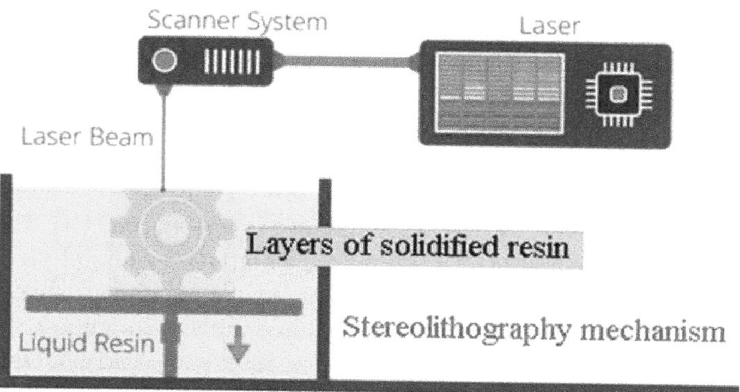

Fig. 9.2 Stereolithography mechanism

9.2 Stereolithography

Stereolithography (SLA) is one of the most commonly used AM processes for rapid prototyping and it is the first commercially available rapid manufacturing technology. SLA was developed by 3D Systems, Inc. in 1986 and it is characterized by photopolymerization of a material. In SLA, a photopolymer resin reacts with the ultraviolet laser and cures to form a solid layer upon layer to produce the final product. The first step in the process is to generate a 3D solid model in a CAD software which is then converted to a STL file in which the model is sliced into different layers containing information regarding each layer. This information will be used later to control the tool path. The part is built on a movable platform which is inside the vat containing the photopolymer resin as shown in Fig. 9.2. The UV laser scans the surface of the vat in X-Y axes per the STL file supplied and the resin solidifies precisely where the UV laser hits the surface. After completion of one layer the built platform descends by one layer thickness in the z axes (25–100 μm) and the built layer is recoated with photopolymer resin. The subsequent layer is then scanned and cured. This process continues until the entire object is complete, the excess is then drained and can be reused.

Once the part is completed, the resin support structure is removed and the cleaned part is put in a UV oven for further curing. After the curing process, finishing is carried out which might require sanding and filing the part to achieve the desired surface finish (Gibson et al. 2014; Melchels et al. 2010).

9.3 Digital Light Processing

Digital light processing (DLP) works on a similar principle as SLA, using a light source to cure photopolymers to manufacture the part. Unlike SLA, which is a bottom-up process, DLP is a top-down process and it is much faster than SLA as an entire layer is cured at once. The DLP printer mechanics are shown in Fig. 9.3.

The DLP printer consists of 3 main components; build platform, resin tank (vat) and the projector. The DLP print process is very simple and is illustrated in Fig. 9.4. As illustrated in Fig. 9.4 the process begins with the build platform lowering down to the vat floor up to one layer thickness (A). The DLP projector which is placed beneath the vat, projects an image slice to cure one layer at a time (B). Then the vat is tilted to separate the cured layer from the vat floor (C). The build platform then rises by one layer height (D). The projector exposes the next image slice and cures the second layer (E). The process repeats until the product is complete (F). One of the advantages of DLP is that it requires a shallow resin vat to carry out the process, so less resin is wasted but similar to SLA the finished parts require post processing like removal of support structures (Wu and Hsu 2015; Bomke 2015; Petrovic et al. 2011).

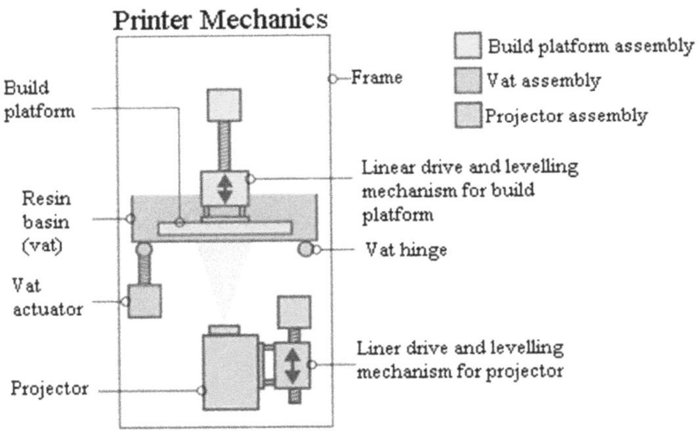

Fig. 9.3 Digital light processing (DLP) mechanism

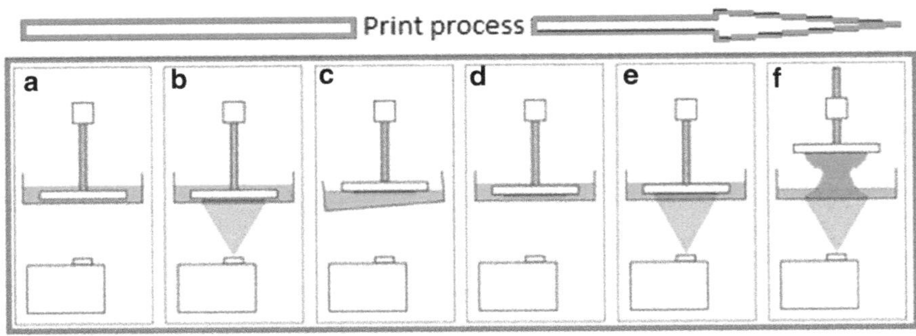

Fig. 9.4 Digital light processing (DLP) print process

Fig. 9.5 Fused deposition modeling (FDM) printer

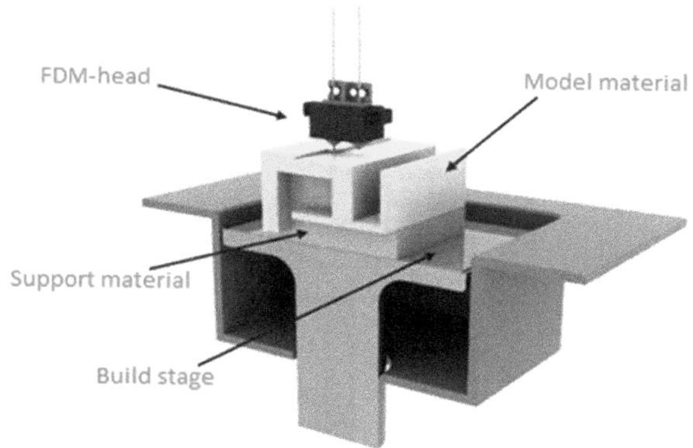

9.4 Fused Deposition Modeling

Fused deposition modeling (FDM) is the most favorable technique for rapid prototyping among all other layer additive manufacturing processes due to its low-cost machinery, ease of operation and durability of final parts. FDM is ideal for concept models, functional prototypes and low volume end-use parts. The FDM process, like other AM processes, begins by slicing the CAD model into layers. The information is then fed to the machine, which constructs the part one layer at a time. A thin thermoplastic filament is fed to the extrusion nozzle, which heats up the material just above its melting point and deposits it on the build platform in a prescribed pattern. The layers harden as soon as they are deposited and bond to the previously deposited layers. The extrusion nozzle continues to move in a horizontal *x-y* plane while the build platform moves down, building the part layer by layer as shown in Fig. 9.5. FDM also consists of a second extruder nozzle, which deposits the support material for overhanging geometries. Most commonly used modeling materials for FDM are ABS and PLA, whereas the support material is usually water soluble and it is easily washed away once the part is completed. Post completion process such as hand sanding can be carried on the completed part to achieve a smooth and even surface (Sood et al. 2012; Sun et al. 2008).

9.5 Selective Laser Sintering

The selective laser sintering (SLS) processes like other AM process begins with creating a 3D CAD model of the part, converting it to a STL file, which can describe the surface geometry of the part by translating it into several small triangles. The model data is then transferred to the build station control software where the part is sliced into thin layers and sent out to the SLS printer. The machine prepares the first layer as the roller spreads a thin layer of the powdered build material across the

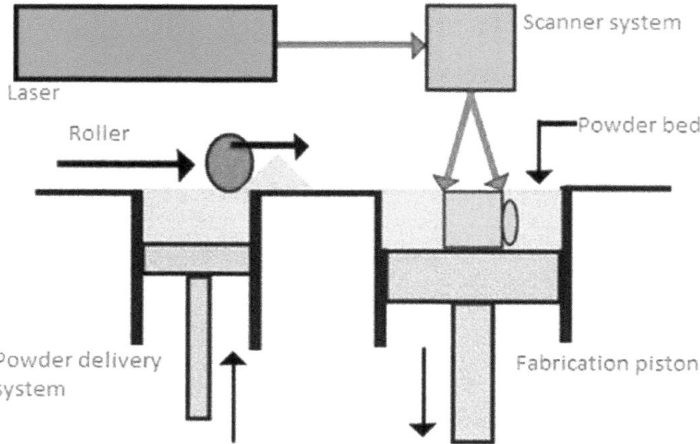

Fig. 9.6 Selective laser sintering (SLS) mechanism

build platform. The laser then scans the cross section on the material in *x*-*y* axes per the 3D data fed to the machine. As the laser scans the surface it heats up the powder material and increases its temperature close to its melting point thereby fusing the powder particles together forming a solid layer. Material which is not part of the model geometry is left unsintered and acts as a support structure. The build platform then lowers by a single layer thickness and the leveling blade sweeps across the build platform and covers it with another layer of powdered material. The laser then selectively sinters the next layer and the process repeats building layer upon layer until the model is completed. Once the part is completed then the powder bed is removed and the part is obtained. The part is then brushed down to remove excess powder surrounding it (Kruth et al. 2003, 2005). The SLS process is illustrated in Fig. 9.6.

9.5.1 Binding Mechanism in SLS

Sintering forms bonds between particles when they are heated. Selective laser sintering and SLS derived technologies use different binding mechanisms as shown in Fig. 9.7. Kruth et al. (2005) classified these binding mechanisms in 4 different categories.

Solid state sintering (SSS) is a thermal process which gives rise to the formation and growth of necks between adjacent powder particles. This process is slow and occurs between the melting temperature of the concerned material (T_M) and $\frac{TM}{2}$. In sintering the atoms of the powder material diffuse across the boundaries of the particles, fusing the particles together and creating a solid piece as illustrated in

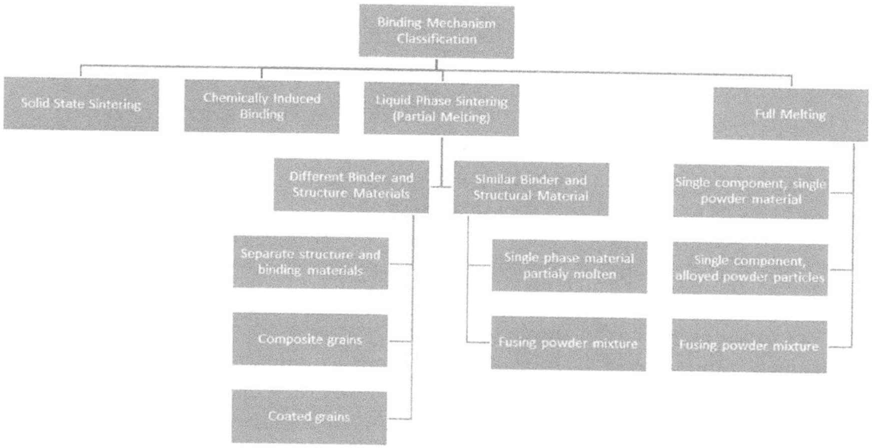

Fig. 9.7 Selective laser sintering binding mechanisms

Fig. 9.8. SSS is a slow process, thus it is necessary to preheat the power material to increase the diffusion rate of atoms (Kruth et al. 2005).

Chemically induced binding is the second consolidation mechanism. Klocke et al. in their study investigated the SLS manufactured SiC ceramic part. During the manufacturing of this part the laser-material interaction time was kept short, which eliminates the possibility of diffusion processes taking place in SSS. Furthermore, no binder element was used during the process. The SIC particles disintegrated into Si and C when heated up at a higher temperature. SiO_2 was formed because of free Si and acted as a binder element between SiC particles. Thus, the parts were composed of a mixture of SiC and SiO_2 (Klocke and Wirtz 1997).

In liquid phase sintering (LPS) partial melting takes place depending on the type of powder used. LPS is further divided into different categories but each consists

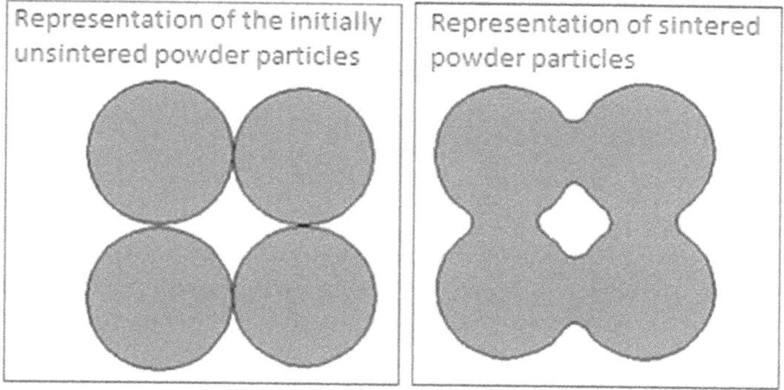

Fig. 9.8 Sintered and unsintered powder particles

of two different materials, one of which is the structural material (unmolten core material) whereas the other one is the binding material. The different categories are as shown in Fig. 9.7.

The importance to increase the density of the final parts and to reduce/eliminate the post processing steps leads to the development of selective laser melting (SLM) which will be discussed further in the following section. In this process the object is built layer by layer by completely melting the powder particles in the built chamber rather than sintering them. SLM achieves full melting of powder particles as compared to direct laser metal sintering (DMLS) which works on the same sintering principle as it is derived from SLS. The group is divided into three subgroups, which are as follows (Kruth et al. 2005); single component, single material powder consists of powder particles made of a single material such as aluminum, titanium, etc. The laser completely melts the powder particles to form the 3D part. Single component, alloyed powder particle consists of powders with alloyed material in each individual grain. Many different kinds of SLM alloys have been tested and are commercially available in the market. In fusing powder mixture the processes are classified into SLM (complete melting of powder particles) or partial melting (some particles melt whereas others do not) based on the degree of melting.

9.6 Direct Metal Laser Sintering

Direct metal laser sintering (DMLS) is derived from SLS and works on the same principle of sintering powder particles layer by layer to manufacture a 3D part. SLS refers to the process being applied to many different materials such as polymers, ceramics, etc., whereas DMLS refers to the sintering process being applied to metals only. A variety of materials are commercially available for DMLS such as aluminum, stainless steel, inconel, etc. (Stratasys Direct Manufacturing: Direct Metal Laser Sintering. Materials. Stratasys Direct Manufacturing). Binding mechanism in DMLS is quite complex and may involve full melting, partial melting, liquid phase sintering or all of the above (Kruth et al. 2005).

9.7 Selective Laser Melting

A lot of research has been carried out on selective laser melting (SLM) and this process has been applied to several different materials such as aluminum, titanium, stainless steel, etc. Rickenbacher et al. (2013) carried out tensile, relaxation and creep test on SLM processed IN738LC material for high temperature application. The tensile test results indicated that SLM processed specimens showed higher strength properties than cast IN738LC. Lore et al. (2010) investigated the effect of SML process parameters such as velocity, hatching space and scanning strategy on the SML processed parts. Aman et al. (2014) studied the effects of two commonly used

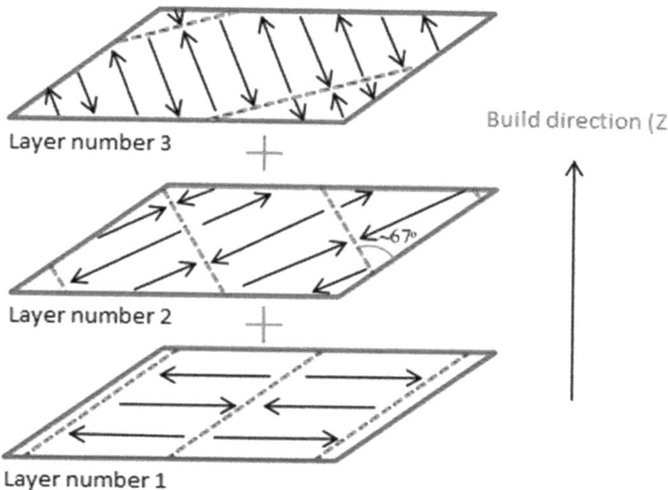

Fig. 9.9 Rotated strip pattern

scan strategies in SLM. The material used was Inconel 625 and the two scanning patterns are as follows.

9.7.1 *Rotated Strip Pattern*

In this pattern, each individual layer is divided into a series of strips (red line in Fig. 9.9), which run across the whole layer (Anam et al. 2014). Raster scan vectors are used within each layer as illustrated in Fig. 9.9. In the rotated strip pattern approach, for each new layer the strips rotate counterclockwise approximately by 67° compared to the previous layer.

9.7.2 *Alternative Block Pattern*

This pattern is similar to a brick stacking sequence. As shown in Fig. 9.10, each layer is divided into rectangular blocks, within each block there are raster scan vectors (Anam et al. 2014). The scan vectors rotate at an angle of 90° between in-line blocks and the blocks themselves shift by half a width between each layer.

Based on the study of different patterns Aman et al. (2014) concluded that SLM is a complex process and the microstructures depend on a variety of different process parameters, but the scanning velocity and scanning strategy significantly affect the orientation of grains. It was also discovered that the elongated grains grow favorably

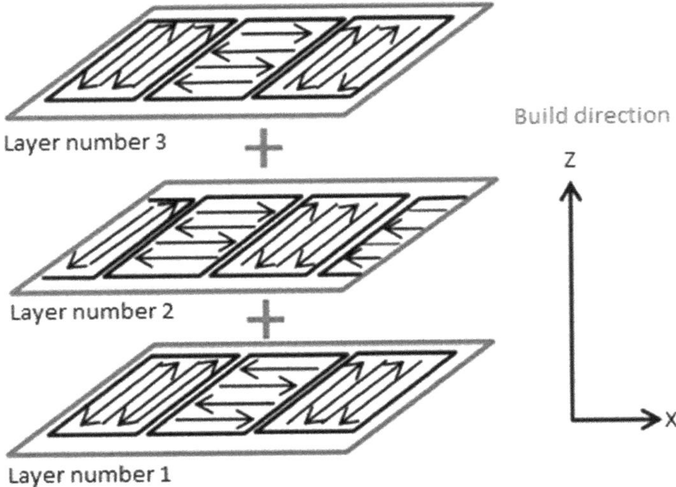

Fig. 9.10 Alternative block pattern

in the build direction. Murr et al. (2009) in his study compared the microstructure and mechanical behavior of SLM and EBM processed parts with conventionally wrought and cast products of Ti-6Al-4 V. The study concluded that EBM and SML processed parts demonstrated similar or superior mechanical properties compared to wrought and cast products of Ti-6Al-4 V. It was also noted that the tensile strength increased by 50% compared to wrought products. Hamza et al. (2016) in his study investigated the effect of different build directions on fracture toughness, tensile strength and density of 316L part processed by SLM. 60% higher ultimate tensile and yield strength was observed for vertically built part as compared to horizontal built part. Even the fracture toughness of 316L stainless steel part built in the vertical direction was 30% higher than the horizontally built part.

9.8 Applications of Additive Manufacturing

Additive manufacturing is fairly new compared to conventional manufacturing techniques, but it has already revolutionized the manufacturing industry. It has significantly changed the designing process, manufacturing and assembly in different industries. Nowadays additive manufacturing is no longer just used for rapid prototyping, but it has ventured successfully into rapid tooling, rapid manufacturing and repair (Azam et al. 2018). AM has been gracefully adopted by aerospace, medical and automotive industry. AM has also made its way into the oil and gas industry with huge success. GE has already begun 3D printing end burners for gas turbine combustion chamber and other companies like Halliburton are actively exploring the possibilities of utilizing additive manufacturing for rapid prototyping and manufacturing fully functional parts. It is estimated that by 2021 the global value for additive manufacturing products would go up to $10.8 billion (SmarTech Publishing 2021; R. A. of Engineering and Engineering Royal Academy 2013). Many different factors have led the transition from rapid prototyping to additive manufacturing. Advancement in technology, reduced cost and broad range of materials have pushed this technology toward manufacturing of consumer products and rapid tooling rather than being just used for rapid prototyping (Campbell et al. 2012; Azam et al. 2018).

9.8.1 Aerospace

Aerospace industry has always been an early adopter and innovator. The innovation in aerospace has always aimed at reducing the cost and weight of parts while maintaining the highest standards without compromising on safety. For the past 20 years, additive manufacturing has been employed by aerospace industry for prototyping, testing concepts and now it is being utilized to manufacture end-use parts (Hiemenz 2013). The forward fuselage of the Boeing F/A-18 comprises of almost 150 parts that have been manufactured through selective laser sintering (Radis 2015). GE Aviation for its passenger jet engine known as "LEAP" is using 3D printed fuel nozzles as shown in Fig. 9.11. LEAP consists of 19 3D printed fuel nozzles which are five times stronger than the previous model. In 3D printing these nozzles allowed engineers to simply design and reduce the number of brazes and welds from 25 to just 5 (Kellner 2014).

NASA is breaking grounds with 3D printing in space to facilitate its astronauts. NASA's space station 3D printer built a ratchet wrench (Fig. 9.12) along with 20 other objects during its first phase of operations. NASA is exploring the opportunity

Fig. 9.11 3D printed fuel nozzle

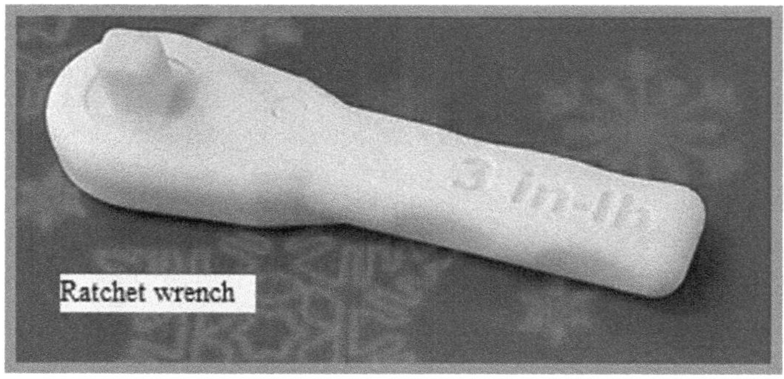

Fig. 9.12 3D printed ratchet wrench

of 3D printing in space for tool manufacturing as well to manufacture objects which previously could not be launched to space (Harbaugh 2015).

TWI employed the LMD process to manufacture a helicopter engine combustion chamber as shown in Fig. 9.13. The component consists of overhanging geometries but it was built without support structures by utilizing the 5-axes of the LMD printer. The thin walled part showed a density of more than 99.5%. The part was built in 7.5 h with 70% powder efficiency (Hauser 2014).

In academic literature Seabra et al. (2016) optimized the topology of an aircraft bracket (Fig. 9.14) to be manufactured using SLM. Compared to the original part, the new part had 54% reduced material volume and weigh 28% less though the material was changed from aluminum to titanium which resulted in increased factor of safety by 2.

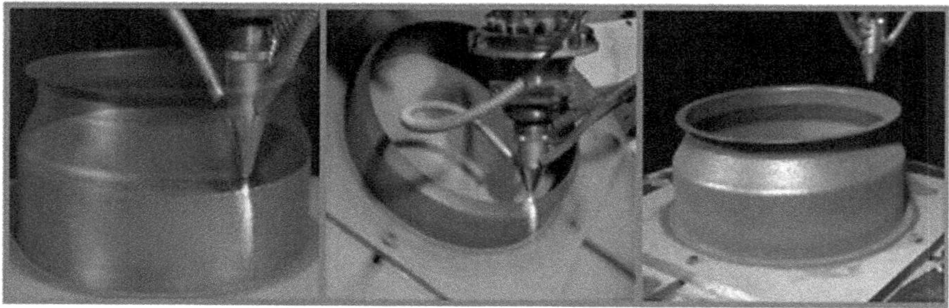

Fig. 9.13 LMD processed helicopter engine combustion chamber

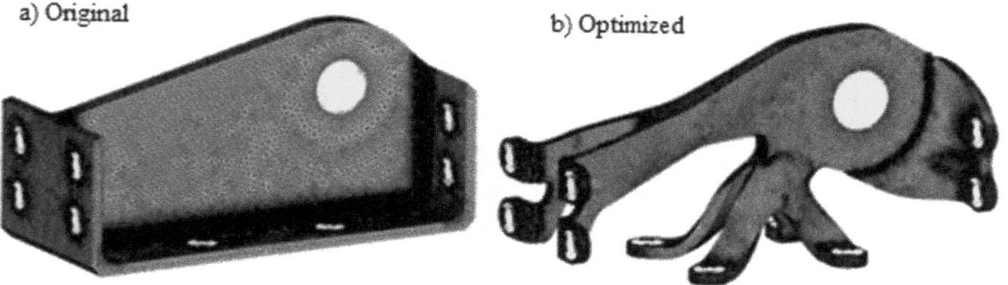

Fig. 9.14 **a** Mesh of original component, **b** mesh of optimized component

9.8.2 Medical

In medical industry customization is really favored as the products must be tailored fitted for each patient, available on demand and at a reasonable price. Additive manufacturing fulfills all these demand, thus it is highly preferred by the medical industry. Additive manufacturing offers patient-specific parts which are strong and lightweight consisting of lattice structures as shown in Fig. 9.15. Such structure was near impossible to manufacture using conventional technologies. Alphaform AG with its partnering firm Novax DMA manufactured a cranial implant for an Argentinian patient. The patient needed a large implant after a stroke-related surgery. A lot of different factors had to be considered such as it should fit perfectly, allow for fusion of bone tissues of the skull, etc. Nova DMA manufactured a lattice structure implant which was capable of meeting all the required characteristics. The implant was 95% porous, which facilitated the flow of fluid though it with least possible resistance and allowed for bone tissue fusion with the outer edges. The implant as shown in Fig. 9.15 was manufactured using EOSINT M 280 (DMLS) (Concept Laser: ConceptLaser—Fast! Direct Components in Vehicle Construction).

In 2014 the hip bone of a 15-year-old patient from Croatia was adversely affected by an aggressive bone cancer. The doctors had no choice but to reconstruct a portion

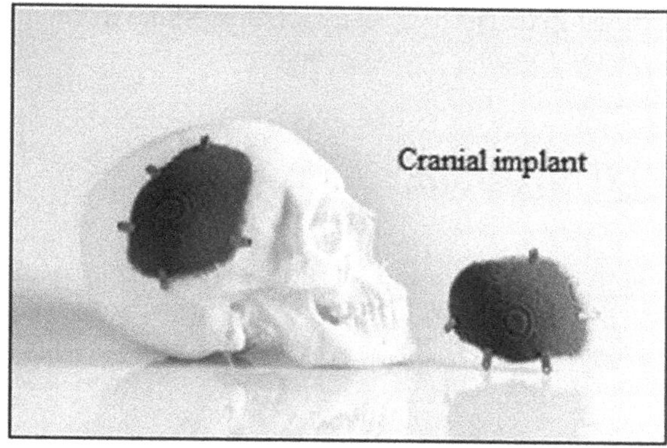

Fig. 9.15 Titanium lattice cranial implant manufactured using additive manufacturing

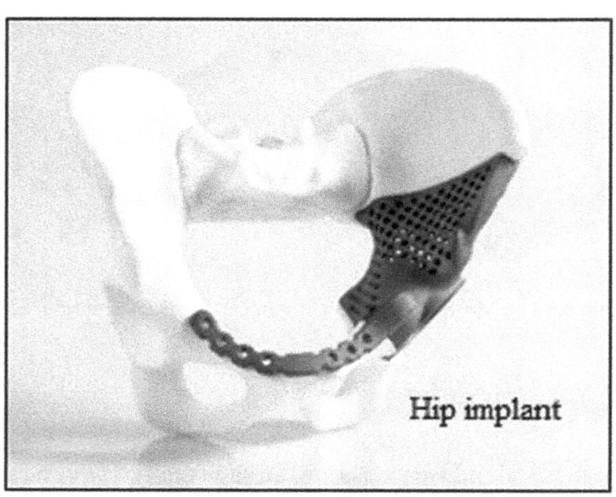

Fig. 9.16 Titanium hip implant

of his hip bone after removing the effected bone tissues. The challenges were that the replacement bone had to be precise and as close to the original as possible so that it perfectly fits and all the angles match one another. Another challenge was to make it strong and lightweight. Alphaform manufactured the replacement bone using additive manufacturing with great success. The replacement hip bone fulfilled all the medical requirements and it is illustrated in Fig. 9.16. The implant consists of a large number of cavities which helped in reducing the weight and this perfect balance of strength and weight was only possible due to additive manufacturing technology (EOS: Medical: alphaform—production of hip implant by using additive manufacturing).

Adler Ortho Group is an Italian manufacturer of orthopedic implants and is widely known for its innovative designs. Utilizing Arcam's EBM technology they manufactured the Fixa Ti-Por acetabular cup shown in Fig. 9.17 and launched it commercially in 2007 after being awarded the CE certification. The new acetabular cup promotes bone ingrowth and after its launch within one year 1000 acetabular cups were implanted with excellent post-op feedback from surgeons (rapidNews: The CE-certified Fixa Ti-Por Acetabular Cup. Manufactured with an Integrated Trabecular Structure for Improved Osseointegration).

9.8.3 Automotive

Additive manufacturing can drastically reduce weight and increase the strength to weight ratio, thus its highly ideal for automotive industry. Automotive industry has been using AM for new design and manufacturing of lighter, stronger and safer products with reduced manufacturing time and cost. AM is highly beneficial for

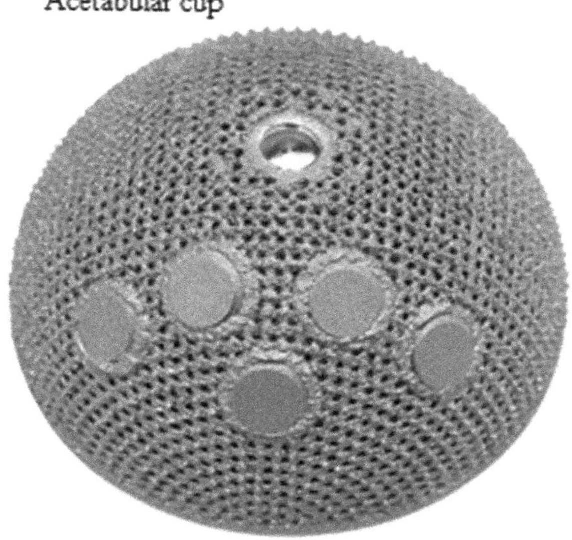

Fig. 9.17 Fixa Ti-Por acetabular cup

manufacturing of small quantities of parts such as gearbox, driveshafts, etc., for luxury or motorsports vehicles (low volume vehicles) (Guo and Leu 2013).

In product development, rapid prototyping has become a standard practice. Designers and engineers at AMP Research printed a prototype of a fuel door using the Dimension 3D Printer by Stratasys. The 3D printer is based on the FDM technology and the fuel door, shown in Fig. 9.18, for General Motors' Hummer H2 sport utility vehicles were manufactured using ABS plastic. The prototype allowed the engineers to test and evaluate the model before finalizing and sending it out for production (Stratasys: AMP Research From Concept to Reality in Record Time. Stratasys).

Ducati is well known for its innovative design, unique engines and advanced engineering in the motorbike industry. By introducing FDM technology by Stratasys into

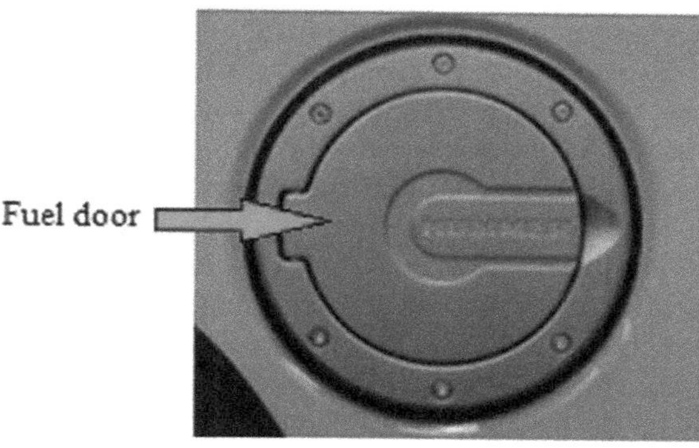

Fig. 9.18 Fuel door prototype for Hummer 2 sport utility vehicle

their design and prototyping phase, Ducati reduced the design and validation cycle by a considerable amount, thus consequentially reducing the time for launching of new products in the market. FDM has enabled Ducati to build functional prototypes and conceptual models from ABS, polycarbonate, etc. During the development process of Ducati's Desmosedici race bike's engine, the design team was able to cut 20 months by utilizing FDM and completed the designing and assembling of the engine in just 8 months. Previously the engines would have taken 28 months from designing to final assembly as most of the prototypes were outsourced (Motorcycle Maker Cuts Development Time with 3D Printing. Stratasys).

Concept Laser manufactures many different automotive parts from materials such as stainless steel, aluminum alloys, titanium alloys, etc. The advantages of using additive manufacturing for automotive parts include reduced manufacturing time, design freedom, no tooling changeover cost and overall cost (Concept Laser 2015; Concept Laser: ConceptLaser—Fast! Direct Components in Vehicle Construction).

9.8.4 Oil and Gas

Additive manufacturing is still in the early stages of adoption in oil and gas industry but it has a great potential for all sectors of this industry (upstream, midstream and downstream). Based on their new report "Additive Manufacturing Opportunities in Oil and Gas Markets" SmarTech Publication believes that AM will cut cost radically for oil and gas industry and it will open up a new era of opulence and opportunities. "I think there is a widespread understanding that the technology has the potential to be just as disruptive for oil and gas companies as it has been for the aerospace, automotive, healthcare and consumer electronics companies of the world. And that's really exciting to me" says 3D Systems CEO Avi Reichental (Peters 2015; Sher 2016). Magma Global and Victrex are taking additive manufacturing for oil and gas industry to a new level with their new flexible m-pipe. Using laser sintering and Victrex's PEEK thermoplastic composite material, they have 3D printed the new flexible m-pipe for use in oil and gas industry. The pipe which will be used for a hydraulic oil and gas pump is capable of being deployed at a depth of 10,000 feet and can handle high pressure and flow rate (Alec 2016; Victrex 2016).

Two companies Furgo and 3D at Depth came together to ensure that an abundant subsea well off the coast Oceania, Australia could fully reach final abundant status. The well in question was drilled decades ago and it was 110 m below the surface of the ocean. Almost no information was available regarding the wellheads, for example measurements, manufacturer, etc., and all this information was essential to acquire the abandoned status. To acquire this information the services of 3D at Depth were hired and using subsea LiDAR scanning 3D data regarding the damaged and outdated wellheads was collected. Using this data, a 1:1 model of the damaged part was printed using a FDM 3D printer. The model helped Furgo to fabricate tools for removing the existing cap and completing the decommissioning project (Goehrke 2016).

GE Oil and Gas at its Kariwa plant in Niigata prefecture, Japan has been utilizing a hybrid metal laser sintering 3D printer (LUMEX Avance-25 metal 3D printer) to manufacture different parts for company's Masoneilan control valve, which is used for several different applications in the energy industry. GE Oil and Gas has been relying on AM for manufacturing these parts because AM offers freedom of design and intricate shapes can be manufactured with ease within a short time. These parts can be designed and fabricated in a matter of weeks compared to months that conventional manufacturing would take (Krassenstein 2015).

Mentioned above are just few examples of additive manufacturing in the oil and gas industry. The oil and gas industry would be the next big adopter of this technology. Companies like GE and Haliburton have already invested in this technology and are actively exploring its possible applications in oil and gas. Furthermore, rising costs, plummeting energy price and long delay in projects are few factors that are shifting attention toward the use of AM in O&G industry.

9.9 Conclusion

With growing interest of AM in various industries, the technology is constantly being developed and commercialized everyday. The purpose of AM technologies in industry ranges from rapid prototyping to producing end user functional products. Many AM processes are available today with the ability to control input parameters to produce a desired final product. The process parameters of the AM process affect the quality and properties of the final product and can be adjusted for a specific application, depending on end-use requirements. Therefore, sound understanding of specifications of the final product is necessary to choose the right AM process.

References

Alec: Magma Global and Victrex 3D print 10,000 foot long PEEK subsea m-pipe for oil and gas production (2016)

Alsalla, H., Hao, L., Smith, C.: Fracture toughness and tensile strength of 316L stainless steel cellular lattice structures manufactured using the selective laser melting technique. Mater. Sci. Eng., A **669**, 1–6 (2016)

Anam, A., Dilip, J.J.S., Pal, D., Stucker, B.: Effect of scan pattern on the microstructural evolution of Inconel 625 during selective laser melting. In: International Solid Freeform Fabrication Symposium—An Additive Manufacturing Conference, pp. 363–376 (2014)

Azam, F.I., Majdi, A., Rani, A., Altaf, K., Rao, T.V.V.L.N.: An in-depth review on direct additive manufacturing of metals. IOP Conf. Ser. Mater. Sci. Eng. (2018)

Azam, F.I., Majdi, A., Rani, A., Altaf, K.: Experimental and numerical investigation of six-bar linkage application to bellow globe valve for compact design. Appl. Sci. (2018)

Bomke, P.: Monkeyprint—an open-source 3d print software (2015)

Campbell, I., Bourell, D., Gibson, I.: Additive manufacturing: rapid prototyping comes of age. Rapid Prototyp. J. **18**(4), 255–258 (2012)

Concept Laser: ConceptLaser—Fast! Direct Components in Vehicle Construction
Concept Laser: What is LaserCUSING?. Concept Laser (2015)
Stratasys Direct Manufacturing: Direct Metal Laser Sintering. Materials. Stratasys Direct Manufacturing
EOS: Medical case study: EOS additive manufacturing for cranial, jaw and facial bone implants
EOS: Medical: alphaform—production of hip implant by using additive manufacturing
Gibson, I., Rosen, D., Stucker, B.: Additive Manufacturing Technologies (2014)
Goehrke, S.A.: In World First, LiDAR Scanning and a 3D Printed Part Help Subsea Well Reach Final Abandoned Status. 3DPrint.com (2016)
Guo, N., Leu, M.C.: Additive manufacturing: technology, applications and research needs. Front. Mech. Eng. **8**(3), 215–243 (2013)
Harbaugh, J.: Space Station 3-D Printer Builds Ratchet Wrench to Complete First Phas (2015)
Hauser, C.: Case Study: Laser Powder Metal Deposition Manufacturing of Complex Real Parts (2014)
Hiemenz, J.: Additive Manufacturing Trends in Aerospace. FOR A 3D WORLD White Paper (2013)
Kellner, T.: Fit to print: new plant will assemble world's first passenger jet engine with 3D printed fuel nozzles, next-gen materials—GE reports. GE Reports (2014)
Klocke, F., Wirtz, H.: Selective laser sintering of ceramics. In: Proceedings of International Conference on Laser Assisted Net Shape Engineering (1997)
Krassenstein, B.: GE Oil and Gas Now 3D Printing End-Use Control Valve Parts in Japan. 3DPrint.com. 3D Printing Business (2015)
Kruth, J.P., Wang, X., Laoui, T., Froyen, L.: Lasers and materials in selective laser sintering. Assem. Autom. **23**(4), 357–371 (2003)
Kruth, J.-P., Mercelis, P., Vaerenbergh, J., Froyen, L., Rombouts, M.: Binding mechanisms in selective laser sintering and selective laser melting. Rapid Prototyp. J. **11**(1), 26–36 (2005)
Melchels, F.P.W., Feijen, J., Grijpma, D.W.: A review on stereolithography and its applications in biomedical engineering. Biomaterials **31**(24), 6121–6130 (2010)
Murr, L.E., et al.: Microstructure and mechanical behavior of Ti-6Al-4V produced by rapid-layer manufacturing, for biomedical applications. J. Mech. Behav. Biomed. Mater. **2**(1), 20–32 (2009)
Peters, G.: 3D printing: futuristic tech set to disrupt oil and gas industry—Offshore Technology (2015)
Petrovic, V., Vicente Haro Gonzalez, J., Jordá Ferrando, O., Delgado Gordillo, J., Ramón Blasco Puchades, J., Portolés Griñan, L.: Additive layered manufacturing: sectors of industrial application shown through case studies. Int. J. Prod. Res. **49**(4), 1061–1079 (2011)
R. A. of Engineering and Engineering Royal Academy of, "Additive manufacturing : opportunities and constraints". R. Acad. Eng. 21 (2013)
Radis, L.: Boeing Files Patent for 3D-printed Aircraft Parts. 3D Printing (2015)
rapidNews: The CE-certified Fixa Ti-Por Acetabular Cup. Manufactured with an Integrated Trabecular Structure for Improved Osseointegration
Rickenbacher, L., Etter, T., Hövel, S., Wegener, K.: High temperature material properties of IN738LC processed by selective laser melting (SLM) technology. Rapid Prototyp. J. **19**(4), 282–290 (2013)
Seabra, M., et al.: Selective laser melting (SLM) and topology optimization for lighter aerospace componentes. Procedia Struct. Integr. **1**, 289–296 (2016)
Sher, D.: The Oil and Gas Industry as a New Opportunity for Additive Manufacturing. Smartech (2016)
SmarTech Publishing: Revenue for 3D Printing From the Oil and Gas Industry to Reach $450 Million by 2021. SmarTech Markets Publishing (2016)
Sood, A.K., Equbal, A., Toppo, V., Ohdar, R.K., Mahapatra, S.S.: An investigation on sliding wear of FDM built parts. CIRP J. Manuf. Sci. Technol. **5**(1), 48–54 (2012)
Motorcycle Maker Cuts Development Time with 3D Printing. Stratasys
Stratasys: AMP Research From Concept to Reality in Record Time. Stratasys

Sun, Q., Rizvi, G.M., Bellehumeur, C.T., Gu, P.: Effect of processing conditions on the bonding quality of FDM polymer filaments. Rapid Prototyp. J. **14**(2), 72–80 (2008)

Thijs, L., Verhaeghe, F., Craeghs, T., Van Humbeeck, J., Kruth, J.P.: A study of the microstructural evolution during selective laser melting of Ti-6Al-4 V. Acta Mater. **58**(9), 3303–3312 (2010)

Victrex: Magma and Victrex partnership breaks record of longest mpipe (2016)

Wu, G.H., Hsu, S.H.: Review: polymeric-based 3D printing for tissue engineering. J. Med. Biol. Eng. **35**(3), 285–292 (2015)

Zhai, Y., Lados, D.A., Lagoy, J.L.: Additive manufacturing: making imagination the major limitation. JOM **66**(5), 808–816 (2014)

Chapter 10
Using a Correction Factor to Remove Machine Compliance in a Tensile Test on DP1000 Steel Validated with 2D Digital Image Correlation Technique

Nurrasyidah Izzati Rohaizat, Khaled Alharbi, Christophe Pinna, Hassan Ghadbeigi, Dave N. Hanlon, and Ishak Abdul Azid

Abstract This paper presents the procedure for the machine compliance removal in the obtained force-displacement data from a tensile test in order to generate a correct stress-strain curve for any material. A tensile test has been conducted on a dual phase steel using an ATSM standard 12.5 mm flat-type tensile specimen geometry and the aim is to produce the correct stress-strain curve without the effect of machine compliance. A correction factor has been developed and is applied to the obtained force-displacement response of the tested DP1000 steel specimen. The strain result obtained from the stress-strain curve with applied machine stiffness correction factor is then validated against the strain result obtained using a virtual strain gauge of digital image correlation (DIC). Strain results obtained from both procedures, through applying correction factor to data and DIC technique are almost similar with a small difference of 0.21%. The implementation of the correction factor procedure in tensile response data are described in this contribution.

N. I. Rohaizat (✉) · I. A. Azid
Universiti Kuala Lumpur, Malaysian Spanish Institute Kulim Hi-Tech Park, 09000 Kulim, Kedah, Malaysia
e-mail: nurrasyidah.izzati@unikl.edu.my

I. A. Azid
e-mail: ishak.abdulazid@unikl.edu.my

N. I. Rohaizat · K. Alharbi · C. Pinna · H. Ghadbeigi
Department of Mechanical Engineering, The University of Sheffield, Mappin Street, Sheffield S1 3JD, UK
e-mail: c.pinna@sheffield.ac.uk

H. Ghadbeigi
e-mail: h.ghadbeigi@sheffield.ac.uk

K. Alharbi
Taibah University, Janadah Bin Umayyah Road, Tayba, Medina 42353, Saudi Arabia
e-mail: kaufe@taibahu.edu.sa

D. N. Hanlon
Tata Steel Research, Development and Technology, 10.000, 1970 CA Ijmuiden, The Netherlands
e-mail: dave.hanlon@tatasteel.com

M. H. Abu Bakar et al. (eds.), *Progress in Engineering Technology III*, Advanced Structured Materials 148, https://doi.org/10.1007/978-3-030-67750-3_10

Keywords Machine compliance · Stress-Strain curve · Tensile test · Dual phase steel · Digital image correlation

10.1 Introduction

The significant effect of machine compliance can be observed when doing a tensile test. Results will be affected in the strain reading of the obtained stress-strain curve. As addressed by a commercial universal testing machine, Instron, when a testing machine is subjected to loading, the components assembled with the system such as specimen, specimen grips, load cell, machine body frame will undergo a small elastic deformation and deflections. The deformation undergone by the machine is called machine compliance. The machine compliance effect may not be noticeable at the machine structure; however, it may greatly affect the tensile response result, which may lead to inaccurate reporting on mechanical properties of tested materials.

There are several methods which can be used to extract only the specimen's elongation, such as applying strain gauges, laser extensometer, digital image correlation techniques, etc. These methods will either be attached directly to the specimen or focused on the specimen gauge length, thus eliminating any external deformations from the machine. However, if any of these devices are not available, another method can be used to generate the correct specimen strain reading for any material with a known Young's modulus. This method will be explained further in this paper and will be called a correction factor.

This paper will later discuss on the procedure in obtaining the correction factor and to apply the correction factor on the obtained force-displacement data of a tensile test. A standard 12.5 mm width geometry of tensile specimen of DP1000 steel has been used for obtaining the force-displacement data. An optical 2D digital image correlation technique has also been set-up during the tensile test to validate the specimen strain result from using the correction factor.

10.2 Literature

The experimental data obtained from the tension testing need to be corrected due to the effect of the machine stiffness. According to Hockett and Gillis, the machine stiffness in mechanical testing is identified as the ratio of the applied load to the deflection of all of the elements of the load train except the specimen gauge length (Hockett and Gillis 1971). It is important to remove the effect of machine stiffness during the elastic interaction between specimen and testing machine to produce a correct result from conducting mechanical testing. It is suggested that in ASTM standards that the frequent approach being employed is to consider that the machine stiffness behaves as a linear spring (Kalidindi et al. 1997). It is also reported that the determination of machine compliance value is important as it has significant influence in the macro-range, thus the uncertainty may affect the reliability in the obtained results (Ullner et al. 2010). In another study involving the shear punch test, it has been found that cross-head displacements are larger compared to the displacement of punch tips at a magnitude greater than an order that has been proven using FEA simulation. The difference in displacement also suggests that it is due to the machine compliance. Their works reported that the compliance of the system (machine) can be measured using an elasticity calculation and is proven to be in an agreement with FEA result (Toloczko et al. 2000).

Digital image correlation is one of the excellent techniques for measuring deformation fields. Full-field displacements up to sub-pixel accuracy can be achieved through processing undeformed and deformed sets of images. Identification of sheet metals mechanical properties has also been obtained using the 2D DIC system (Nguyen et al. 2017). Evolution of deformation and strain in a uniaxial test to study single lap joints up to failure has also been carried out using a 2D and 3D DIC system and has been found to have a good correlation between data from strain gauge for higher level of deformation (Comer et al. 2013). It has been reported that the results from the 2D DIC system can be affected by a small out-of-plane deformation of the specimen during loading, distortion due to imaging lens, etc., (Pan et al. 2013) non-perpendicular alignment of cameras (Lava et al. 2011), speckle pattern, illumination, subset size, interpolation scheme, etc. (Pan et al. 2009).

10.3 Methodology

10.3.1 Material

In this study, the material being tested is a Dual Phase (DP) grade steel with a UTS of 1000 MPa or commercially known as DP1000. Mechanical properties of the DP steel can be found in Table 10.1. The suitable way to test this DP sheet-type steel is by using the 12.5 mm width geometry according to the ASTM Standard E8/E8M Standard Test Methods for Tension Testing of Metallic Materials (ASTM Standard 2011). The DP steel has then undergone EDM wire cut to produce the tensile test specimens.

Table 10.1 Properties of DP1000 Steel

Material	Thickness	Yield strength, $YS_{0.2\,\%}$ (MPa)	Ultimate tensile strength, UTS (MPa)
DP1000	1.5 mm	729	1051

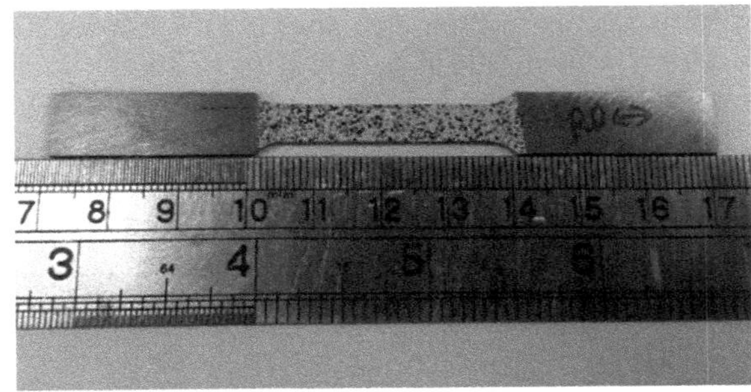

Fig. 10.1 Tensile specimen with matte black and white uniformly distributed speckles

10.3.2 Tensile Test with Optical 2D DIC

The tensile test is then set-up on a 25 kN Tinius Olsen electric machine, using a pair of wedge-type grips to hold the flat tensile specimen. For this experiment, the tensile test set-up is equipped with an optical 2D digital image correlation, where the strain measurements will be later used to be validated against the data subjected to the correction factor. In order to enable strain measurements to be captured using the optical 2D DIC system, the tensile specimen needs to be painted along the gauge length with random black speckled pattern with matte white background. An example of a speckled specimen is shown in Fig. 10.1. The speckling pattern on the specimen should have a good contrast, fine size, uniform size, and distribution to allow accurate strain measurement using DIC. Lighting is also important; thus, a good and steady light source is required to avoid any errors.

The final set-up of the experiment is shown in Fig. 10.2. The machine's cross-head speed is set to 2 mm/min. For the optical 2D DIC, the rate of image capture is set at one image per second. The output signal from the Tinius Olsen machine is connected to a DIC encoder so that the force and displacement obtained from the machine are synchronized with the captured DIC data. Once the tensile test is performed, two sets of data will be obtained. The DIC system provides the true strain measurement of the tensile specimen, while the machine provides force-displacement data which will later be corrected to remove the effect of the machine stiffness.

For the image processing to obtain the strain measurements for the DIC result, the interrogation window of 64 by 64 pixel has been chosen with an overlap of 50%. The system being utilized in this study is the 2D La Vision Strainmaster software.

10.3.3 Machine Compliance Correction

The machine compliance correction will focus on removing the external deformations (machine frame, fastenings, grips, etc.) from the acquired force-displacement data from the machine. This method can be used for any material with known Young's modulus. The concept being used to derive the correction factor can found in Eq. (1).

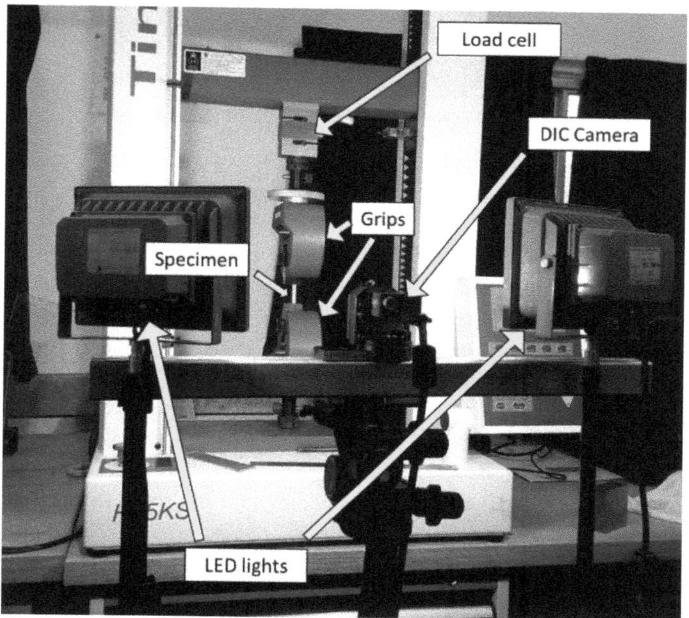

Fig. 10.2 Tensile test with an optical 2D digital image correlation set-up

Where the change of length of the specimen, $\Delta L_s(\text{mm})$ can be obtained by deducting the product of force, $F(N)$ and correction factor constant, M_C from the total displacement, $\Delta L(\text{mm})$. Force, $F(N)$ and total displacement, $\Delta L(\text{mm})$ are the data obtained from the machine. Note that the subscript *data* in L_{data} and F_{data} are the raw data taken from the machine and that Eq. (1) can be applied directly in an excel file.

$$\Delta L_s = \Delta L_{\text{data}} - F_{\text{data}} M_C \tag{1}$$

The correction factor constant, M_C is derived from the equation of stress, σ. Where the stress, σ is equal to the product of the material's Young's modulus, E and strain, ε.

$$\sigma = E\varepsilon \tag{2}$$

Since the raw data obtained from the machine are in terms of force, $F(N)$ and total displacement, $\Delta L(\text{mm})$; the terms stress, σ and strain, ε can be expressed as shown in Eq. (3). Where stress, σ is equal to force, $F(N)$ divided by cross-sectional area of the tensile specimen, $A(\text{mm}^2)$. The term strain, ε can also be replaced with the change in length, ΔL over the original length, L_o. Equation (3) is then, rearranged to get the formula for the change in length, ΔL, as shown in Eq. (4).

$$\frac{F}{A} = E\left(\frac{\Delta L}{L_o}\right) \tag{3}$$

$$\Delta L = F\frac{L_o}{AE} \tag{4}$$

Previously, it is mentioned that the total displacement, $\Delta L_{\text{data}}(mm)$ of the force-displacement data obtained from the machine is the sum of the deformation in specimen and the deformation in machine. This can be expressed as shown in Eq. (5). Here in Eq. (5), the change in length, ΔL can be substitute with Eq. (4). After substituting Eq. (5) with Eq. (4), the new equation will look like Eq. (6). Note that the subscript m and s in the brackets in Eq. (6) refer to machine and specimen, respectively. For this equation, the specimen's original length, $L_{o,s}$, cross-sectional area, A_s and Young's modulus, E_s is known. The loading, F_{data} is also known, which is from the raw data obtained from the machine. The only unknowns are the terms with subscript m, which are the properties of the machine. The unknown terms of the machine can be grouped and expressed as Eq. (7), which in this paper is named as the correction factor, M_C.

$$\Delta L_{\text{data}} = \Delta L_{m,\text{machine}} + \Delta L_{s,\text{specimen}} \tag{5}$$

$$\Delta L_{\text{data}} = F_{\text{data}}\left(\frac{L_{o,m}}{A_m E_m}\right) + F_{\text{data}}\left(\frac{L_{o,s}}{A_s E_s}\right) \tag{6}$$

$$M_C = \left(\frac{L_{o,m}}{A_m E_m}\right) \tag{7}$$

Rearranging Eq. (6) will allow the correction factor, M_C to be solved using the readily known terms of specimen as well as force-displacement data. This is shown in Eq. (8). Once the constant value of correction factor, M_C is obtained, this constant will then be applied in Eq. (8) to produce the change of length of specimen, ΔL_s. Hence, by obtaining the values of change of length of specimen, ΔL_s, the correct stress-strain curve of the tested DP1000 steel can be produced. Equation (1) shows that the deformation due to the machine is removed from the raw displacement data, ΔL_{data}.

$$M_C = \Delta L_{\text{data}} - \left(\frac{L_{o,s}}{A_s E_s}\right) \tag{8}$$

10.4 Result and Discussion

Once the tensile test is performed on the DP1000 material, the data retrieved from the machine can be plotted into a force-displacement curve and is shown in Fig. 10.3a. The curve being shown here is the result of both deformation due to specimen and machine. The maximum displacement recorded before the specimen fails is 10.2 mm with a force of 16.8 kN. The maximum force measured on this curve is at 20.7 kN during 7.06–7.09 mm extension (or displacement). Due to the fact that the force is not affected by the machine compliance, the force values will remain the same before and after applying the correction factor.

As observed in Fig. 10.3a, the beginning of the graph from point (0, 0) is not linear, where it should be theoretically linear. When the test is started and the loading begins to take place, at the range of a very small load around below 15 N, this may be due to the extensions happening in gaps in grips fastenings, load cells, etc., and is not due to the deformation of the specimen. The deformation in the specimen will begin when

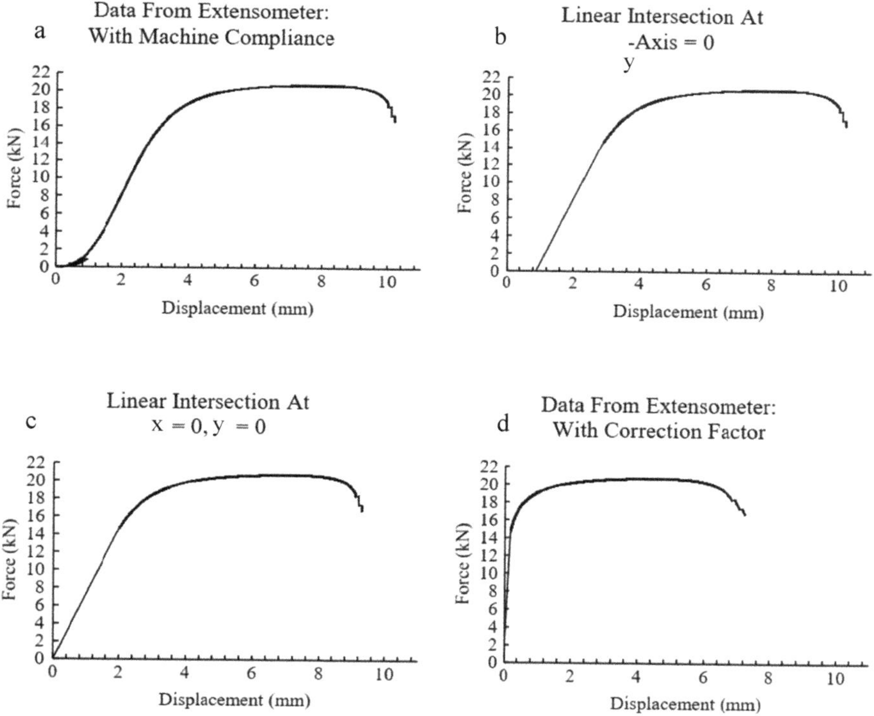

Fig. 10.3 The four stages in removing machine compliance from the force-displacement response of the tensile test. Figure **a** is the force-displacement response with the effect of machine compliance. Figure **b** is to correct the linear-elastic region of the curve. Figure **c** is to bring the starting position to zero. Figure **d** is the final curve for the specimen elongation with applied correction factor

the force increases linearly, hence from observing the curve it can be declared that the specimen is undergoing an elastic deformation. Therefore, the nonlinear extension happening during the beginning of the test can be removed by extending the linear part of the graph to the *y*-axis at 0 as shown in Fig. 10.3b. The starting position of this graph can be brought back to (0, 0) by shifting the position of the *x*-axis to 0, as shown in Fig. 10.3c.

In Fig. 10.3c, the maximum displacement before the specimen failed is now reduced to 9.31 mm with a similar force of 16.8 kN as seen in the previous two figures, (a) and (b). At this stage, Fig. 10.3c still needs to be corrected to remove the displacement due to the machine stiffness. After applying Eq. (1) to the force-displacement data of Fig. 10.3c, the final corrected force-displacement curve on DP1000 can be plotted as shown in Fig. 10.3d. Figure 10.3d shows that the specimen's maximum displacement is now reduced and measured at 7.25 mm before fracture occurs at the same force of 16.8 kN. The extension due to machine stiffness is found to be around 2.06 mm. The comparison between data obtained before applying the correction factor to remove machine stiffness and the data after correction to produce the correct specimen deformation can be found in Fig. 10.4.

By applying the correction factor, the correct displacement of the specimen can be obtained. This will allow the correct stress-strain curve for the studied DP1000 steel to be produced. Figure 10.5 shows the comparison between before and after applying the correction factor to the force-displacement data. For the data set with the effect of machine stiffness, the maximum strain obtained is 17.9% at a stress value of 896 MPa. Meanwhile, the data set with applied correction factor has a maximum strain of 12.72% before fracture, with the same stress value of 896 MPa. A total reduction of 5.18% in strain is recorded and is due to the effect of machine stiffness.

From the stress-strain curve, the ultimate tensile strength of the DP1000 steel tested for this experiment can also be measured and is recorded at 1104 MPa.

The final step is to validate the reliability of the correction factor being used in this investigation, which is to compare to the virtual strain gauge using the 2D DIC. The result from 2D DIC is shown in Fig. 10.6. By using the virtual strain

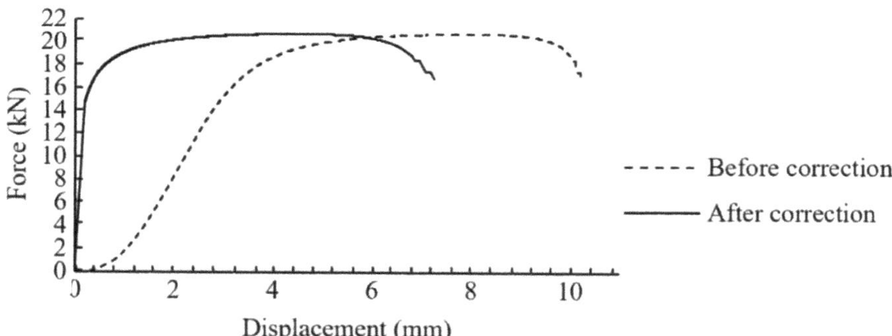

Fig. 10.4 Force-displacement curves comparisons between before correction (with the effect of machine stiffness) and after correction curve (specimen only)

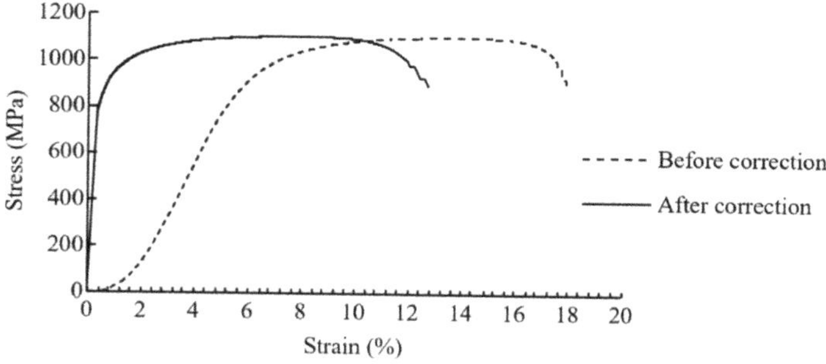

Fig. 10.5 Stress-strain curves comparison between before and after applying the correction factor

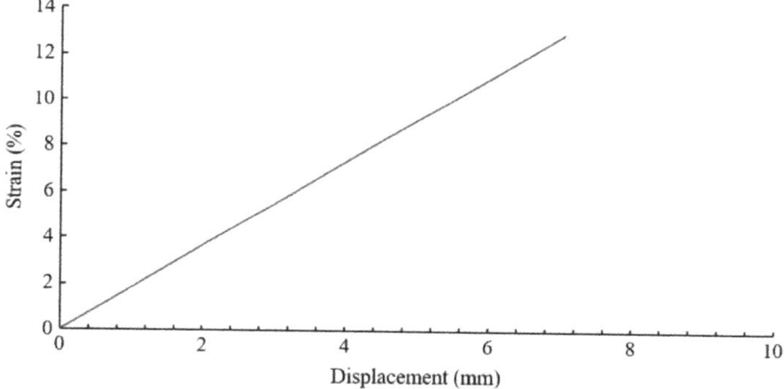

Fig. 10.6 Strain and displacement measurements obtained using optical 2D DIC virtual strain gauge

gauge, the maximum strain measured just before the specimen fails is 12.93% compared to the correction factor method which is 12.72%. The comparison between strain and displacement results obtained for raw data sets before correction and data after applying the correction factor for removing the machine stiffness and DIC are tabulated and shown in Table 10.2.

Table 10.2 Maximum strain and displacement recorded prior to specimen failure for data sets before correction, after applying correction factor and data from DIC

	Before correction	Correction factor (CF)	DIC strain gauge	Difference, CF versus DIC
Maximum displacement (mm)	10.2	7.2486	7.0534	0.1952
Maximum strain (%)	17.8947	12.716	12.93	0.214

The relatively small difference of around 0.21% found in strain measurements between both correction factor method and DIC virtual strain gauge, provides a good reliability of the developed correction factor. There can be various factors that can attribute to the difference found between both measurements. Such factors can be due to the parameters being used to process the DIC results, e.g., window size, algorithm, the selection of area of interest along the gauge length for DIC, etc. These factors can be further investigated and validated by comparing the correction factor method against strain gauges and extensometers. The difference between the two methods, correction factor and DIC, however is relatively very small. Therefore, this technique is recommended to be used for a quick validation and whenever there is no access to strain gauges or other strain measurement equipment.

10.5 Conclusions

A correction factor to remove the machine stiffness in a tensile test experiment has been presented in this paper. The developed correction factor has successfully produced the correct strain measurements for the tested DP1000 steel. Strain measurement result using the correction factor has been validated against strain measurements captured using an optical 2D DIC technique and it is found that the maximum elongation-to-fracture for DP1000 steel to be 12.72% after using the correction factor meanwhile, strain result from DIC is measured at 12.93%. The result difference is relatively small at around 0.21% and proves confidence that using the correction factor method can be used as an alternative, for cases in tensile testing when the elastic (Young's) modulus is known.

Acknowledgments The authors gratefully thank Tata Steel Europe Ijmuiden, The Netherlands for providing the materials used for this research. We would also like to express our profound thanks to The University of Sheffield for the accommodating facilities and the assistance from the technical staff. The first author would also like to thank Majlis Amanah Rakyat (MARA) Malaysia and Universiti Kuala Lumpur for the financial support to make this study possible.

References

Comer, A.J., Katnam, K.B., Stanley, W.F., Young, T.M.: Characterising the behaviour of composite single lap bonded joints using digital image correlation. Int. J. Adhes. Adhes. **40**, 215–223 (2013)

Hockett, J.E., Gillis, P.P.: Mechanical testing machine stiffness: part I—theory and calculations. Int. J. Mech. Sci. **13**(3), 251–264 (1971)

Kalidindi, S.R., Abusafieh, A., El-Danaf, E.: Accurate characterization of machine compliance for simple compression testing. Exp. Mech. **37**(2), 210–215 (1997)

Lava, P., Coppieters, S., Wang, Y., Van Houtte, P., Debruyne, D.: Error estimation in measuring strain fields with DIC on planar sheet metal specimens with a non-perpendicular camera alignment. Opt. Lasers Eng. **49**(1), 57–65 (2011)

Nguyen, V.-T., Kwon, S.-J., Kwon, O.-H., Kim, Y.-S.: Mechanical properties identification of sheet metals by 2D-digital image correlation method. Procedia Eng. **184**, 381–389 (2017)

Pan, B., Qian, K., Xie, H., Asundi, A.: Two-dimensional digital image correlation for in-plane displacement and strain measurement: a review. Meas. Sci. Technol. **20**(6), 062001 (2009). https://doi.org/10.1088/0957-0233/20/6/062001

Pan, B., Yu, L., Wu, D.: High-accuracy 2D digital image correlation measurements with bilateral telecentric lenses: error analysis and experimental verification. Exp. Mech. **53**(9), 1719–1733 (2013)

Toloczko, M.B., Abe, K., Hamilton, M.L., Garner, F.A., Kurtz, R.J.: The effect of test machine compliance on the measured shear punch yield stress as predicted using finite element analysis. Mater. Trans., JIM **41**(10), 1356–1359 (2000)

Ullner, C., Reimann, E., Kohlhoff, H., Subaric-Leitis, A.: Effect and measurement of the machine compliance in the macro range of instrumented indentation test. Meas. J. Int. Meas. Confed. **43**(2), 216–222 (2010). https://doi.org/10.1016/j.measurement.2009.09.009

Chapter 11
Productivity Improvement Through Improving the WorkStation of Manual Assembly in Production Systems

Muhammad Nazirul Syahmi Nazri, Mohd Norzaimi Che Ani, and Ishak Abdul Azid

Abstract The design of a production layout directly impacts the process cycle time of any manufacturing process. The purpose of this research is to design and improve the workstation of manual assembly in the production system. This research aims to improve the line balancing of a manual process, to improve the ergonomic issues, and ultimately improve the productivity of the production system. The research started observing the manual calculation before converting into line balancing percentage and at the same time measuring the current process time. Then, the data were analyzed to identify the potential issues that affected productivity. The results of this research successfully improved the workstation of the manual process and improved productivity. The obtained results found that the percentage of line balancing drastically improve from 30 to 45.2%, while the process cycle time was improved from 15.161 to 9.313 s.

Keywords Manual WorkStation · Production system · Cycle time · Line balancing

11.1 Introduction

The operation layout of the production system normally impacts the performance of a process assembly or in terms of efficiency, effectiveness, or productivity. Effectiveness of the operation layout not only for the manufacturing industry but also in the services industry required efficient flow of the operating system because every operation layout was designed or created usually solving multiple problems. In the

M. N. S. Nazri · M. N. C. Ani (✉) · I. A. Azid
Universiti Kuala Lumpur—Malaysian Spanish Institute, Kulim Hi-Tech Park, 09000 Kulim, Kedah, Malaysia
e-mail: mnorzaimi@unikl.edu.my

M. N. S. Nazri
e-mail: mnazirul.nazri@s.unikl.edu.my

I. A. Azid
e-mail: ishak.abdulazid@unikl.edu.my

M. H. Abu Bakar et al. (eds.), *Progress in Engineering Technology III*,
Advanced Structured Materials 148, https://doi.org/10.1007/978-3-030-67750-3_11

manufacturing industry, multiple approaches were applied to achieve several objectives such as reduced manufacturing costs, improved quality, and many more. The production system requires several processes which are from raw materials until the components are gathered into a finished-good product. The semi-finished good products in between the process flow are connected between the processes and they are assembled through a sub-assembly process. The production process is normally designed based on a set of workstations and each of the workstations requires a specific task that is carried out in a restricted sequence. Every production flow consists of several employees and thousands of bundles of sub-assemblies producing different styles simultaneously (Bahadir 2011). The joining together of components in the interconnection processes is the most labor-intensive part of manufacturing processes especially in manual assembly processes because some of the processes requires a complex process (Jalil et al. 2015).

The production flow also known as assembly line in the production system requires a set of specific tasks which is assigned to a set of workstations and every workstation is linked together by using a special transportation system. The transportation system mechanism under detailed assembling sequences is specifying how the assembling process flows from one station to another (Boysen et al. 2009). One of the ways in measuring the effectiveness of production flow is through measuring the assembly line balancing (ALB). ALB is measured based on several parameters such as the allocation of tasks to workstations and the timely completion of each task. The objective of measuring the ALB is to minimize the workflow among the operators, reducing the throughput time as well as the work in progress and thus increasing the productivity. Some of the workstations requires sharing tasks between several workers and this is called the division of labor. The division of labor should be balanced equally by ensuring the completion of the processing time at each station being approximately the same. Each step in the assembly process in every workstation is analyzed carefully and ensures that the process cycle time is balanced equally. The dedicated worker of each workstation is carrying out tasks properly in ensuring the workflow is smooth. In a detailed workflow, the synchronized process includes transferring product between workstations, inventory level, production downtime, quality issues, and predictable production quantity (Yemane et al. 2020).

An inefficient ALB of the production process is concerned with finding the most efficient arrangement of the tasks in any workstation which is considered as unequal time spent on each process. The problem of the inefficient production layout between workstations is influenced by several factors such as long-distance travel and the number of inventory. In this research, the main problem was identified that the worker has many movements to transfer the materials. The unnecessary movement required higher time and workers easy to get tired very fast. This situation also affected late delivery to customers. Thus, this research has been conducted in measuring and improving the effectiveness of the workstation design, especially in eliminating or minimizing the unnecessary movement. Three objectives have been set-up in this research which are to identify the effectiveness of the current production flow, to design the effectiveness of the production flow, and to conduct the verification and validation through implementation in the selected case study industry.

Five main sections have been divided in this article to discuss the findings of this research. This research starts with the elaborations on the effectiveness of production flow in the first section, and then the following with the second section which presents the techniques employed to solve the ineffective workstation based on the findings from published literature. Completion of both sections, then it was driven into the next section which is the discussion on the research methodology in optimizing the effectiveness of the workstation design. From the created methodology, then data collection results analysis and discussions are performed in the following section which is the fourth section. The conclusion of the overall findings and achievement of this research article will be concluded in the final section of this article.

11.2 Literature

In the manufacturing industry, facility layout will be the main consideration in the control system to maintain and improve the production efficiency in the production line. It also is known as factory layout or plan layout and it depends on the organization of company, consideration, and focusing. The facility layout considered as a main element of the production system which is contributing to the manufacturing performance, and it has been studied by many researchers many times over the past decades. Drira et al. (2007) was concluding that an effective placement of facilities contributes to the overall efficiency of the operation management and it reduced up to 50% of the overall operating costs. Tompkins et al. (2010) also stated that the good placement of facilities is reducing between 20 and 50% of the overall operating overheads in terms of reducing costs. The operating expenses are reduced by about 10–30% annually by improving the material handling activities. From the previous literature review, it can be stated that an effective facility layout requires an arrangement of everything including the shape of production flow, inventory control, batch size, movement, and material handling.

Generally, in the manufacturing system, there are four main types of layout which are; fixed-position layout, process layout, product layout, and cellular layout (Bennett 2015). The fixed-position layout concept differs from the other three in the fact that material is being brought to the workstation. It is used in aircraft assembly, shipbuilding, and heavy machinery. The process layout obtains machines of a similar type and they are arranged together in one place. Normally, the process layout is applied in the toy factory industry and hospitals. The product layout is based on machines and equipment that are arranged in one line depending on the sequence of operations required for the product. This is used for the realization of one product in the high volume of production and low variety of products such as in this research. The machines, and equipment are arranged in a line and the product one focuses on the job-shop production system.

According to Tompkins et al. (2010) the distinction between different types of production layout is made based on system characteristics such as production capacity and product mixing. The production capacity and product mixing is the first consideration in layout design among researchers to determine the type of layout that is suitable in the literature. ElMaraghy et al. (2013) states that the production layout is associated with the capacity of the production process and variety of the product mixing will be key in determining the type of production systems, i.e., either product-based or process-based production systems.

Nowadays, in a competitive market, Edmondson and Nembhard (2009) found differences suggesting that the effectiveness and efficiency of the manufacturing system is measures based on several factors such as multi-variances of the product mixing, quality issues, and optimum time of the delivery process. For instance, the international competition also influences the product mixing and production capacity. Nunes et al. (2006) suggest that manufacturing companies are needed to work extremely to be knowledge-intensive and highly creative to develop a new product. In deciding design layout, the manufacturing companies have to identify the production volume and product variety before allocating the type of layout in the factory. It will encourage worker production efficiency and expenditure cost of manufacturing companies.

An effective production layout is contributing to lower costs of the production expenses and maintaining the smoothness of the product flow. Optimizing the production layout also will increase the utilization and at the same time enhances the production capacity. Continuous improvement is one of the techniques for the increase in the effectiveness and efficiency of their production system. One of the reasons for designing an effective production layout is increasing the performance of the existing production process (Ani and Hamid 2014). The most crucial element that affects the efficiency and effectiveness of a production process is the design of the facility layout. Following the changes in the technology evolution, some old machines are upgraded or replaced. This situation is considered to ensure that the improvement is identified for the continuous improvement activity.

The effectiveness of the production layout design is not only impacting large industries but small and medium enterprises (SMEs) are also affected if their production layout is not efficient. SMEs can be defined based on the overall company annual sales turnover or the number of full-time employees. Most of the workers in SMEs are invariably semi-skilled and required to work with multi-task jobs. Some of the SMEs facing the problem of unsafe movement of the workers because limited space of operation floor and arrangement of the production layout (Ani and Chin 2016). This situation requires improving the production layout to get better utilization of the production process and the process of redesigning an existing production layout is considered as a critical consideration of several key factors toward the effectiveness of the production system (Khan et al. 2013).

Furthermore, the improvement of the production layout design must be compared with the existing layout, to evaluate the effectiveness of the production system. Other factors influencing SMEs are the nature of production demand such as if there is frequent change and fluctuation demand, which creates complicated production scheduling. This situation requires a careful study of the production layout in SMEs by understanding the overall situation prior to changing the production layout. The commonly available method of production layout development and improvement is the selection of the best among proposed layouts based on the evaluation methods. Each management has its own and different evaluation methods and normally a single performance measure which is the effectiveness. The main objective of evaluating the production layout design has to be very clear which helps in deciding the effective method of process flow.

Since SMEs have limited space and resources the planning of the production layout should solve the priority issues on the production floor. The concept of production layout and assignable tasks of each workstation might be different in terms of shape of the production layout, shift pattern, inventory control, and other factors (Rekiek et al. 2000). The bottleneck is one of the major problems faced by most SMEs in their production floor due to improper design of the material movement and transportation process. The reason for the highest production cycle time in the bottleneck workstation is mainly due to improper task distribution. Bottlenecks should be eliminated or minimized in any operation because they severely affect the smoothness of production flow.

11.3 Methodology

In the completion of this research, the objective is the main part that should be a concern. This is because the objective of the research is to finalize the effectiveness of the production sequence in the production floor. To achieve the objective, all type of research has been done to make the improvement. The methodology shows rough ideas on how to conduct the project successfully. Also it will describe the flow chart and identify resources and tools. Quality tools such as flowcharts will be applied to study the sequences of the production flow and they are applied to represent the flow ideas in a graphical manner. Graphical analysis using flowcharts makes it easier to see and clearly identify the complexity of the process.

This section gives a review of the research methodology that was implemented in this study. It provides information on the criteria for each phase as a developed methodology in order to achieve the research objectives. The section describes the developed research methodology and purpose of each phase of the study. The tools and application of equations will be highlighted and included to ensure the successful implementation based on data collection in the selected case study industry of this research. Lastly, the graphical presentation of the developed methodology that was followed in the process is also discussed.

Figure 11.1 shows the flowchart of the researcher to complete this research by following the flow. First, it starts with a literature review whereby the topic was chosen to analyze and study to make sure no research was done yet for this selected topic and to identify the current production layout. The procedure goes to the facility layout of the company. Once completion of stage 1, this study was continued by observing the actual layout of the company. Study the effectiveness of the current layout and redesigning the layout by calculating the cycle time and line balancing. In the third stage, this research will be continued by studying the effectiveness

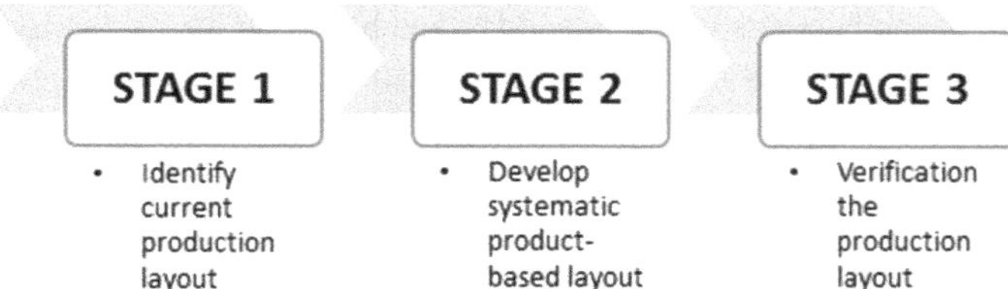

Fig. 11.1 Research methodology

of a newly designed layout. Next, designing the requirements to fulfill the needs. Finally, the layout was verified and validated in the selected case study industry. This methodology was designed to meet the defined research objectives as discussed in the previous section.

11.4 Result and Discussion

This section discussed the result that has been obtained from the research and the data collection. Based on the result that occurred, it will be discussed the performance of production layout improvement for the job-shop production system. This discussion will be made based on the results obtained. The main objective of the implementation of the new layout is to fulfill the demands. This shows the current layout and proposed layout. To achieve the defined objectives, the line balancing should meet a certain target as decided by the management.

In the selected case study, the current required a minimum of five workers to operate the production system. As shown in Fig. 11.2, the function of worker 1 is to

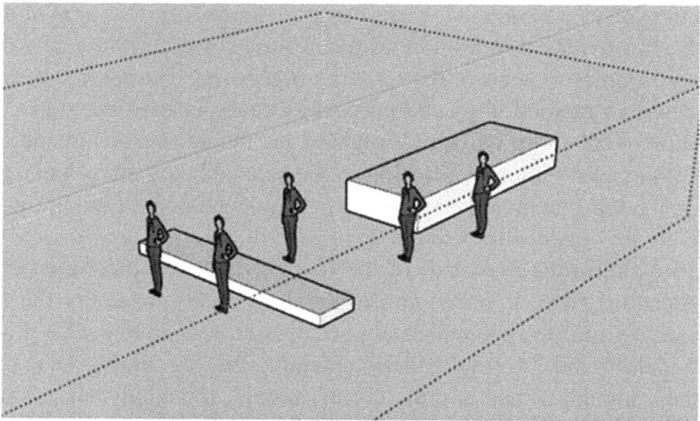

Fig. 11.2 Current layout

transfer the part from conveyor A to worker 2. Then, worker 2 transfers the part to worker 3. Worker 3 will place the part into the trolley B, while worker 4 stamps the product's logo to the part. Finally, worker 5 transfers the part into a pallet for the next workstation. To measure the effectiveness of production system, the assembly line balancing (ALB) will be calculated using the Eq. (1). ALB is a production strategy that sets an intended rate of production to produce a particular product within a particular time frame. The time required and line balancing for this workstation is as follows:

Tare weight (s) = 10.12 s
Logo stamp (s) = 5.041 s
Worker = 5

$$
\begin{aligned}
\text{Line balancing} &= \frac{\text{Total cycle time} \times 100\%}{(\text{bottleneck time}) \times (\text{total no. of worker})} \\
\text{Line balancing} &= \frac{(10.12 + 5.041) \times 100\%}{(10.12) \times (5)} \\
\text{Line balancing} &= \frac{(1516)}{(50.6)} \\
\text{Line balancing} &= 30\%
\end{aligned}
\tag{1}
$$

Based on the observation results as discussed, the current design on the selected workstation seems to require several wastes of motions and movements. Thus, the workstation has been redesigned to ensure the effectiveness of the production system. In this proposed layout the number of workers is reduced to three persons to operate the production system. This proposed layout has reduced the workers and the material's trolley has been redesigned and changed to a location which is to near the conveyor A. The function of worker 1 is to transfer the part from conveyor A direct to the material's trolley. Then, worker 2 is stamping the product's logo to the part. While worker 3 requires to transfer the part to pallet into the next workstation. The time required and line balancing for this proposed workstation is as follows (Fig. 11.3):

Tare weight (s) = 6.862 s
Logo stamp (s) = 2.451 s
Worker = 3

$$
\begin{aligned}
\text{Line balancing} &= \frac{(6.862 + 2.451) \times 100\%}{(6.862) \times (3)} \\
\text{Line balancing} &= \frac{(931.3)}{(20.6)} \\
\text{Line balancing} &= 45.2\%
\end{aligned}
$$

A comparison of the results before and after the improvement process is shown in Fig. 11.4. It compares from the existing layout of the company and the proposed layout in this research. Figure 11.4 indicates that improvements occur at bottleneck

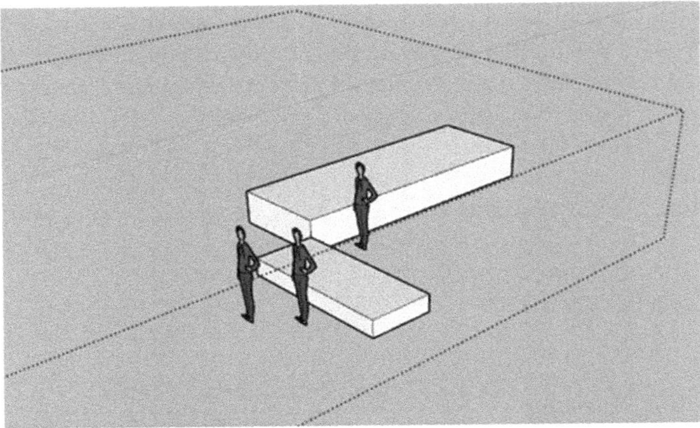

Fig. 11.3 Proposed layout

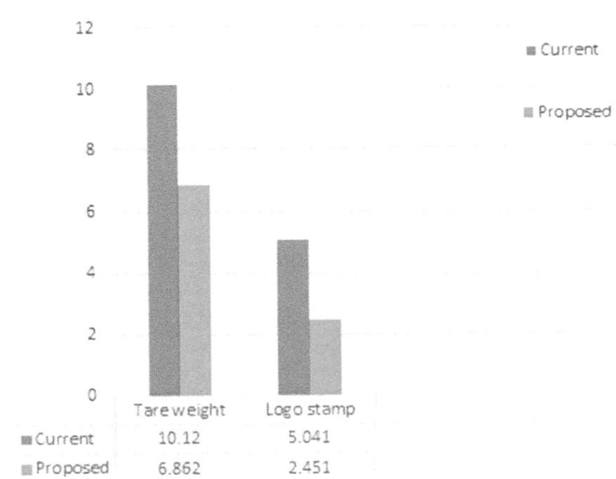

Fig. 11.4 Graph of time cycle of current and proposed layout

stations which is the tare weight station. By that, the objective of the research has been achieved by minimizing the cycle time. Moreover, the other factor contributing to the reduction of the cycle time is the redesigned material's trolley which is more ergonomic.

Based on the observation results, the function of the material's trolley is to place all the semi-finished good parts and stamping the logo. The capacity of the material's trolley is only for 8 units each time (as shown in Fig. 11.5). Currently, this material's trolley is so heavy to pull or push to another place. So, the new material's trolley has been designed and fabricated to ensure comfort for workers. The current design of the material's trolley was designed based on four multi-direction wheels. That is the reason that the current design is heavy once it is loaded by the semi-finished good

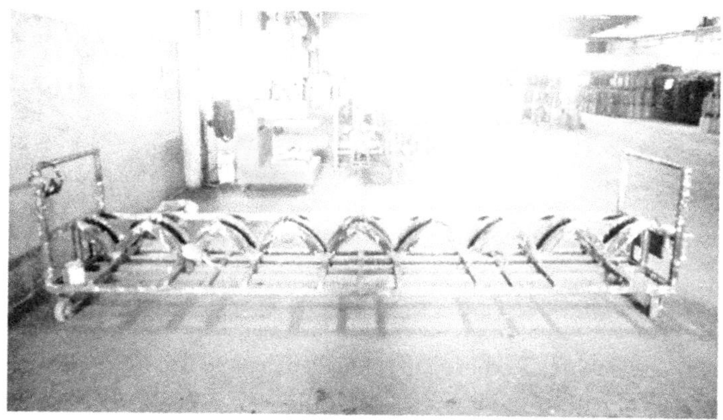

Fig. 11.5 Current design of material trolley

part. Furthermore, it will affect the body of workers and at the same time reduce the productivity of the production process.

The newly designed material's trolley applied the "Pipe & Joint" concept to ensure more comfort for the transferring process as shown in Fig. 11.6. Furthermore, the cost of this newly designed material's trolley is cheaper and it is easily maintained because it just used a single Allen-key size for the maintenance process (shown in Fig. 11.7). Moreover, this new design of this material's trolley is lighter. This pipe and joint system is considered as a smart system, dynamic, and flexible modular assembly system consisting of plastic-coated steel pipes and metal joints. The flexibility of

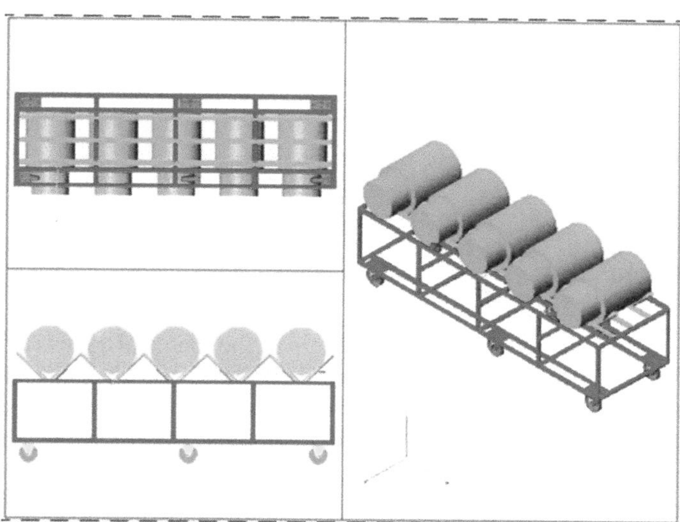

Fig. 11.6 Proposed design of material trolley

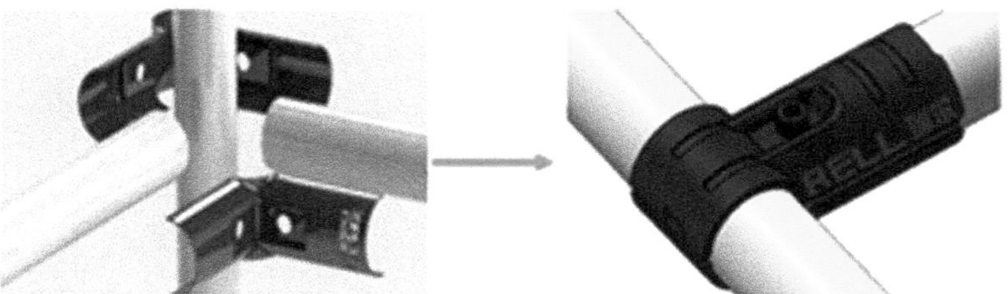

Fig. 11.7 Structure of proposed design material trolley

this system to tailor with several requirements such as racks, trolleys, workstations, gravity flow racks, and even light-duty machine structures was considered during the selection of this system. The wide varieties of available designs make the system flexible to be quickly molded into any required shape suitable for any production layout and workstations.

Based on the results of the before mentioned improvement, many workers are needed to operate the production system and required many manual handling. The new design of the material's trolley and redesigned the process tasks drastically reduced the workers and also reduced the manual handling process. Based on the observation results of the after improvement, the layout has operated smoothly with minimum manual handling processes and the process cycle time drastically reduced.

For further research, it is highly recommended that the implementation of a similar developed methodology is done in another workstation such as the semi-automation process or fully automation process. Eliminating or minimizing the waste of movement will increase the production performance and at the same time optimize the effectiveness and efficiency of the production layout. Since this research is composed of various independent activities, carrying out those activities simultaneously might not be possible because limited available resources may not allow performing those activities at the same time. Moreover, since implementing into other workstations will affect the overall production system, its implementation should be carried out with minimum disruption of the production process.

11.5 Conclusions

This study was conducted to identify and eliminate the source of defects or problems that occur at the production line. It is structured to focus on production planning. This study can help the selected case study industry to identify the problem, analyze and understand the causes of the problem and thus eliminate the problem. This research deals with production planning and scheduling and customer demands. This type of production process is susceptible to demand fluctuation. Therefore, it is quite difficult to implement methods for production planning improvement in the standard way. One part of this research is a case study where current production planning and scheduling in the real factory is improved just through the use of simulations. The purpose of the current study was to propose the new production layout of a selected workstation in the case study industry by increasing their line balancing and productivity is successfully achieved. Moreover, productivity is increased by 15.2% for the proposed layout which from 30 to 45.2%. Apart from that, there is a reduction of workers from 5 to 3 workers.

Acknowledgements The authors of this article acknowledge the Malaysian Spanish Institute, Universiti Kuala Lumpur (UniKL MSI) for funding this research that resulted in the publishing of this article. Also supported by the selected case study industry and anonymous reviewers to improve the quality of this research article are highly appreciated.

References

Ani, M.N.C., Chin, J.F.: Self-reinforcing mechanisms for cellularisation: a longitudinal case study. Int. J. Prod. Res. **54**(3), 696–711 (2016)

Ani, M.N.C., Hamid, S.A.A.: Analysis and reduction of the waste in the work process using time study analysis: a case study. Appl. Mech. Mat. **660**(1), 971–975 (2014)

Bahadir, S.K.: Assembly line balancing in garment production by simulation. In: Grzechca, W. (ed.) Assembly Line. IntechOpen, Rijeka (2011)

Bennett, D.: Product Layout. Wiley Encyclopedia of Management, Online edn. Wiley (2015). https://doi.org/10.1002/9781118785317.weom100061. Accessed 25 May 2020

Boysen, N., Fliedner, M., Scholl, A.: Sequencing mixed-model assembly lines: survey, classification and model critique. Eur. J. Oper. Res. **192**(2), 349–373 (2009)

Drira, A., Pierreval, H., Hajri-Gabouj, S.: Facility layout problems: a survey. Annu. Rev. Control **31**(2), 255–267 (2007)

Edmondson, A.C., Nembhard, I.M.: Product development and learning in project teams: the challenges are the benefits. J. Prod. Innov. Manag. **26**(2), 123–138 (2009)

ElMaraghy, H., Schuh, G., ElMaraghy, W., et al.: Product variety management. Cirp. Annals **62**(2), 629–652 (2013)

Jalil, M.A., Hossain, M.T., Islam, M.M., Rahman, M.M. et al.: To estimate the standard minute value of a polo-shirt by work study. Glob. J. Res. Eng. (2-G), 25–31 (2015)

Khan, A.J., Tidke, D., Scholar, M.: Designing facilities layout for small and medium enterprises. Int. J. Eng. Res. Gen. Sci. **1**(2), 1–8 (2013)

Nunes, M.B., Annansingh, F., Eaglestone, B., et al.: Knowledge management issues in knowledge-intensive SMEs. J. Doc. **62**(1), 101–119 (2006)

Rekiek, B., De Lit, P., Delchambre, A.: Designing mixed-product assembly lines. IEEE Trans. Robot. Autom. **16**(3), 268–280 (2000)

Tompkins, J.A., White, J.A., Bozer, Y.A., et al.: Facilities Planning. Wiley, United State (2010)

Yemane, A., Gebremicheal, G., Hailemicheal, M., Meraha, T.: Productivity improvement through line balancing by using simulation modeling. J. Optim. Ind. Eng. **13**(1), 153–165 (2020)

Chapter 12
Effect of a Gap Between Electrodes by Using the Coplanar Copper Plate in Capacitive Sensing

Norhalimatul Saadiah Kamaruddin, Muhammad Amir Mohammad Zohir, Ahmad Kamal Ismail, and Nor Haslina Ibrahim

Abstract Capacitive sensing is a technology that can detect and measure anything that has a dielectric property different from its surrounding by measuring the capacitance between the electrodes. The use of this technology enables better solution in measurement such as detecting the height of liquids, finding properties of composite materials, acting as human machine interface and many more. The gap between the electrodes plays a major role in detection and measurement. In this work, a capacitive water level sensor has been fabricated and tested to evaluate the effect of the gap between electrodes by using a coplanar copper plate. The sensor measurement has been done on four different gaps of the electrodes which are divided into 15, 35, 55 and 75 mm gap. Arduino Mega 2560 is used to capture the voltage reading from the sensor. The results from the sensor reading show that the voltage is proportional to the gap between electrodes. The bigger the gap between electrodes, the smaller is voltage reading. The results were also supported by analysis of variance for better reliability.

Keywords Capacitive sensor · Copper plate · Electrode gap · Analysis of variance

12.1 Introduction

Recently, capacitive sensing is well known for its robust application. This type of sensor has many applications and is customizable (Wang 2014). The orientation of

N. S. Kamaruddin · M. A. M. Zohir · A. K. Ismail (✉) · N. H. Ibrahim
Universiti Kuala Lumpur, Malaysian Spanish Institute Kulim Hi-Tech Park, 09000 Kulim, Kedah, Malaysia
e-mail: ahmadkamal@unikl.edu.my

N. S. Kamaruddin
e-mail: norhalimatul@unikl.edu.my

M. A. M. Zohir
e-mail: muhammadamir1248@gmail.com

N. H. Ibrahim
e-mail: norhalimatul@unikl.edu.my

M. H. Abu Bakar et al. (eds.), *Progress in Engineering Technology III*,
Advanced Structured Materials 148, https://doi.org/10.1007/978-3-030-67750-3_12

the capacitor is normally made by two electrodes at with the insulator acts as the separator. The capacitor material must be an electric conductor while the insulator is a non-conducting material called dielectric. More explanation regarding capacitor for measurement and liquid gauging can be found from previous literatures (Axt et al. 2016; Maltby 1975; Nguyen 2016; Nawrocki 1991). The accuracy of the capacitive sensing can be determined by the distance or gap between the conducting materials or electrodes used. The gap normally can be controlled according to the surrounding condition as well as the type of substances to be measured. Some errors may exist in this type of measurement such as inaccurate level of liquids due to sloshing which effects the electrodes response (Terzic et al. 2012). In this work, capacitance sensor circuit was constructed, validated and tested. The sensor is used for measurement of water level with statistical analysis to validate the results. Effect of gap between electrodes using coplanar copper plate is being focused in this study.

12.2 Literature

In general, a capacitive sensor is made out of a couple of adjacent terminals or electrodes. At the point when a human being (or some other conductive object) comes in closeness to these electrodes, there is an extra capacitance between the electrodes and the object which can be measured to the object's presence. Utilizing this technology, it is easy to fabricate touch sensors acting as buttons, sliders, trackpads and many more. Capacitive sensing is progressively used to supplant mechanical button with touch sensitive buttons for infotainment systems. The work by Arora (Aurora and Varma 2011) represents the utilization of a programmable mixed signal controller to accomplish intelligent capacitive detecting. In designing the capacitive sensor system, the size and orientation between electrodes, the surrounding and the substances to be measured must be properly analyzed. There are guidelines to guarantee maximum performance from the sensor system (Wang 2015). The sensor electrode can be in the shape of plate or rod with different thickness or diameter. The signal may be in the form of pico-farad (pF). The noise to the sensor can be reduced by shielding (Kuttruff 2012). This sensor works by creating an electric field around the electrodes when supplied with certain amount of voltages. Therefore, the electrodes must be a good conductor and placed at the correct arrangement (Alam 2014). At the higher voltage potential charged plate, the electric field lines begin, and it ends at the lower side of the plate. Due to the intricacy of modeling the behavior, the parallel plate equation neglects the fringing effect. Hence, more study must be done to encounter the problem.

12.3 Methodology

A capacitance water level sensor circuit has been developed by an Acondiniv 547 which has two combined circuits. It contains the ultrasonic water level sensor circuit and the capacitance water level sensor circuit (Fig. 12.1). The circuit is then simulated and validated using the NI Multisim software.

Validation was made by connecting the positive probe of the oscilloscope to the drive electrode (in the NI Multisim software) where it has been found that the voltage signal obtained has the same value as compared to the actual signal from the real drive electrode. The analog circuit was applied to the test as shown in Fig. 12.2. Adhesive copper tape with 10 mm width was used as electrode for the sensor. Five electrodes were pasted on the test section. One of the electrodes acted as drive electrode which was E1 while the other four electrodes acted as sense electrodes arranged with four different gaps between electrode named as G1 (15 mm gap), G2 (35 mm gap), G3 (55 mm gap) and G4 (75 mm gap) as shown in Fig. 12.2. The drive electrode energized at 1 V. The electrodes were placed 26 mm above the floor inside the test section to reduce the liquid sloshing effect during the water was pumped into the test section so that the water rose steadily to avoid inconsistency in sensor reading.

Figure 12.3 shows details of the experimental setup labeled by 1-laptop with Arduino software (IDE), 2-Arduino board, 3-analog circuit board, 4-the auto shut-off pump circuit board, 5-the test section with sensor electrode and 6-the water reservoir with water pump. The experiment has been done at room temperature to maintain the water temperature. Any changes of the water temperature may result

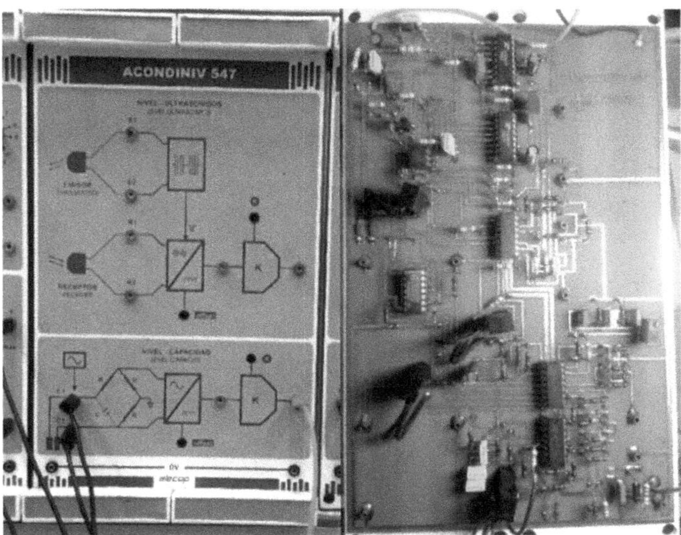

Fig. 12.1 Acondiniv 547 with two types of sensor application (analog circuit). (Left) outside view. (Right) two combined circuits inside Acondiniv 547

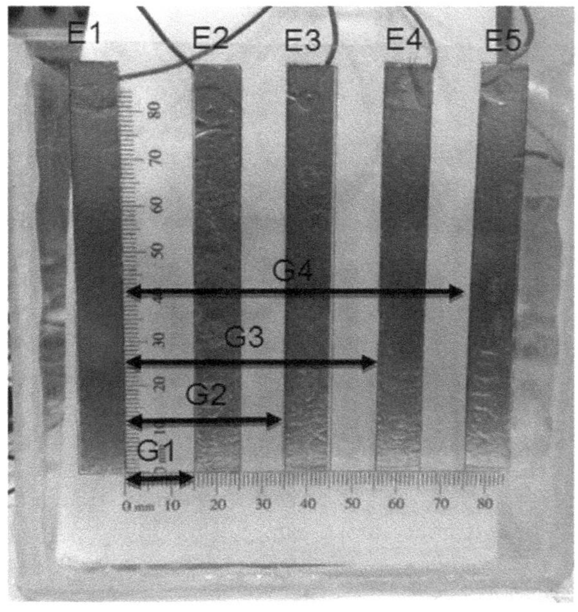

Fig. 12.2 Electrodes arrangement with four different distance of the gap

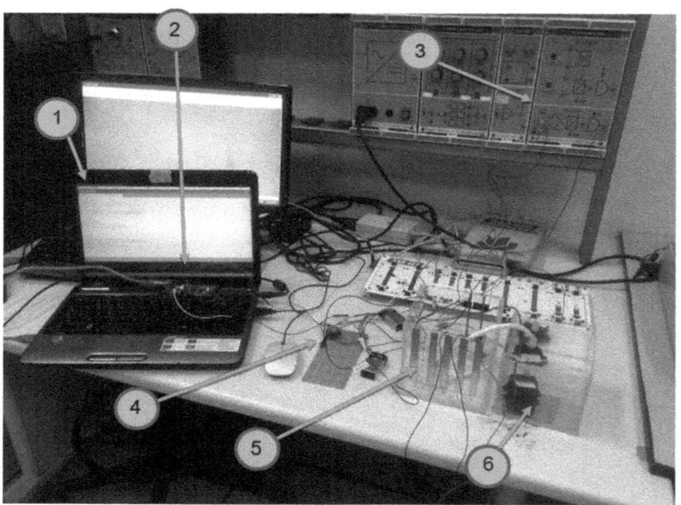

Fig. 12.3 Full apparatus setup for analysis

in changes ofin water dielectric properties and reduce the accuracy of measurement. The maximum level of water was fixed at 56 mm (30 mm of electrode height will be in contact with water) from the test section floor by using an automatic shut-off water pump so that the differences in sensor reading can be compared and analyzed. Arduino Mega 2560 was used to capture the voltage reading from the sensor. The data was sent through the Arduino's UART over USB to the computer. This data was continuously sent every 0.1 s. The PLX-DAQ software is used to transfer the data to Excel.

The capacitance sensor works by identifying the changes of the capacitance value that can be observed from the changes of voltage on different electrodes. The voltage differences have been correlated with the water level to measure the level of water. The sensor reading from Arduino is validated with a multimeter. Before the analysis started, the circuit need was calibrated so that the reading of the water level is the same as its actual level. Theoretically, the capacitance value will have a direct proportional relation to the liquid level height. However, error may still exist. Gain and offset compensation must be applied to the system measured to get the actual level. A linear correction algorithm as given in Eq. (12.1) below can be used to solve the problem (Wang 2015).

$$\text{Level}' = \text{Level} \times \text{Gain} + \text{Offset} \tag{12.1}$$

The calibration process was made by setting-up the offset potentiometer to 0 V while the water level is at the lowest height and then setting-up the gain potentiometer (K) to a certain volt while the water level is at a certain height.

12.4 Result and Discussion

The results of the sensor reading for four difference electrode gaps from 15, 35, 55 and 75 mm were displayed. The results were compared and discussed. Analysis of variance (ANOVA) technique was used and discussed to observe significant differences of the sensor reading from three times sensor readings for each electrode gap. The effect of the electrode gap on the sensor reading is discussed below.

12.4.1 Sensor Reading

The results of four different gaps between electrodes are shown in Figs. 12.4, 12.5, 12.6 and 12.7. The water pump stops at a water level of 56 mm. From the results in these four graphs, it can be seen that there was a small disturbance along the line when the voltage started increasing. This might be due to the liquid sloshing during pumping water into the test section that caused the disturbance.

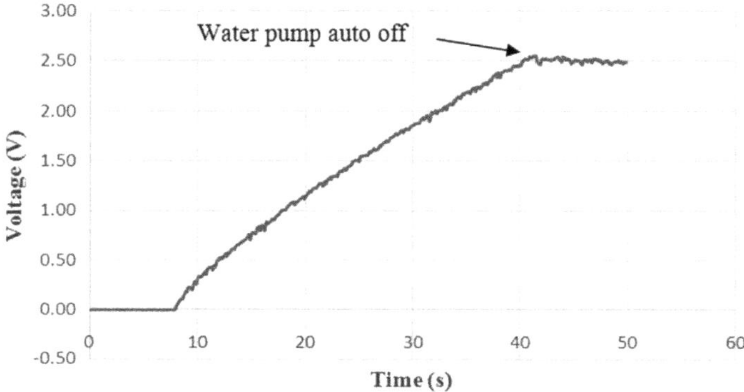

Fig. 12.4 Voltage reading of electrode gap of 15 mm

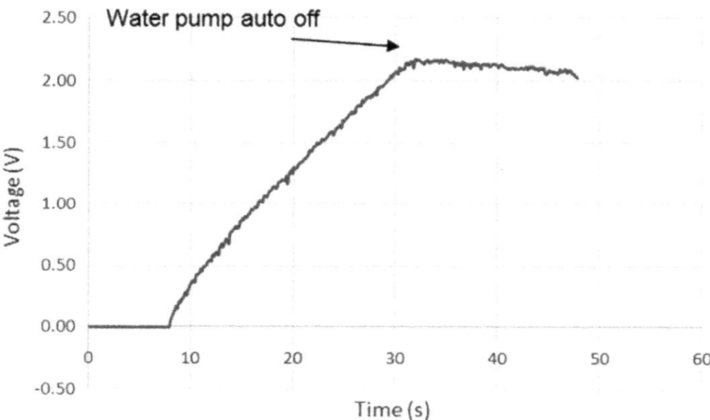

Fig. 12.5 Voltage reading of electrode gap of 35 mm

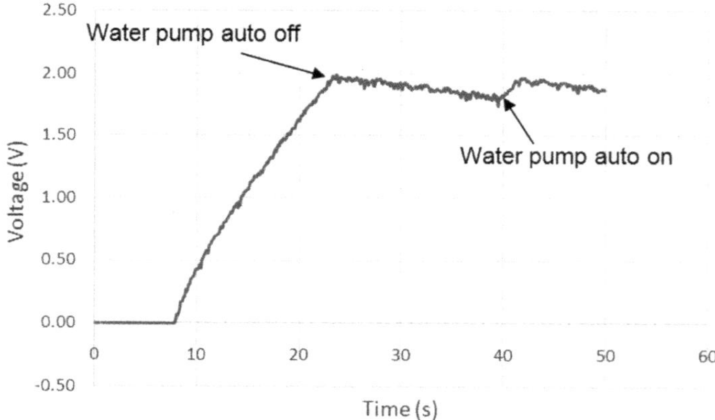

Fig. 12.6 Voltage reading of electrode gap of 55 mm

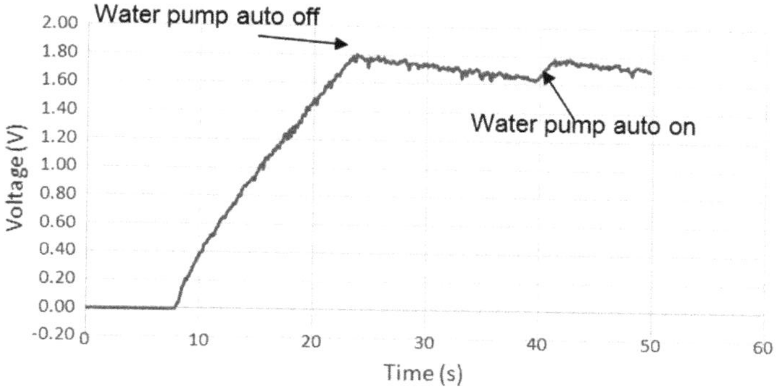

Fig. 12.7 Voltage reading of electrode gap of 75 mm

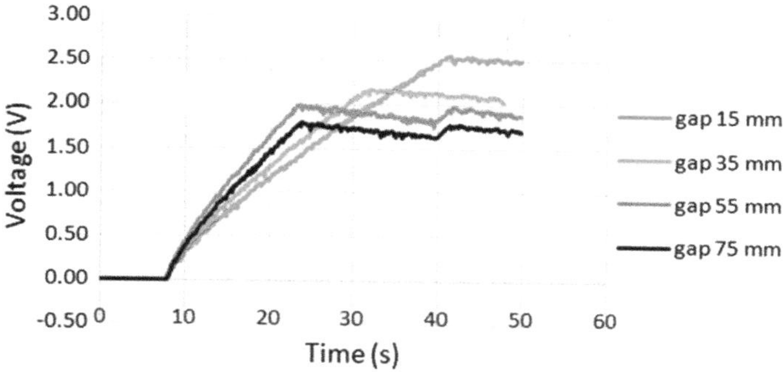

Fig. 12.8 The differences of voltage reading with four different electrode gaps

Figure 12.8 shows the comparison of sensor readings of the four different electrode gaps. It can be seen that the gradient of these 4 readings is different. The reason was that when the water was pumped into the test section, the starting water level was different from each analysis thus the time to reach the water level of 56 mm and the water pump to stop pumping water was different. The differences in gradient have no significant meaning in this case. Note that the starting water level for all analyses is always below the sensor electrodes before the sensor can detect the water and reads the voltage reading.

12.4.2 Analysis of Variance (ANOVA)

The analyses for each four electrode gaps were done three times to see if there is a significant difference in the sensor reading for one electrode gap. The analysis of variance (ANOVA) technique is used to see if there is a significant difference on three times sensor reading taken on one electrode gap.

Table 12.1 ANOVA result for electrode gap of 15 mm

Source of variation	F-value	P-value	F-critical value
Between three voltage reading	0.260489	0.770711	3.002012

Table 12.1 shows the ANOVA result for the electrode gap of 15 mm. The null hypothesis of ANOVA is that there is no significant difference between three voltage readings. Based on the result in Table 12.1, it can be found that the null hypothesis is accepted. For the hypothesis to be validated, the F-value must be lower than the F-critical value. In addition, the P-value is higher than the α value ($\alpha = 0.05$). Obviously, a level of 0.05 indicates a 5% risk of difference value exists. The null hypothesis is accepted since the F-value < F-critical value and P-value > α. It can be concluded that there are no significant differences between the three sensor readings of the electrode gap of 15 mm and the sensor reading is acceptable. The ANOVA technique also can be applied on the other three sensor readings for each of the three electrode gaps (35, 55 and 75 mm). The result of ANOVA for the other three electrode gaps shows no significant differences between the three sensor readings thus those sensor readings are acceptable.

12.4.3 *Effect of the Gap Between Electrodes on the Sensor Reading*

From the graph in Fig. 12.9, for the 15 mm gap result, the maximum voltage at a water level of 30 mm (height of electrodes in contact with water) is 2.56 V, while the maximum voltage for 35, 55 and 75 mm gaps is 2.18 V, 1.99 V and 1.8 V, respectively. Figure 12.9 shows the graph of the maximum voltage reading against the gap between

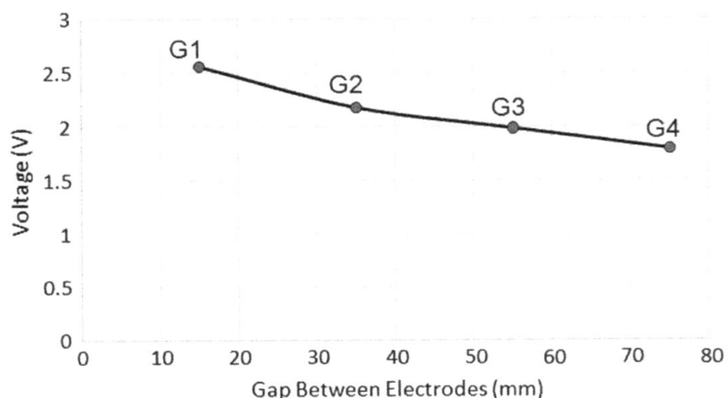

Fig. 12.9 Graph of maximum voltage reading at water level of 30 mm (height of electodes in contact with water) against the gap between electrodes

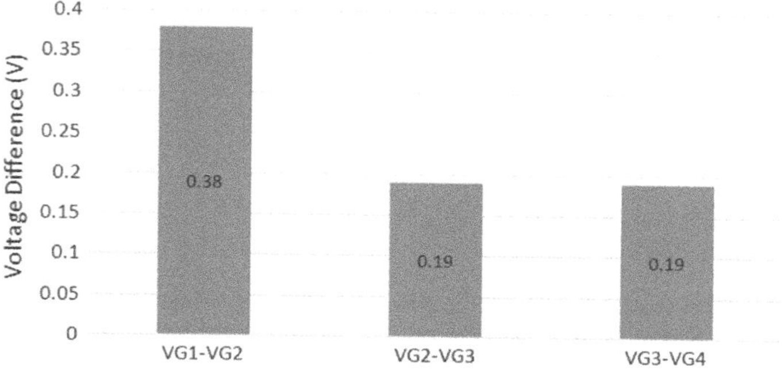

Fig. 12.10 Voltage difference between the maximum sensor reading at water level of 56 mm

electrodes. The voltage is proportional to the gap between the electrodes. The bigger the gap between the electrodes, the lower the voltage value across the electrodes.

The voltage difference between the maximum sensor reading at water level of 56 mm was shown in Fig. 12.10. The comparison has been plotted as the VG1-VG2 was the difference between the maximum sensor reading of G1 and G2 (2.56-2.18 V). VG2-VG3 was the difference between the maximum sensor reading of G2 and G3 (2.18-1.99 V). While VG3-VG4 was the difference between the maximum sensor reading of G3 and G4 (1.99-1.80 V). From Fig. 12.10, it can be noticed that VG2-VG3 and VG3-VG4 are equal while the VG1-VG2 is exactly two-times higher compared toVG2-VG3 and VG3-VG4. VG1-VG2 is higher because the G1 is smaller compare to G2 (15 mm difference in electrode gap) thus the voltage difference is higher since the smaller the distance between the electrodes, the larger the voltage. VG2-VG3 is equal with VG3-VG4 because the electrode gap difference between G3-G2 (55-35 mm) and G4-G3 (75-55 mm) is equal to 20 mm. Based on the voltage difference as shown in Fig. 12.10, it is indicated that the voltage difference value is maintained as the electrode gap difference was kept constant.

12.5 Conclusions

From this study, it can be concluded that the capacitive sensor for water level measurement had been successfully developed and tested. The sensor reading from four different electrode gap has been collected and observed. The result shows that the voltage is proportional to the gap between electrodes. The bigger the gap between electrodes, the lower the voltage across the electrodes. Simulation of the electric field around the electrodes to investigate how the electric field affects the capacitive sensor reading in the small test section can be done in the future. Comprehensively, capacitive sensing technology is good and better over other detection approaches. This was due to its capability to detect different kinds of materials. This technology is contactless and wear-free with a large sensing range. The design of this sensor can be compacted by using low-cost PCB sensors at low-power solution. In capacitive liquid level sensing, it can detect the presence and level of liquid in a container without any physical contact.

References

Alam, M.N.: Applications of electromagnetic principles in the design and development of proximity wireless sensors. Thesis. University of South Carolina, Columbia, United States (2014). https://scholarcommons.sc.edu/etd/2784/. Accessed 30 June 2020

Aurora, P.K., Varma, K.: Revolutionizing Automotive HMI Design Through Capacitive Sensing. EE Times (2011). https://www.eetimes.com/revolutionizing-automotive-hmi-design-through-capacitive-sensing/. Accessed 30 June 2020

Axt, B., Zhang, S., Rajamani, R.: Wearable coplanar capacitive sensor for measurement of water content—a preliminary endeavor. ASME J. Med. Dev. **10**(2) (2016). https://doi.org/10.1115/1.4033149

Kuttruff, H.: Ultrasonics: Fundamentals and applications. Elsevier Science Publishers Ltd, Springer, Netherlands (2012)

Maltby, F.L.: System for Measuring Fluid Levels in a Vehicle Transmission. US3918306A. USA (1975). https://patents.google.com/patent/US3918306. Accessed 25 May 2020

Nawrocki, R.: Apparatus and method for gauging the amount of fuel in a vehicle fuel tank subject to tilt. US5072615A. USA (1991). https://patents.google.com/patent/US5072615A. Accessed 20 June 2020

Nguyen, T.: Capacitive sensing: water level application. Thesis. Metropolia University of Applied Sciences, Finland (2016). https://www.theseus.fi/handle/10024/113153. Accessed 20 June 2020

Terzic, E., Terzic, J., Nagarajah, R., Alamgir, M.: A Neural Network Approach to Fluid Quantity Measurement in Dynamic Environments. Springer, London (2012)

Wang, D.: FDC1004: Basics of Capacitive Sensing and Application. Application Report SNOA927. Texas Instruments, Dallas (2014). https://www.ti.com/lit/an/snoa927/snoa927.pdf. Accessed 25 May 2020

Wang, D.: Capacitive-Based Liquid Level Sensing Sensor Reference Design. Application Report TIDU736A. Texas Instruments, Dallas (2015). https://www.ti.com/lit/ug/tidu736a/tidu736a.pdf. Accessed 30 June 2020

Chapter 13
Optimization of a Production Layout Model to Increase Production Efficiency in Small Medium Enterprises

Muhamad Hazrul Alif Abdul Halim, Baizura Zubir, and Ahmad Fauzie Abdul Rahman

Abstract Considering sustainability in current manufacturing industry under the worldwide situation of aggressive competition, a productive action measure needs to be done in order to reduce or eliminate the idle and/or down time of an operation. Therefore, it will improve the current production efficiency and working method in the company itself. A case study has been carried on facility and process layout analysis of a small and medium sized enterprise (SME) in a manufacturing company that produces coffee product in Malaysia. The whole production layout suffers due to the absence of established standard time for activities carried out by workers, the non-value added (NVA) activities involved and the inefficient methods. The goal of this case study is to improve the production efficiency of the current production layout by identifying and minimizing unnecessary activities at the bottleneck in production line. This study is conducted through time study technique to estimate the time allowed to a qualified and well-trained worker on normal situation to complete a specific task. The new layout has been designed and compared with the current layout. Subsequently, the WITNESS simulation software was applied also to verify and validate all proposals to improve the production layout. The result shows through improving the working method and rearranging the layout. Thus, it will possible to well balancing the process flow as well as ensuring better financial benefits to the company.

Keywords Production efficiency · Production layout · Time study · Witness simulation

---

M. H. A. A. Halim · B. Zubir (✉)
Universiti Kuala Lumpur, Malaysian Spanish Institute Kulim Hi-Tech Park, 09000 Kulim, Kedah, Malaysia
e-mail: baizura@unikl.edu.my

M. H. A. A. Halim
e-mail: hazrulalif92@gmail.com

A. F. Abdul Rahman
Politeknik Tuanku Sultanah Bahiyah, Kulim Hi-Tech Park, 09000 Kulim, Kedah, Malaysia
e-mail: fauzie@ptsb.edu.my

M. H. Abu Bakar et al. (eds.), *Progress in Engineering Technology III*,
Advanced Structured Materials 148, https://doi.org/10.1007/978-3-030-67750-3_13

13.1 Introduction

A good layout design is important to maximize the productivity of manufacturing process in a production line. It depends on several factors such as the product made, the quality of raw material, method of manufacturing process and arrangement of workstation that contributes to production line. The production layout is known as facility layout that can be divided by four processes that include planning location, process flow, floor layout and material handling system. The challenge in determining the best solution arrangement of the workstation is one of the elements that will make a huge change in the manufacturing system performance. The objective of each manufacturing industry is to ensure the smooth and efficient execution of the production process from the first process of material until the end of the process based on the production layout design that will reduce the waste activities in the production line and also can improve the overall effectiveness in the production. In this project, the company KILANG KOPI MOHD CHIN DAN ANAK-ANAK SDN. BHD. has been selected to perform the verification and validation of the production layout performance in terms of improving production efficiency. KILANG KOPI MOHD CHIN DAN ANAK-ANAK SDN. BHD. is located at Kampung Badelishah, Terap, Kedah. This company produces original coffee powder and mixed coffee powder. The company's vision is to become one of the competitive Bumiputera companies in Kedah and to produce high-quality food products on top to penetrate the international market. The company's mission is to produce the highest standards of coffee and not to compromise on the quality and keeping the cost price economical for the benefit of our coffee lovers.

13.2 Literature

13.2.1 Facility Layout

The layout of facilities is the most crucial element affecting the efficiency of a production process. A good layout keeps costs low, while keeping the product flow through the facility, reducing unnecessary material handling (Tompkins et al. 1996). Improving the layout also increases the use of machines which enhances the shop floors machining capacity. Management often feels the need for redesign. An enterprise is considered as an SME based on the annual sales turnover or the number of full-time employees of a small unit that does not exceed 50 employees. Most workers are semi-skilled and required to work on more than one machine. Most SMEs are faced with the problem of unsafe operator movement from machine to machine. It is desired to arrange the machines in such a way that single operator can move to number of machines easily and safely. Improving the layout for better machine and operator utilization requires lean thinking. The common thing about all the plant layout development methods available is that they develop several alternative layouts. Choosing the best among these layouts is based on some methods of evaluation. Each layout technique of the plant has its own and different method of evaluation. Each method of evaluation is based on a single measure of performance. To design a new plant layout, it is therefore very necessary to decide and fix the performance measure. The layout design objective must be very clear as it helps to determine the best method to implement for improved layout design. In addition, the proposed layout must be evaluated and compared to the existing layout based on the performance criterion chosen. The nature of plant dynamics in SME is that demand changes frequently and scheduling is very complicated. Direct mathematical calculations make it very difficult to measure performance. Some researchers suggest the use of simulation tools to measure performance. It is seen that knowing the problems faced by SMEs, the available techniques of layout design, the use of simulation in layout design and evaluation methods is very important for designing a new layout.

13.2.2 Witness Simulation

Witness has been introduced as a pedigree software package and has been used successfully by thousands of models over the past 20 years (Waller 2012). It was a group with a modern platform for software development and interfaced design as witnesses of Microsoft Gold Partner status and Window 7 from Lanner. In the research activity of layout design, numerous studies have attempted to demonstrate witness simulation as a simulation technique tool. In making a flexible simulation tool suite (Markt and Mayer 1997), the model of witness simulation was used, complete with the outline of the AutoCAD facilities and the exact location of the machine. The schematic layout to simulate the actual production process and provide additional information on system operation such as the location of possible bottlenecks and the requirement for buffer size. A schematic layout in the working data file was created from the data. The machines were placed in the spreadsheet strictly according to order without any technical or economic restrictions being considered. While the material flow was graphically represented as arrow, these arrows can also represent the individual product flow or a summation of all product flows. The arrows thickness and color were varied to show the volume of flow graphically. Manually and automatically, the layout optimization was done (Markt and Mayer 1997). The manual method, on the other hand, involved exchanging the position of one or more pairs of machines or buffers and evaluating the resulting control number to see if any improvements were made. After complete production runs, the result using witness simulation shows a good way to predict and solve any problems and inefficiencies such as production bottlenecks, overly idle resources and storage areas that were too small or too large.

13.2.3 Lean Production Principle

To become lean requires a way of thinking, philosophy and management system that follows Toyota's "4P" model. In his book "The Toyota Way," Liker (2004) describes fourteen principles that provide the basis for lean at Toyota Production System (TPS). These principles are divided into four categories, all starting with the letter "P"—Philosophy, Process, People and Partners and Problem Solving. This is known as Toyota's so-called "4P" model (Liker 2004).

In the lean production principle, there were several key elements that need to be understood in order to implement lean in facility layout system (De Carlo et al. 2013). The lean production principle was important as a guide before it was applied in the company. One of the most critical principles in lean production was the elimination of waste of any material, processes and features that were not required from the customer point of view in creating value. This type of waste should be eliminated in the facility layout that would interrupt production line operations. Seven basic types of waste were applied in the manufacture of a study for existing layout design (De Carlo et al. 2013).

13.3 Methodology

The case study was done at a food industry to improve the production efficiency on one of their coffee production lines. The first step was doing a survey and taking all necessary data from the company which includes working hour per day, working shift, working procedure for each job, cycle time for each procedure, operator number, daily output target. Then, investigating the current layout and studying the effectiveness of the actual layout practice. After that, generate a few other possible production layout design models. Then for each production layout model, the simulation was conducted using Witness software and the result was recorded and analyzed. After the witness simulation has been completed, verification and validation of the proposed production layout was done.

13.4 Result and Discussion

13.4.1 First Proposal

In first proposal, the focus is on the reduction of time at the bottleneck workstation which is the cooking station. In this part, they need to use two kitchens to reduce the time. Table 13.1 shows the proposed cycle time for the production layout.

In this workstation, when two kitchens are used, it can reduce half of the time in cooking process because the mixture of coffee beans can be divided into two portions

Table 13.1 Proposed cycle time for the first proposal

Workstation	Current practice (min)	Proposed change (min)
Cooking	37.95	22.95

in the kitchen port and run at the same time than it can decrease the time to cook the mixture, more over this method also can reduce the time in the mixing process in workstation two because it uses the same port. The calculation is done in Fig. 13.1 that shows more details and explanation for the first proposal.

So for workstation 2, time is decreased by 15.00 min and the bottleneck time was reduced from 37.95 to 22.95 min which gives a reduction of 39.52% in time.

$$\begin{aligned}\text{\% Bottleneck Time Reduction} &= (15.00/37.95) \times 100 \\ &= 39.52\%\end{aligned}$$

<table>
<tr>
<td>

Workstation 2
In Cooking Process

Data from Kilang Kopi Mohd Chin dan Anak-Anak Sdn. Bhd.
70 kg = 10 minute
Total Current Input = 140 kg per packaging
Current Time Recorded = 19.96 minutes
If we use two machine:
140/2 = 70 kg per packaging

New Time = 19.96 - 10.00 (minutes)
= 9.96 minutes for two kitchen

We can reduce 10 minutes in cooking process for workstation 2

</td>
<td>

Workstation 2
In Mixing Process

Hence, this process uses the same place as the cooking process, so we can assume the time can reduce by half if using two machines :
70 kg = 5 minutes
Current time recorded = 10.00 minutes

New Time = 10.00/2
= 5.00 minutes/per Kitchen

So we can reduce 5.00 minutes in mixing process for workstation 2

</td>
</tr>
<tr>
<td colspan="2">

Total new time for workstation 2 :
In Cooking Process = 9.96 minutes
Mixture = 5.00 minutes
Cooling = 7.99 minutes
Total = 22.95 minutes

</td>
</tr>
</table>

Fig. 13.1 Calculations in first proposal

Table 13.2 Proposed cycle time for second proposal

Workstation	Current practice (min)	Proposed change (min)
Roasting	34.58	20.53

13.4.2 Second Proposal

The recommendation in the second proposal is to add the roasting machine to workstation 1. This is because the bottleneck occurred during the roaster process. When two roasting machines operate simultaneously, the time for this phase can be decreased. Table 13.2 shows the proposed cycle time for the production layout in the second proposal.

In this workstation, the coffee beans from workstation 1 can also be divided into two portions in a roasting machine. Hence, this process uses the same machine in the roaster process, so we can divide it by two and run at the same time. We can assume the time can be reduced by half if we use two machines. The calculation is done in Fig. 13.2 that shows more details and explanations for the second proposal.

So for workstation 1 time is decreased by 14.05 min and the bottleneck time was reduced about 40.63%.

$$\begin{aligned}\% \text{ Bottleneck Time Reduction} &= (14.05/34.58) \times 100 \\ &= 40.63\%\end{aligned}$$

13.4.3 Witness Simulation

Witness software is used to simulate the layout with the objective of producing the part by meeting the expectations of the customer in terms of output production. The input parameters for this simulation are the cycle time and machine orientation and while the output parameters are the number of enter and number of shipped. The result of all the simulations in this research is presented in Fig. 13.3.

(a) Current Layout

Figure 13.3 for the current layout-based process design shows the structure of the witness simulation after run for the production line. The simulation result based on process design shows that the current layout meets the daily target of 16 units per day. Therefore, the output based on the process design of the existing production layout will be compared for further improvement in the proposed new production layout.

Workstation 1

Roasting Process

Data from **Kilang Kopi Mohd Chin dan Anak-Anak Sdn. Bhd.**

70 kg = 14.05 minutes

Total Current Input = 140 kg per kitchen

Current Time Recorded = 29.58 minutes

If we use two machines:

140/2 = 70 kg per kitchen

New Time = 29.58 – 14.05 (minutes)

= 15.53 minutes for two machines

We can reduce 15.53 minutes in roasting process for workstation 1.

Total new time for workstation 1 :

Roaster Process	**: 15.53 minutes**
Cooling	**: 5.00 minutes**
Total	**: 20.53 minutes**

Fig. 13.2 Simple calculation in second proposal

(b) First Proposal

The simulation has proven that the proposed layout will meet the objective of the daily target of production output. In the first proposal, one more kitchen is add to workstation 2 to make the bottleneck reduced as the proposal. The difference is from the current layout, the cooking station is now located in a new position to reduce the time in the production line. More over this type of plan layout follows the I-line plan layout in the manufacturing industry. The framework of the witness simulation structure after run for the proposed is shown in Fig. 13.4.

The result after running the simulation shows that the proposed design meets the target and increases the input by 9 units per day. The simulation result of the proposed one layout shows that the number of shipped is 25 units per day. This layout is very efficient in terms of production output compared to the existing layout. In addition, the time in the bottleneck workstation will be reduced. In terms of efficiency, 56.25% of the production output was increased in the first proposal compared with current practice.

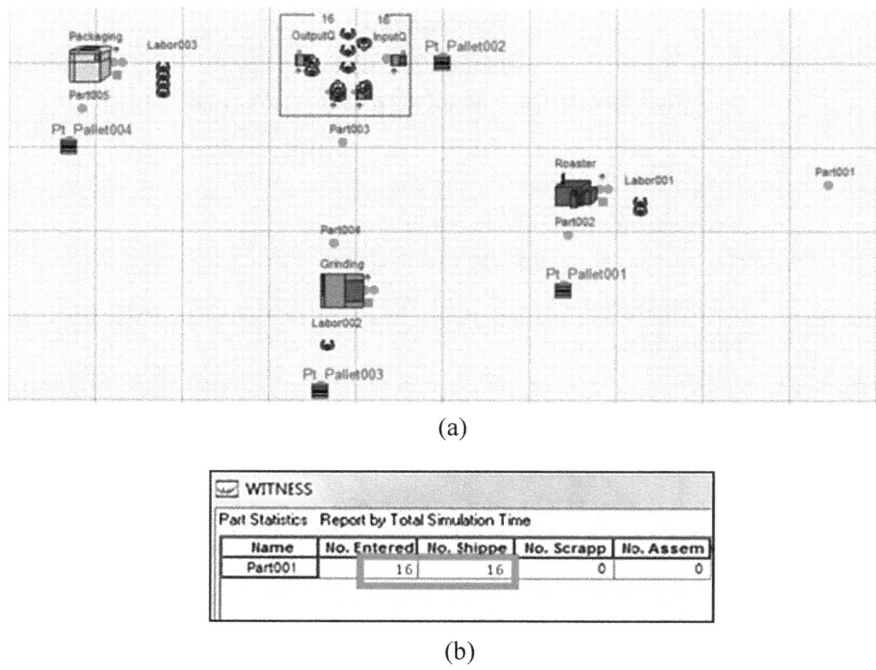

Name	No. Entered	No. Shippe	No. Scrapp	No. Assem
Part001	16	16	0	0

(b)

Fig. 13.3 **a** The simulation structure of the current layout, **b** the simulation of daily production for the current layout

(c) Second Proposal

For the second proposal, the roasting machine was added to workstation 1. The difference from the current layout is the roasting station is now located at a new position to reduce time in the production line. This plan layout follows the L-line in the manufacturing industry. The framework of witness simulation structure after run, and the proposed design is shown in Fig. 13.5.

The simulation result shows that the second proposal is able to meet the daily target of 20 units per day. The simulation result for the second proposal layout shows that the number of shipped is 20 units per day. This layout is more efficient in terms of production output compared to the existing layout. In addition, it decreases the workstation time in the bottleneck and also increases the performance by 4 units compared to the current design. In terms of efficiency, a 25% increase in production output was given to the second proposal compared with current practice.

From the results, it is shown that the first proposal has better production output compared to the current one and the second proposal. The cycle time is increased from 141.67 to 126.67 min with 73.61% more efficiency for the production line as a result of the validation. To conclude, a possible way to improve the efficiency of

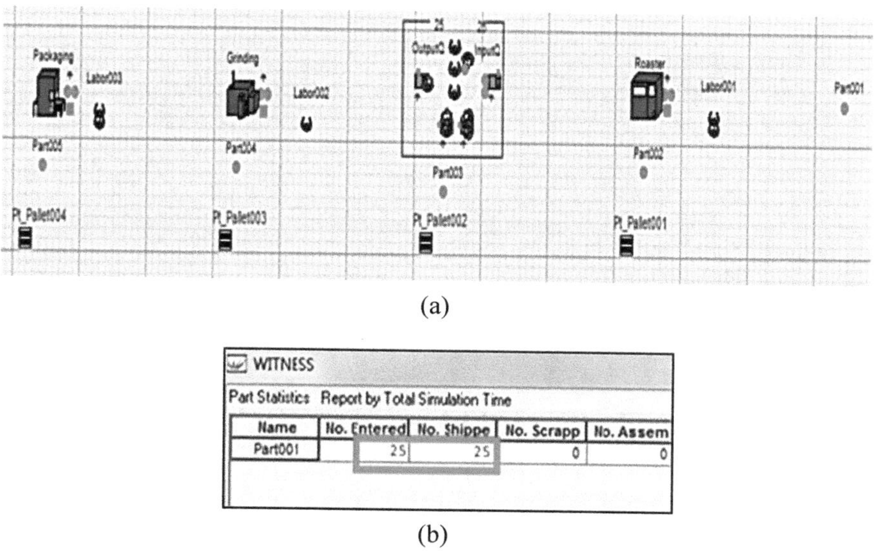

(a)

WITNESS

Part Statistics Report by Total Simulation Time

Name	No. Entered	No. Shippe	No. Scrapp	No. Assem
Part001	25	25	0	0

(b)

Fig. 13.4 **a** The simulation structure of the new layout, **b** the simulation of daily production for new layout for first proposal using I-Line

production was undertaken by a case study presented in this research. The result shows that the equipment and machines are improved.

13.5 Conclusions

The result shows that the efficiency of production has improved. The cycle time increased from 141.67 to 126.67 min with 73.61% more efficiency for the production line as a result of the validation. For future work, it is recommended to place conveyors from the workstation to other workstations and stores. In addition, this conveyor can make it easier for workers to move sacks of coffee beans to store, it can also make them work faster. Moreover this step can give more efficiency in the production line. Next, it is necessary to reduce the work in progress (WIP) store and make it in one place only because it can cause limited production line space. Workers have to be comfortable in the production line to do their job. Therefore, in the production line limited space occurs, it can impact the production line operation. Next, make a complete automation system at the cooking workstation, because workers at the cooking workstation are exposed to high temperatures when the cooking process is running, they can easily be injured such us scalding and can also have long-term effects on their health. In this system worker only controls the process in the center room, this system can also save a number of cooking workstation employees.

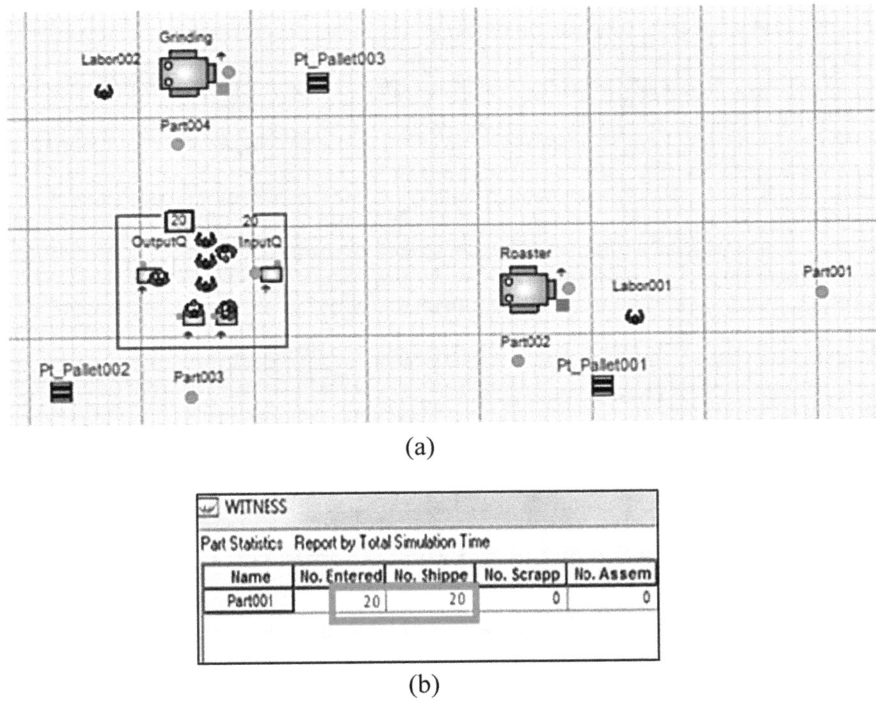

Name	No. Entered	No. Shippe	No. Scrapp	No. Assem
Part001	20	20	0	0

Fig. 13.5 **a** The simulation structure of the new layout, **b** the simulation of daily production for new layout for second proposal using L-Line

References

De Carlo, F., Arleo, M.A., Borgia, O., Tucci, M.: Layout design for a low capacity manufacturing line: a case study. Int. J. Eng. Bus. Manag. **5**(SPL. ISSUE), 1–10 (2013). https://doi.org/10.5772/56883

Liker, J.K.: The Toyota Way. McGraw Hill, NY (2004)

Markt, P.L., Mayer, M.H.: Witness simulation software a flexible suite of simulation tools. In: Winter Simulation Conference Proceedings, Atlanta, GA, USA, pp. 711-717 (1997). https://doi.org/10.1109/wsc.1997.640943

Tompkins, J.A., White, J.A., Bozer, Y.A., Frazelle, E.H., Tanchoco, J.M., Trevino, J.: Facilities Planning. Wiley, New York (1996)

Waller, A.: In: Proceedings of the 2012 Winter Simulation Conference, pp. 1–12 (2012)

Chapter 14
Study on Series Motor Four Quadrants DC Chopper Operation Controlled by an Expert System

Saharul Arof, A. R. Amir Shauqee, M. K. Zaman, N. H. Nur Diyanah, Philip Mawby, Hamzah Arof, and Emilia Noorsal

Abstract DC drive electric vehicles will have several driving conditions such as driving on a flat surface, climbing a steep hill, or cruising downhill with different types of loads, i.e., passengers and luggage. A four quadrants DC chopper (FQDC) has been developed to work with a series motor for the application of DC Drive electric vehicle in previous work, capable of operating in such driving conditions and offers several types of chopper operation modes, such as driving and reverse, regenerative and resistive braking, generator mode, field weakening, and series–parallel drive. For the FQDC to operate in the methods as mentioned earlier, it requires a control algorithm that can process input signals such as signals from the accelerator pedal, brake, speed, torque, voltage, current, load, SOC, etc., and choose the most suitable chopper operation mode. Hence, this paper describes an expert system control algorithm as the chopper operation controller. The control algorithm

S. Arof (✉) · A. R. Amir Shauqee · M. K. Zaman · N. H. Nur Diyanah
Universiti Kuala Lumpur, Malaysian Spanish Institute Kulim Hi-Tech Park, 09000 Kulim, Kedah, Malaysia
e-mail: saharul@unikl.edu.my

A. R. Amir Shauqee
e-mail: amirshauqee@unikl.edu.my

M. K. Zaman
e-mail: mkzak@unikl.edu.my

N. H. Nur Diyanah
e-mail: diyanahhisham94@gmail.com

S. Arof · P. Mawby
University of Warwick School of Engineering, Coventry CV47AL, United Kingdom
e-mail: p.a.mawby@warwick.ac.uk

H. Arof
Department of Engineering, Universiti Malaya, Jalan Universiti, 50603 Kuala Lumpur, Malaysia
e-mail: ahamzah@um.edu.my

E. Noorsal
Faculty of Electrical Engineering, Universiti Teknologi MARA, Cawangan Pulau Pinang, Permatang Pauh Campus, 13500 Permatang Pauh, Penang, Malaysia
e-mail: emilia.noorsal@uitm.edu.my

M. H. Abu Bakar et al. (eds.), *Progress in Engineering Technology III*,
Advanced Structured Materials 148, https://doi.org/10.1007/978-3-030-67750-3_14

has been simulated in MATLAB/SIMULINK. Results showed that the controller could handle several modes of operation, for different types of driving patterns, battery SOC, and loads.

Keywords Electric vehicle · Expert system · Four quadrants DC chopper · Series motor · Chopper operation

14.1 Introduction

The emission of hydrocarbons not only pollutes the environment but also contributes to global warming, which melts the icebergs and increases the sea level. Using efficient electric vehicles (EV) for transportation is one of the solutions in reducing global hydrocarbon emissions. Unfortunately, the price of EVs is still high, making it not famous for many people, especially those living in developing countries. Thus, there is a need for an efficient, compact drive system for EVs that can reduce the cost and thus making them economical and affordable.

14.2 Review Stage

Interest to embark research in DC drive EC started when Oak Research National Laboratory (ORNL) (Oak Ridge National Laboratory 2009), United States in 2009, had successfully designed a DC brushed motor with high power output (55 kW), high efficiency (92%) that can operate at low operating voltages (13 V). A new series motor four quadrants DC chopper, such as shown in Fig. 14.1, was designed, and the proposed chopper has multiple operations (Arof et al. 2014). Several other studies related to DC drive EV led to research on EC battery chargers (Arof et al. 2019a) and different types of DC drive motors such as separately excited (Arof et al. 2019b). Detailed investigations on the chopper operation modes led to the establishment of a simulation model to test the chopper operations for the application of electric car and light rail transit (LRT) (Arof et al. 2017a). This led to further investigation on each of the chopper operations in detail on the specific pattern of voltage, current, torque, speed, of the FQDC running DC series motor (Arof et al. 2017b). For the DC series motor traction of EC application, the speed and torque control has been successfully done and implemented with direct current control (Arof et al. 2016).

For power regeneration, the FQDC offers the generator mode, and several techniques to regenerate power are studied and discussed in (Arof et al. 2019c). For the FQDC to be applied in the real world; it needs controllers running control algorithm in the embedded system. The controller and its control algorithm are studied and tested using a processor in the loop technique (Arof et al. 2019d). The close-loop controller using the PID tuned using pole placement (Arof et al. 2020a) and Fuzzy logic (Arof et al. 2020b) are applied to improve the performance and stability. These

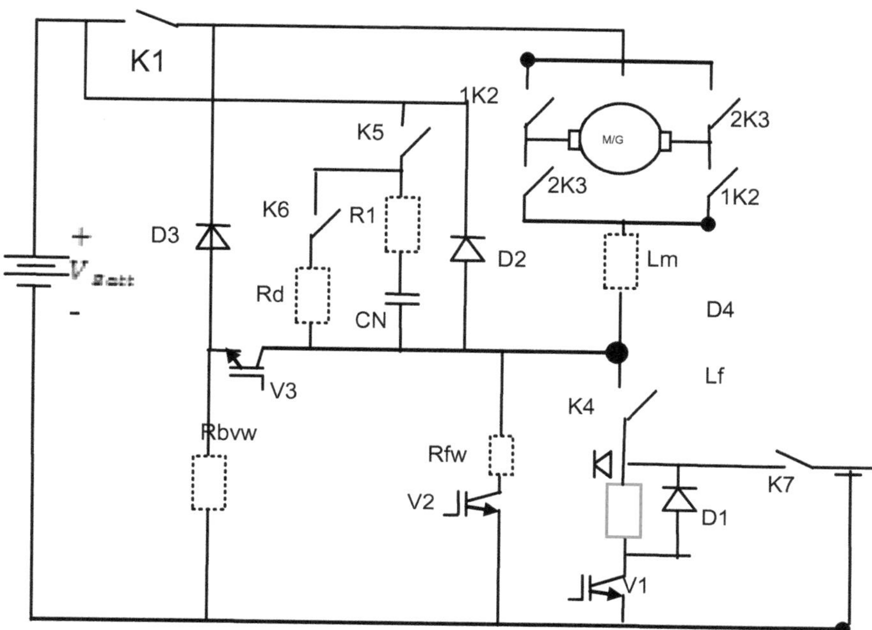

Fig. 14.1 Four quadrants DC chopper (FQDC)

methods and direct current control have improved performance of torque fluctuations and after improvisation made and using cascade PID associated with steering (Arof et al. 2020c) and vehicle movement control (Arof et al. 2020d), it was used for autonomous EV automatic reverse parking. For fault diagnosis and online system tuning and optimization, a numerical representation using Taylor's series (Arof et al. 2020e) was studied and tried for the driving mode. To improve the performance and optimizations tool of a new FQDC such as artificial intelligence with self-tuning Fuzzy (Arof et al. 2015a), neural networks as well as ANFIS was introduced to control all the chopper operations of the proposed FQDC chopper (Arof et al. 2015a, b, 2019e). Each process of FQDC modes performance can be improved using AI optimization tools such as genetics algorithm to set up a specific look-up table (Arof et al. 2019f).

In this paper, a study on a four quadrants drive DC chopper (FQDC), chopper modes of operations controlled by an expert system controller for electric vehicles (EVs) application, is carried out as shown in Fig. 14.1.

14.2.1 Chopper Hopper Operation Controller

The chopper operation controller has expected to choose the most appropriate operation by processing signals obtained from the accelerator pedal, brake pedal, battery voltage, motor speed, etc. This paper will focus on a chopper operation controller controlling the proposed FQDC to drive a DC series motor. Upon receiving these signals, the controller will generate other signals such as error and rate of speed. The vehicle traction operation of FQDC can be divided into three modes, such as driving, generator, and braking. Meanwhile, the FQDC EV driving operation has three general categories which are (1) driving mode for slow driving, (2) field weakening mode for fast driving but at low load, and (3) parallel mode for climbing steep hills.

Additionally, the FQDC braking operation can be divided into two categories such as (1) generative braking for braking action at high speed and (2) resistive braking for braking at low speed. The regenerative braking mode is to slow down the vehicle and charge batteries, while the resistive braking mode is used to charge batteries and halt the vehicle. In contrary, the generator mode is only used to charge batteries like a standard generator. The EV will experience three driving conditions according to earth profile which is driving at a flat surface, downhill, and hill-climbing. It also carries different kinds of loads which fall into three load categories: low load, medium load, and high load. The loads mainly referred to as passengers or luggage. Additionally, the EV must consider the battery's remaining state of charge (RSOC) and the remaining distance traversed (RDT) which provides the ratio RSOC/RDT. Commonly, an EV will experience driving condition running at high SOC while the remaining distance travelled is high, or at medium. In contrast, the distance traversed is medium, or low while the SOC is low.

Hence, the performance of the chopper operation controller is determined and measured by three performance criteria. The first criterion is whether the chopper controller can select the appropriate operation mode. For instance, when the EV is supposed to be in driving mode, the controller should not select braking mode. The second criterion is whether the chopper controller can select the most appropriate operation mode. The controller should be able to differentiate the most appropriate operation for a driving mode which consists of normal driving, field weakening, or parallel mode. The ability to select the most appropriate function depends on the signal inputs. For example, the controller will first select the driving mode and then picks the parallel mode instead of field weakening mode for better performance in climbing a steep hill.

The third criterion is to consider the EV maximum speed and distance traversed concerning the selected operation mode. Hence, would the chopper controller be able to maintain the ratio of RSOC/RDT? To accelerate at top speed, the EV consumes more battery power. On the contrary, to achieve maximum distance traversed, the EV consumes low battery power. The maximum distance travelled is not directly proportional to EV speed because there are losses such as heat losses, IGBT firing losses, chopper losses, etc. These losses were due to operating the FQDC at its maximum power by firing the IGBTs at their maximum limit. For example, the parallel mode used for climbing the steep hill cause the EV consumed high battery power. However, the increase in speed is much lower than in normal driving mode or field weakening mode.

Accordingly, the FQDC chopper operation controller must balance these criteria at its optimum, not only to achieve better performance but also to prevent chopper failure. Failing to pick the most appropriate operation could cause chopper traction failure.

14.2.2 Expert System

An expert system is an interactive computer-based decision tool that uses both facts and heuristics to solve difficult decision problems based on knowledge acquired from an expert (Negnevitsky 2011). It is a computer program that simulates the thought process of a human expert to solve complex decision problems in a specific domain. The characteristics of expert systems make them different from conventional programming and traditional decision support tools.

14.2.2.1 Characteristics of an Expert System

Successful expert systems will be those that mix facts and algorithms and in solving issues, integrate human understanding with computer power. To be efficiently mentioned below, an expert system must concentrate on a specific issue area. The method selected to use in this project is the expert system. An expert system is defined as an interactive and reliable computer-based decision-making system that uses both facts and algorithms to solve complex decision-making problems. The expert system can resolve many issues which generally would require a human expert. It based on knowledge acquired from an expert.

(a) Domain specificity
Expert systems are typically very domain specific. For example, a diagnostic expert system for troubleshooting computers must perform all the necessary data manipulation as a human expert would.
(b) Special programming languages
Expert systems have typically written in special programming languages such as LISP, PROLOG, and C++.
(c) Operates as an interactive system
The expert system responds to questions, ask clarifications, makes recommendations, and aid the decision-making process.
(d) Tools could filter knowledge
Expert systems have mechanisms to expand and update knowledge continuously.

(e) Make logical inferences based on knowledge stored
The knowledge base must have means of exploiting the knowledge stored; else it is useless, e.g., learning all the words in a language, without how to combine those words to form a meaningful sentence.

(f) Ability to explain the reasoning
The expert system remembers the logical chain of reasoning.

14.3 Methodology

The chopper operation controller must be able to choose the most appropriate chopper operation. For this purpose, it must be tested and pass the earth profile test. The EV has tested on three different earth profiles that are flat driving, downhill, and uphill as shown in Fig. 14.2. The expected chopper operations tested are driving, field weakening, regenerative braking, resistive braking generator, and parallel modes. Form this test, the EV load has set to a minimum, and the battery SOC has set to 100%, or the battery is fully charging. The accelerator and brake pedal has been set-up as input signals. The error difference between accelerator signals and speed and rate of speed are extracted from the test and are used as input signals in designing the chopper operation of the controller control algorithm. A simulation model shown in Fig. 14.3 is used to test and simulate the FQDC. The chopper operation modes have been determined and preset in this model.

The rule-based expert system proposed by Newell and Simon in 1970 has used as the basis for the control algorithm in this work. Backward chaining is the goal for driven reasoning. In backwards chaining, an expert system has the goal (a hypothetical solution), and the inference engine attempts to find the evidence to prove

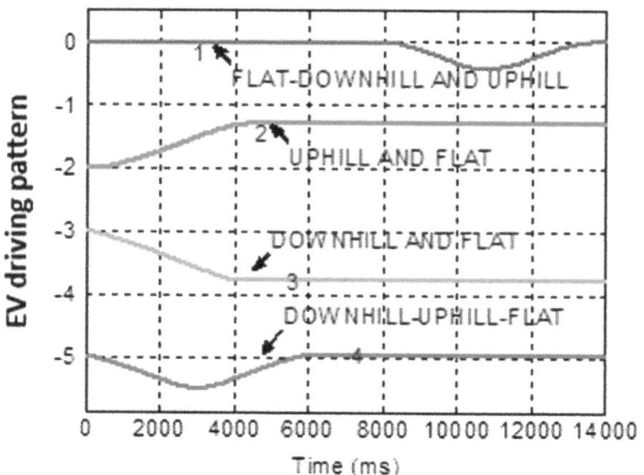

Fig. 14.2 Driving pattern

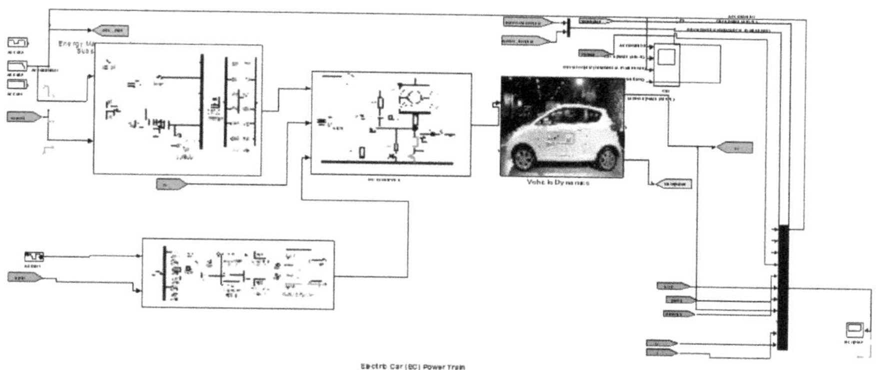

Fig. 14.3 MATLAB/SIMULINK model

it. First, the knowledge base has searched to find rules that might have the desired solution. Such rules must have a goal in their *THEN* (action). If such as a rule is found and the *IF* condition matches data in the database, then the rule is fired, and the goal is proven (Arof et al. 2020b). In this work, the goal refers to the chopper operation which consists of driving (DRV), field weakening (FW), generator (GEN), regenerative braking (RGB), resistive braking (RSB), and parallel mode (PAR) while the *IF* part refers to signals from the accelerator pedal, brake, speed, error, state of charge, and load as described in Table 14.1.

First, establish the basic rule for the first layer of *IF–THEN* rules for the general chopper operation mode:

IF the accelerator pedal > 0 *and brake* ≤ 0 *THEN the action is driving mode.*

IF the brake > 0 *THEN is a braking mode.*

IF error ≤ -0.3 *THEN is generator mode.*

The previous three basic *IF–THEN* rules can use to set new rules:

IF driving mode is selected and the speed > 2.5 *THEN field weakening mode.*

Table 14.1 Rule selection for input signals

	DRV	FW	GEN	RGB	RSB	PAR
Acc_pedal	>0	>2.5	>0	N/A	N/A	>2.5
Brake	0	0	0	>0	>0	0
Speed	> 0 < 2.5	>2.5	>2.0	>2.5	<2.5 and >1.0	>2.5
Rate of speed	N/A	>0	<5 >−0.2	<0	<0	<0
Error	>0	>0	>−0.3 <0	N/A	N/A	>1.5 <3
SOC	>0	>1.5 <5	>1 <5	>0	>0	>1.5 <5
Load	>0	>0 and <2.5	N/A	>0	>0	>0 and <2.5

IF driving mode is selected and the speed > 0 and speed < 2.5 THEN driving mode.

IF driving mode selected and the speed > 2.5 and the ROS < -0.5 THEN its parallel mode.

IF braking mode is selected and the speed > 3.5 THEN regenerative braking is selected.

IF braking mode is selected and the speed is below 3.5, THEN resistive braking mode is selected.

14.4 Results and Discussion

The performance of the proposed expert system-based strategy applied on a controller can be evaluated on the real hardware system or simulated systems. A simulated system is easily developed, time-saving, and cost-effective. It provides a convenient platform to test the algorithms and to stimulate the process of chopper and EV operations. If the algorithm fails, it not effected too many risks in terms of the costs spent and losses brought to the system. Furthermore, various parameters can be simply adjusted and optimized. It is hence speeding up the simulation process toward achieving the desired results. In this work, a computer simulation has performed as a method to test the expert system algorithm and to study its performance and effectiveness.

For this purpose, MATLAB/Simulink is used to develop the controller and the system. The user provides some parameter values required by the simulation software to simulate the system. The car has tested according to the accelerator signal (signal 9), brake signal (signal 7), and also according to earth profile (signal 8) as shown in Fig. 14.4. The expected operation for the four-quadrant DC chopper concerning the test signals given during start-up, the driving mode is selected. As the vehicle speed increases and due to accelerator demand, the field weakening mode is chosed. When the accelerator signal is set low, while the speed is high, the generator mode is selected. But when the accelerator signal is set maximum and the speed at medium or high, the field weakening mode is selected. When the brake command is activated, and the vehicle speed is still high due to inertia, the regenerative mode is operated. As the vehicle speed drops further, the resistive braking mode is expected. If the brake command is released and replaced with a low driving command while the vehicle speed is low, driving mode operated. As the vehicle moves downhill while the driving command is low, the generator modes are picked. However, if the driving command is high, while vehicle speed is increases, the field weakening mode is selected. If the vehicle drives on a steep hill, the speeds expected to drop, hence, the parallel mode is expected. Finally, as the vehicle regained its speed, the field weakening mode is selected.

The performance of the expert system (ES) as a chopper operation controller depicted in Fig. 14.5. The ES controller is expected to select the correct operation for the four-quadrant DC chopper concerning the test signals given. During start-up, the

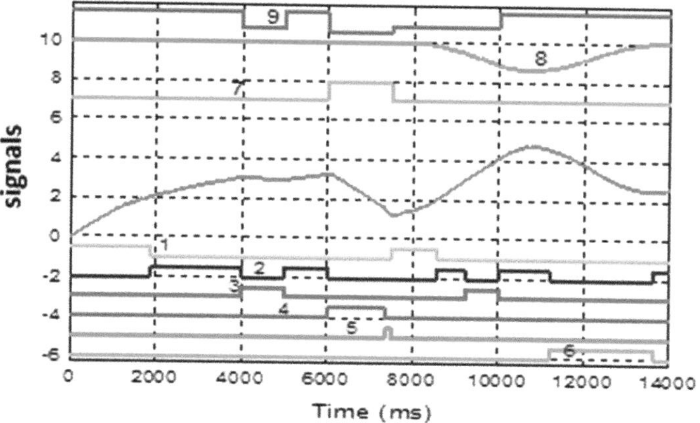

Legend:

1. Driving
2. Field Weakening
3. Generator
4. Regenerative Braking
5. Resistive Braking
6. Parallel Mode
7. Brake Signal
8. Earth Profile
9. Accelerator Signal

Fig. 14.4 FQDC operation results

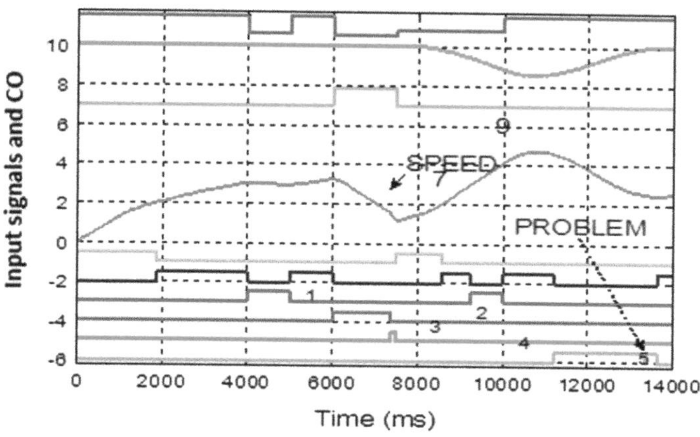

Fig. 14.5 Results of ES as FQDC controller

ES selected the driving mode. As the vehicle speed increases and due to accelerator demand, the ES decided the field weakening mode. However, when the accelerator signal is set low, while vehicle speed is at high or medium, the generator mode is selected. In contrary, when the accelerator signal is maximum and the speed still high, the field weakening mode is selected. When the brake command is activated, while the vehicle speed is still high due to inertia and the regenerative mode is

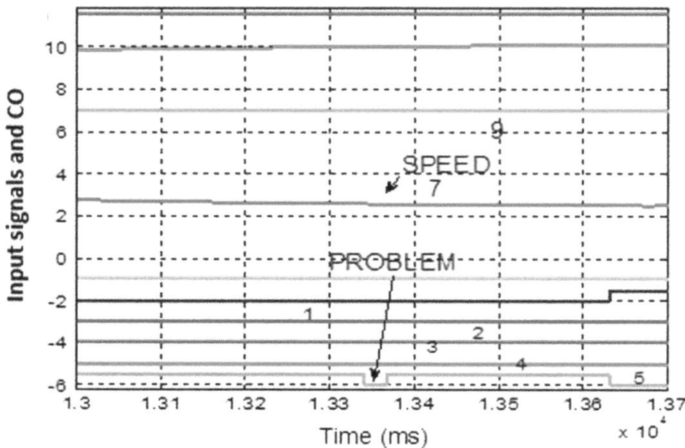

Fig. 14.6 ES controller output

operated. As the vehicle speed drops further, the resistive braking mode is expected. When the brake command is released and replaced with a low driving command, while the vehicle speed is low, the ES should select driving mode. As the vehicle moves downhill while the driving command is low, the ES should select the generator mode. However, if the driving command is high, the controller should then select the field weakening mode. As the vehicle drives on a steep hill, the speed has expected to drop. Hence, the parallel mode need to be selected. Finally, as the vehicle regained its speed the field weakening mode is selected.

Overall, ES picks the right mode of operation most of the time except where glitches occur at the end of the test. Figure 14.6 shows an enlarged view of the part where the glitches are. It is due to a slight confusion when the ANN controller is not sure whether to choose the field weakening or parallel mode. But eventually, it picks the parallel mode, which is the correct decision.

14.5 Conclusion

The "Expert System If Then Rules" offers the simplest method. But plenty of rules must be written depending on each type of chopper operation, and input signals cause the drawback. Additional sensors of car position and weight sensors are needed to facilitate the control algorithm and to differentiate the selection mode of the chopper operations such as driving, field weakening, regenerative braking, resistive braking, etc. The signal is also required to ensure that changes from field weakening to parallel mode are easily distinguishable.

Commonly used IF–THEN rules in the expert system are the fundamental and most straightforward way of controlling the chopper operation of FQDC for the application of EV. The ES controller has the potential to be used as the four-quadrant chopper operation controller. If the appropriate tuning of the controller can be done, and this could result in the improvement of the EV performance. The DC drive series motor and four quadrants DC chopper has high potential to be utilized in EV. It is due to a simple design, low cost, and excellent controllability.

References

Arof, S., Yaakop, N.M., Jalil, J.A., Mawby, P.A., Arof, H.: Series motor four quadrants drive DC chopper for low cost, electric car: Part 1: Overall. In: 2014 IEEE International Conference on Power and Energy (PECon), pp. 342–347 (2014)

Arof, S., Jalil, J.A., Kamaruddin, N.S., Yaakop, N.M., Mawby, P.A., Hamzah, A.: Series motor four quadrants drive DC Chopper part2: driving and reverse mode with direct current control. In: 2016 IEEE International Conference on Power and Energy (PECon), pp. 775–780 (2016)

Arof, S., Hassan, H., Rosyidi, M., et al.: Implementation of series motor four quadrants drive DC chopper for dc drive electric car and LRT via simulation model. J. Appl. Environ. Biol. Sci. **7**(3S), 73–82 (2017a)

Arof, S., Noor, N.M., Alias, M.F., et al.: Investigation of chopper operation of series motor four quadrants DC chopper. J Appl. Environ. Biol. Sci. **7**(3S), 49–56 (2017b)

Arof, S., Diyanah, N.H.N., Mawby, P.A. et al.: Low harmonics plug-in home charging electric vehicle battery charger utilizing multi-level rectifier, zero crossing and buck chopper: part 1: general overview. In: Bakar, M.H.A., Sidik, M.S.M., Öchsner, A. (eds.) Progress in Engineering Technology. Advanced Structures and Materials, 119. Springer, Cham (2019a)

Arof, S., Diyanah, N.H.N., Noor, N.M. et al.: A new four quadrants drive chopper for separately excited DC motor in low-cost electric vehicle. In: Bakar, M.H.A., Sidik, M.S.M., Öchsner, A. (eds.) Progress in Engineering Technology. Advanced Structures and Materials, 119. Springer, Cham (2019b)

Arof, S., Diyanah, N.H.N., Noor, N.M.N. et al.: Series motor four quadrants drive DC chopper: Part 4: Generator mode. In: Bakar, M.H.A., Sidik, M.S.M., Öchsner, A. (eds.) Progress in Engineering Technology. Advanced Structures and Materials, 119. Springer, Cham (2019c)

Arof, S., Diyanah, N.H.N., Yaakop, M.A. et al.: Processor in the loop for testing series motor four quadrants drive DC chopper for series motor driven electric car: part 1: Chopper operation modes testing. In: Ismail, A., Bakar, M.H.A., Öchsner, A. (eds.) Advanced Engineering for Processes and Technologies. Advanced Structures and Materials, 102. Springer, Cham (2019d)

Arof, S., Diyanah, N.H.N., Mawby, P.A. et al.: Study on implementation of neural network controlling four quadrants direct current chopper: part1: using single neural network controller with binary data output. In: Ismail, A., Bakar, M.H.A., Öchsner, A. (eds.) Advanced Engineering for Processes and Technologies. Advanced Structures and Materials, 102. Springer, Cham (2019e)

Arof, S., Diyanah, N.H.N., Noor, N.M. et al.: Genetics algorithm for setting up look up table for parallel mode of new series motor four quadrants DC chopper. . In: Bakar, M.H.A., Sidik, M.S.M., Öchsner, A. (eds.) Progress in Engineering Technology. Advanced Structures and Materials, 119. Springer, Cham (2019f)

Arof, S., Noor, N.M., Alias, M.F. et al.: Digital proportional integral derivative (PID) controller for closed-loop direct current control of an electric vehicle traction tuned using pole placement. In: Bakar, M.H.A., Zamri, F.A., Öchsner, A. (eds.) Progress in Engineering Technology II. Advanced Structures and Materials, 131. Springer, Cham (2020a)

Arof, S., Noor, N.M., Mohamad, R. et al.: Close loop feedback direct current control in driving mode of a four quadrants drive direct current chopper for electric vehicle traction controlled using fuzzy logic. In: Bakar, M.H.A., Zamri, F.A., Öchsner, A. (eds.) Progress in Engineering Technology II. Advanced Structures and Materials, 131. Springer, Cham (2020b)

Arof, S., Said, M.S., Diyanah, N.H.N. et al.: Series motor four-quadrant direct current chopper: reverse mode, steering position control with double-circle path tracking and control for autonomous reverse parking of direct current drive electric car. In: Bakar, M.H.A., Zamri, F.A., Öchsner, A. (eds.) Progress in Engineering Technology II. Advanced Structures and Materials, 131. Springer, Cham (2020c)

Arof, S., Said, M.S., Diyanah, N.H.N. et al.: Series motor four-quadrant DC chopper: reverse mode, direct current control, triple cascade PIDs, and ascend-descend algorithm with feedback optimization for automatic reverse parking. In: Bakar, M.H.A., Zamri, F.A., Öchsner, A. (eds.) Progress in Engineering Technology II. Advanced Structures and Materials, 131. Springer, Cham (2020d)

Arof, S., Sukiman, E.D., Diyanah, N.H.N. et al.: Discrete-time linear system of New series motor four-quadrant drive direct current chopper numerically represented by Taylor series. In: Bakar, M.H.A., Zamri, F.A., Öchsner, A. (eds.) Progress in Engineering Technology II. Advanced Structures and Materials, 131. Springer, Cham (2020e)

Arof, S., Muhd Khairulzaman, A.K., Jalil, J.A., Arof, H., Mawby, P.A.: Self tuning fuzzy logic controlling chopper operation of four quadrants drive DC chopper for low-cost electric vehicle. In: 2015 6th International Conference on Intelligent Systems, Modelling and Simulation, pp. 40–45 (2015a)

Arof, S., Zaman, M.K., Jalil, J.A., et al.: Artificial intelligence controlling chopper operation of four quadrants drive DC chopper for low-cost electric vehicle. Int. J. Simul. Syst. Sci. Technol. **16**(4), 1–10 (2015b)

Negnevitsky, M.: Artificial Intelligence, a Guide to Intelligent System, 3rd edn. Pearson Education, Addison Wessley (2011)

Oak Ridge National Laboratory: Advanced brush technology for DC Motors. https://peemrc.ornl.gov/projects/emdc3.jpg. Accessed 15 May 2009

Chapter 15
Fault Diagnose of DC Drive EV Utilizing a New Series Motor Four Quadrants DC Chopper Using an Expert System and Quadratic Solver Running in Embedded

Part 1: During Start Up

Saharul Arof, M. R. Faiz, N. H. N. Diyanah, N. M. Yaakop, Philip Mawby, H. Arof, and Emilia Noorsal

Abstract This paper proposes a fault diagnose method in finding problems related to four quadrants DC chopper and the propulsion motor during start up while vehicle is still at standstill. The check is done during start up of the DC drive electric vehicle (EV). If the problems exist and if it is major, the EV will not be allowed to operate. The expert system and quadratic solver are implemented as part of the control algorithm to search for the fault. MATLAB/Simulink is used to test and verify the checking control algorithm. The simulation results using MATLAB/Simulink with the expert

S. Arof (✉) · M. R. Faiz · N. H. N. Diyanah · N. M. Yaakop
Universiti Kuala Lumpur, Malaysian Spanish Institute, Kulim Hi-Tech Park, 09000 Kulim, Kedah, Malaysia
e-mail: saharul@unikl.edu.my

M. R. Faiz
e-mail: faizroslan966@gmail.com

N. H. N. Diyanah
e-mail: diyanahhisham94@gmail.com

N. M. Yaakop
e-mail: nurazlin@unikl.edu.my

S. Arof · P. Mawby
University of Warwick School of Engineering, Coventry CV47AL, UK
e-mail: p.a.mawby@warwick.ac.uk

H. Arof
Department of Engineering, Universiti Malaya, Jalan Universiti, 50603 Kuala Lumpur, Malaysia
e-mail: ahamzah@um.edu.my

E. Noorsal
Faculty of Electrical Engineering, Universiti Teknologi MARA (UiTM), Cawangan P. Pinang, Permatang Pauh Campus, 13500 Permatang Pauh, P. Pinang, Malaysia
e-mail: emilia.noorsal@uitm.edu.my

M. H. Abu Bakar et al. (eds.), *Progress in Engineering Technology III*,
Advanced Structured Materials 148, https://doi.org/10.1007/978-3-030-67750-3_15

system and quadratic solver algorithm techniques show that it can successfully detect if problems exist.

Keywords DC drive · Four quadrants chopper · Fault diagnose · Expert system · Quadratic solver · EV

15.1 Introduction

Problems associated with hydrocarbon-fueled vehicles as well as improvements in electric vehicle technology has increased the interest in electric vehicles. Using efficient electric vehicles (EV) and hybrid electric vehicles (HEV) for transportation is one of the solutions to reducing global hydrocarbon emission. Unfortunately, the price of EV and HEV is expensive making it unattainable to many people, especially those living in poor countries resulting in research for economical DC drive EV is carried out.

15.2 Literature Review

Oak Research National Laboratory (ORNL) (2009), United States in 2009, had successfully designed a DC brushed motor with high power output (55 kW), high efficiency (92%) that can operate at low operating voltages (13 V) and this has initiated the interest to embark research in DC drive EV. Many attempts to improve the conventional H-bridge chopper by increasing the operations or the motor reverse action have been continuously carried out. But ever since the development of of the new DC motor by ORNL, a new series motor four quadrants DC chopper as shown in Fig. 15.1, was designed and the proposed chopper has more operations compared to the conventionl version (Arof et al. 2009a). Other studies related to DC drive EV led to research on different types of DC drive brushed motor such as the separately excited DC motor that can be used for motor traction for DC EV has been done (Arof et al. 2014). Detailed investigations on the chopper operation modes led to the establishment of a simulation model to test the chopper operations for the application as electric car and light rail transit (LRT) have been done (Arof et al. 2019a). This simulation model led to further detailed investigations on each of the chopper operations and on the specific pattern of motor voltage, current, torque, speed, of the series motor and FQDC running for DC drive EV application have been continuously carried out (Arof et al. 2017a). For DC series motor traction of EV application, the speed and torque control for the series motor in an attempt to reduce jerk and tire slip have been successfully done and implemented with the direct current control technique (Arof et al. 2017b). For power regeneration, the FQDC offers the generator mode with several techniques of starting the regenerated power and voltage control as studied and discussed in (Arof et al. 2016). For the FQDC to be applied in real world it needs

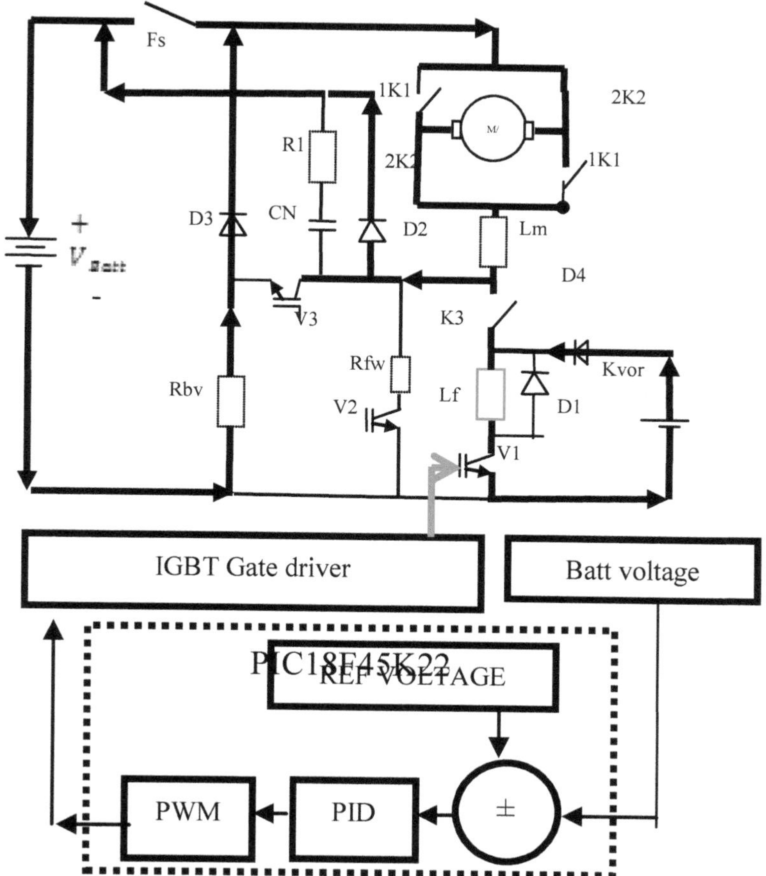

Fig. 15.1 New four quadrants DC chopper

controllers running control algorithms in the embedded system. The controller and its control algorithm are studied and tested using the processor in the loop (PIL) technique (Arof et al. 2019b). To improve the new FQDC performance, an optimization tool such as artificial intelligence (AI) is introduced to control all the chopper operations of the proposed FQDC chopper (Arof et al. 2015a, 2019c). Among the three AI controllers, ANFIS shows the best performance followed by neural network and self-tuning Fuzzy logic controller. A study on neural network controller to uncover the proper method of tuning has been carried out such as using single controller with binary output (Arof et al. 2015b). For each specific FQDC chopper operation mode the performance can be further improved using AI optimization tools such as a genetics algorithm to set up a specific look up table for the field current in parallel mode operation (Arof et al. 2019d). Several other studies related to DC drive EV led to research on EV battery chargers such as shown in Fig. 15.1, that offer low harmonics (Arof et al. 2019e). This battery charger uses zero cross, multilevel rectifier, and buck chopper. Close loop controller using PID tuned using pole placement (Arof et al. 2019f) and Fuzzy Logic (Arof et al. 2020a) are applied to improve the performance and stability. This method has undergone improvements using cascade PID and ascend descend algorithm and after improvisation made and associated with steering (Arof et al. 2020b) and vehicle movement control (Arof et al. 2020c) it was used for autonomous EV automatic reverse parking. For fault diagnose and online system tuning and optimization, a numerical representation using Taylor series (Arof et al. 2020d) is studied and tried for driving mode.

To make the EV efficient it requires a fault diagnose controller that associated to Battery Management system (BMS) to make the EV safe, reliable, and a cost-efficient solution. The system controls the operational conditions of the FQDC, traction motor, battery, etc. to prolong its life and guarantee its safety. This paper extends the FQDC research specifically and the focus is for fault diagnoses of FQDC and propulsion motor using expert system and quadratic solver during the Electric Vehicle start up.

15.2.1 Faultly in Electric Vehicle and fault Diagnosis System

An EV is a fragile system as the system is still under the relatively weak circumstance (Sobhani-Tehrani and Khorasani 2009). If one of the core components of EVs is broken down, the whole system might be affected. The EVs system may have various problems, and any failure could further expand thereby affecting the safe operation of the vehicle (Lin et al. 2013). The fault type varies depending on the location of the failure and can be specified into five which are the motor failure, motor controller failure, a failure between motor controller and motor lines, motor power supply failure, and the motor system communication failure (Miljković 2011).

A fault diagnosis system should ideally meet some general requirements. Some of the most important desirable attributes of a diagnostic system are explained in the following:

- **Early detection and diagnosis**: This refers to the potential of a diagnostic system in detecting early faults. Early detection of faults before their full manifestation into a failure is of utmost importance for fault-tolerant control of safety–critical systems.
- **Isolability**: This is the capability of a diagnostic system in distinguishing the origins of a fault from other potential fault sources or to locate a faulty component among various components of a system.
- **Fault identifiability**: To estimate the severity, type, or nature of the fault. While being useful for fault accommodation purposes, fault identifiability is a definitive requirement for fault prognosis. Accurate fault identification is usually very difficult to achieve due to the presence of measurement noise, system disturbances, modeling uncertainties, and many more.

- **Robustness**: Uncertainties are inevitable in practical settings. Therefore, robustness to measurements noise, system disturbances, and modeling uncertainties is one of the foremost extremely fascinating attributes of a diagnostic system meant for practical implementations.

15.2.2 Fault Detection Methods

Faults detection methods exist in several overlapping taxonomies of the field. Some are more oriented toward control engineering approaches, other to mathematical/statistical/AI approaches. The following division is the methods of fault detection (Miljković 2011):

i. *Data Methods and Signal Models*
- Limit checking and trend checking
- Data analysis (PCA)
- Spectrum analysis and parametric models
- Pattern recognition (neural nets)

ii. *Process Model-Based Methods*
- Parity equations
- State observers
- Parameter estimation
- Nonlinear models (neural nets)

iii. *Knowledge-Based Methods*
- Expert systems
- Fuzzy logic

15.2.3 Expert System

An expert system is an interactive computer-based decision tool that uses both facts and heuristics to solve difficult decision problems based on knowledge acquired from an expert. It is a computer program that simulates the thought process of a human expert to solve complex decision problems in a specific domain. The characteristics of expert systems that make them different from the conventional programming and traditional decision support tools (Gupta and Singhal 2013).

15.3 Characteristics of Expert Systems

Successful expert systems will be those that mix facts and algorithms and in solving issues, integrate human understanding with computer power. To be efficient, as mentioned below, an expert system must concentrate on a specific issue area. The

```
switch state

state 1

        if(condition)
                {......................
                 statements
                 ......................
                }
        else
                {......................
                 statements
                 ......................
                }
state 2
        if(condition)
                {......................
                 statements
                 ......................
                }
        else
                {......................
                 statements
                 ......................
                }|
endstate
```

Fig. 15.2 Example expert system with if then rules algorithm

ES also needs to have ability to explain the reasoning in which the expert system remembers the logical chain of reasoning (Fig. 15.2).

15.3.1 Quadratic Equation

In algebra, a quadratic equation is an equation having the form where x represents an unknown, and a, b, and c represent known numbers, with $a \neq 0$. If $a = 0$, then the equation is linear, not quadratic, as there is no term. The values of x that satisfy the equation are called solutions of the equation, and roots or zeros of its left-hand side.

A quadratic equation has at most two solutions. If there is no real solution, there are two complex solutions. If there is only one solution, one says that it is a double root.

15.3.2 Embedded System

An embedded system is a combination of computer hardware and software, either fixed in capability or programmable, designed for a specific function or functions within a larger system. Industrial machines, agricultural and process industry devices, automobiles, medical equipment, cameras, household appliances, airplanes, vending machines, and toys, as well as mobile devices, are possible locations for an embedded system. Such example are Microcontroler and Digital signal Processing controller.

15.4 Methodology

In this paper the mathematical representation of the four quadrants chopper which needed to be solved using quadratic regulator running in embedded systems is as shown in Eq. (15.1).

$$I_a = \frac{V_{\text{batt}} - I_a(R_a + R_f)}{L_a + L_f} \tag{15.1}$$

The embedded system such as system, such as the PIC microcontroller, which receives an input signal from voltage and current sensors and produces digital and PWM output. The objectives of this research are to develop a fault diagnose algorithm to detect faults in the four quadrants chopper, the propulsion motor and all the associated devices such as contactors, diode, propulsion motor, etc. The error examples could be due to components failure, short circuit, and broken parts. The fault-finding only operates during the start-up of the electric vehicle. As soon as the EV started, this fault diagnosis system will be operated. Expert system is used based on the knowledge of the researcher to conclude what is the problem in the circuit with respect to reading of the current feedback signals and voltage. In this project, the expert system is used to set the parameters of voltage and current to a certain value being able for the system to recognize if any error in the FQDC circuit arises. The quadratic equation and quadratic solver are used to solve and define the expected current that flows to the circuit and propulsion motor due to changes in EV batteries voltage. Equation (15.1) is the most commonly used to determine the line current and voltage. This includes solving and finding the expected voltage and current due to resistance of the propulsion motor. At high battery voltage if contactor is make contact and the igbt is fired, the computation and actual value (from sensors) of voltage and current is high and at the low battery voltages the voltage and current is low. Two embedded controllers are required for this research, i.e., the firing sequence controller and the fault diagnose controller. The sequence controller provides the sequence of firing of contactors and IGBT while the fault diagnoses controller, which is running the algebraic solver, and the expert system controller are the ones who decide what problems pertained to the system. The sequence controller provides the delay time so that the allocated time is enough so that voltage and current sensors feedback can be read by the fault diagnose controller. There is intercommunication between the controllers where the sequence controllers pass information such as sequence number while the fault diagnoses controller also pass crucial information that decides whether the sequence of checking should be continued or stopped to avoid further damage to the propulsion motor or FQDC as a result of continuous testing the sequence.

The FQDC has 7 contactors and 3 IGBT. It has 5 current sensors and 4 voltage sensors. The fault diagnoses test starts by checking the battery voltage, and then starts to operate main contactors, forward contactors, parallel contactor, field excite contactor, filed weakening contactor, and continue to fire IGBT1, IGBT2, and IGBT3. The delay time is guaranteed to allow the associated voltage and current sensors to have enough time to be read by the controller. Details on the firing sequence and the expert system algorithm and quadratic solver will be discussed in detail in another chapter.

Fault Diagnose of FQDC for EV Algorithm Flowchart.

The flow chart of control algorithms for fault diagnose controller is as shown in Fig. 15.3 and the step of operation for firing sequence controller is as shown in Fig.15.4. In the Fig. 15.4, on each of the state of operation, FQDC contactors and IGBTs are turn on and off for the purpose of FQDC circuit checking. When necessary the quadratic solver which is running in embedded system is used to solve Eq. 15.1. This is to compare the expected current and voltage resulting from computation algorithm with actual reading from sensors and if there is obvious difference a decision to allow or stop the EV from operating will be done by the fault diagnose controller.

15.4.1 Simulation Model Development

MATLAB and Simulink Simulation model as shown in Fig. 15.5 is established for fault diagnose and MATLAB/Simulink matlab function is used as the sequence controller that represents the embedded system controllers. The fault diagnoses controllers are running the quadratic solver and expert system algorithm (Fig. 15.6).

15.5 Results and Discussion

Two tests were conducted, i.e., one is a normal circuit without error and the second is with a short circuit of the FQDC circuit.

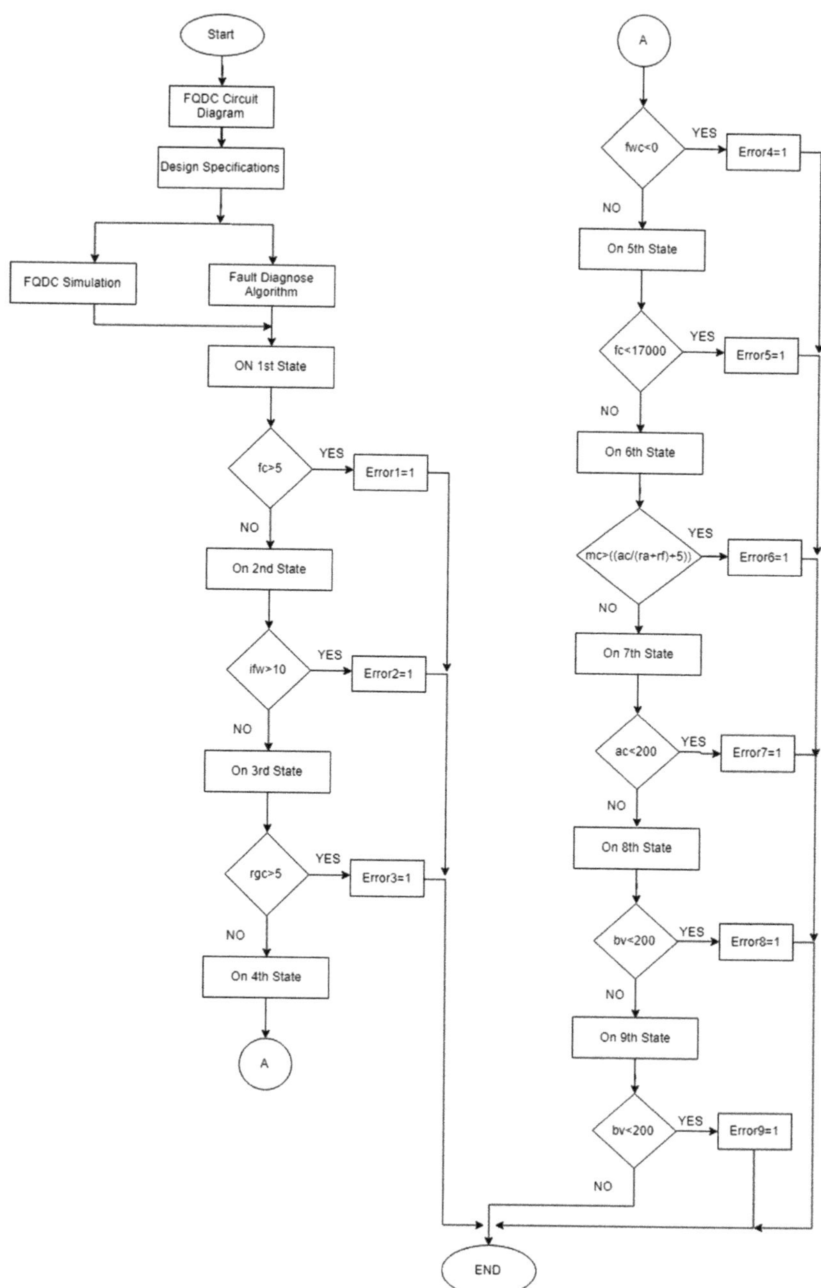

Fig. 15.3 Block diagram of sequence signals test

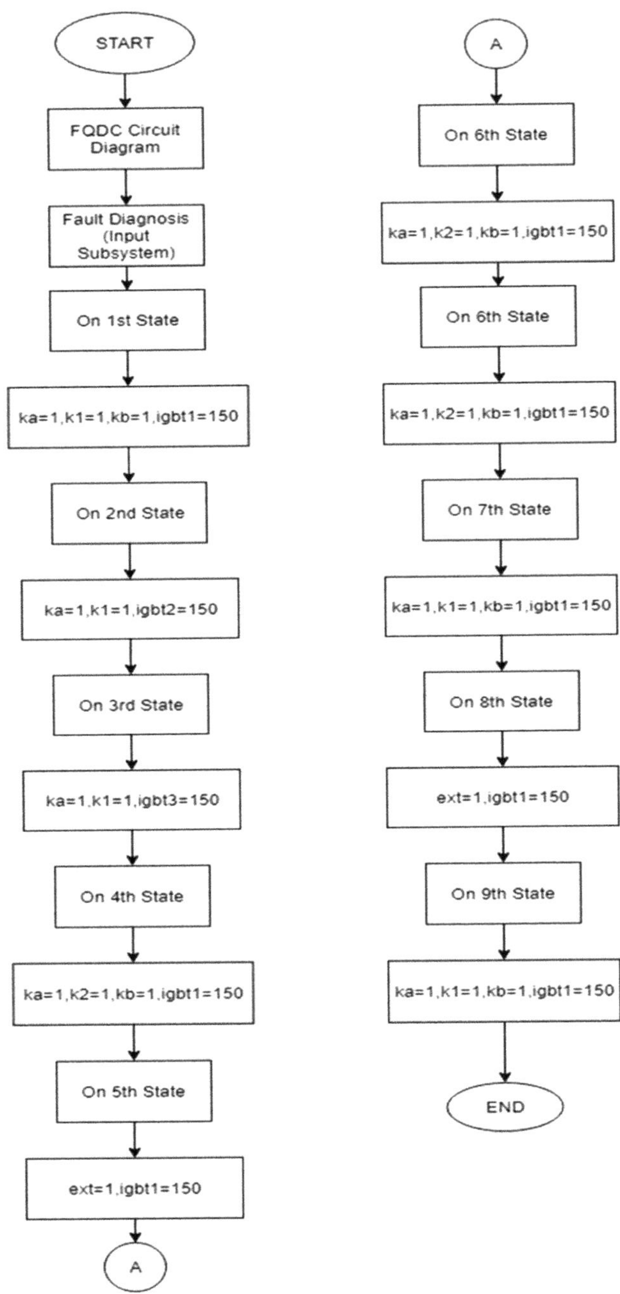

Fig. 15.4 Block diagram of the sequences

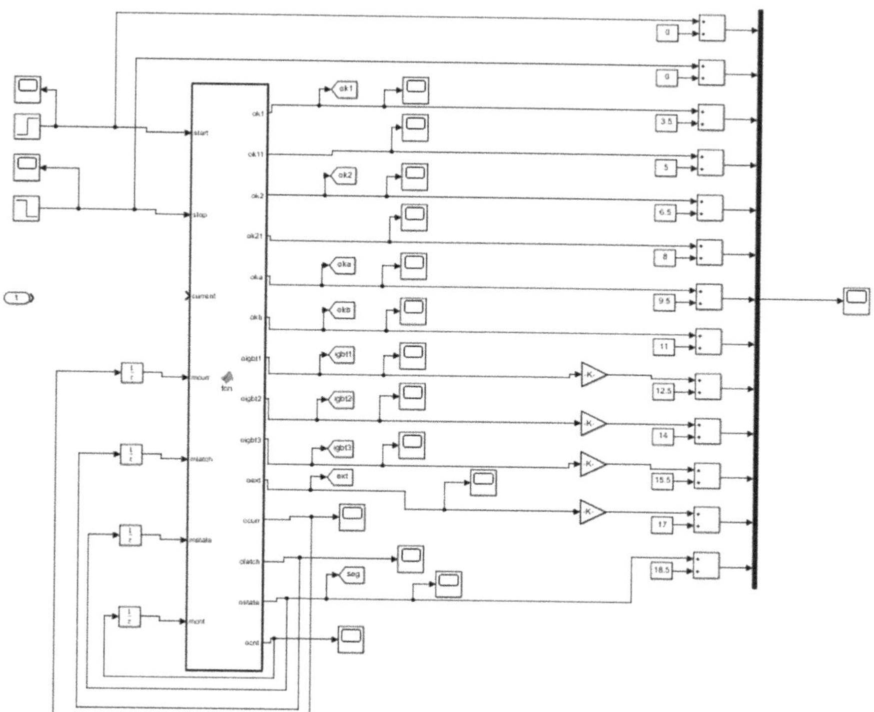

Fig. 15.5 Sequence controller

The graphs in Figs. 15.7 and 15.8 indicate no fault error, so there is no signal that appears as the error signal. From graphs 9 and 10 the error is created by creating a short circuit in Fig. 15.9, the error signal appears at the graph the FS contactor is only switched on for a while and then reset. Figure 15.10 shows the Err1 happens. The Err1 is a series error problem such as short circuit at FQDC. This type of error could damage so many components of the FQDC. With this error, the rest of the sequence test as normal will be prohibited.

15.6 Conclusion

The proposed sequence and fault diagnose controller successfully performed the expected task to detect problems or failure resulting from faulty FQDC devices or propulsion motor.

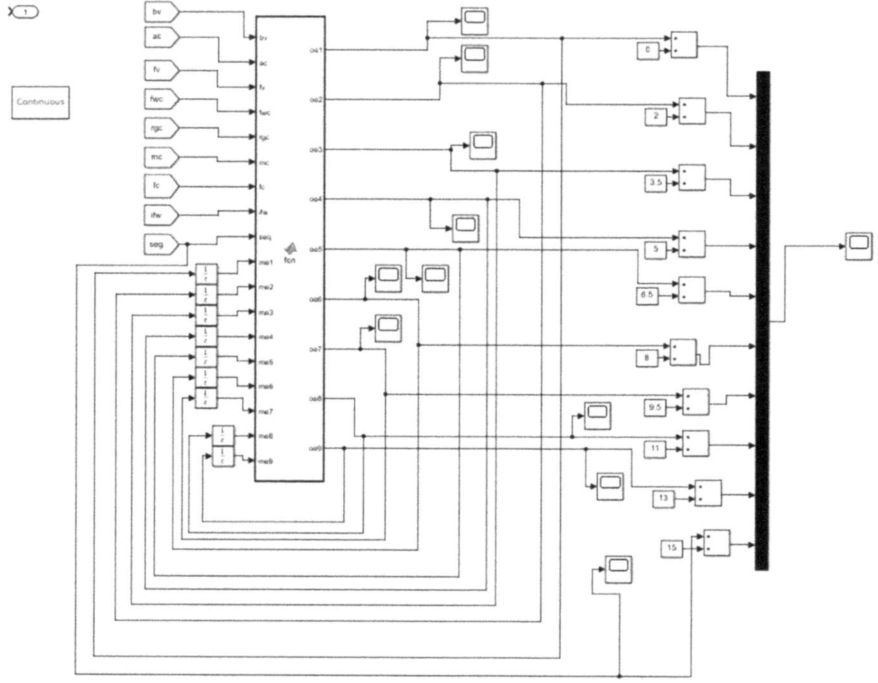

Fig. 15.6 Fault diagnose controller

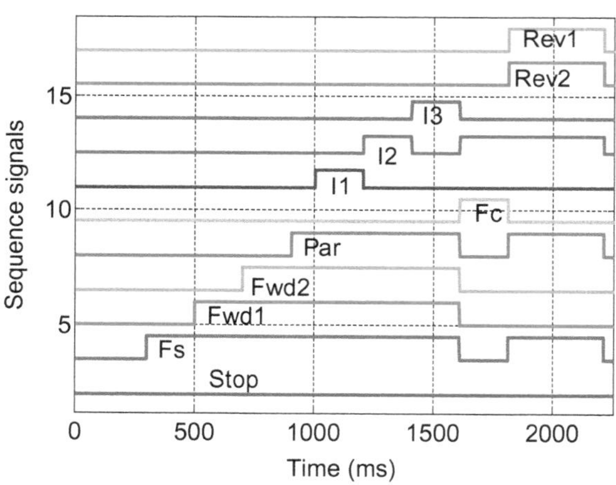

Fig. 15.7 Sequence results without error

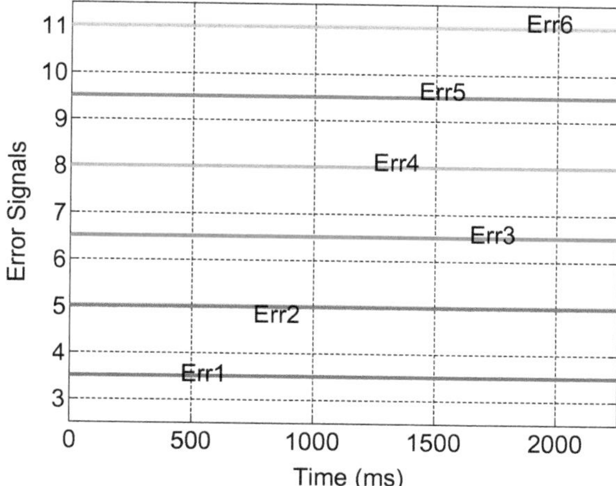

Fig. 15.8 Fault error signals without error

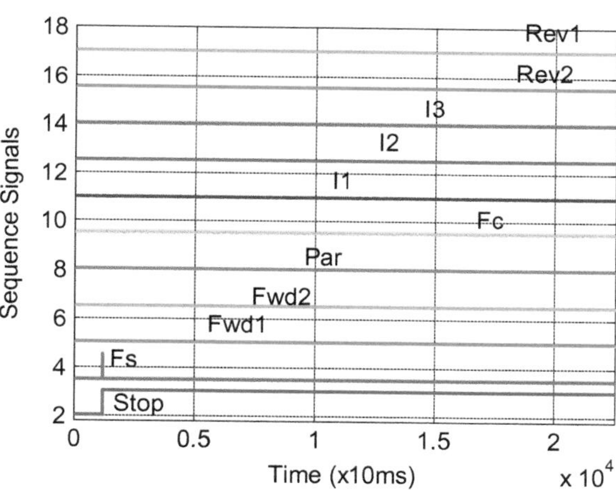

Fig. 15.9 Sequence signals with error

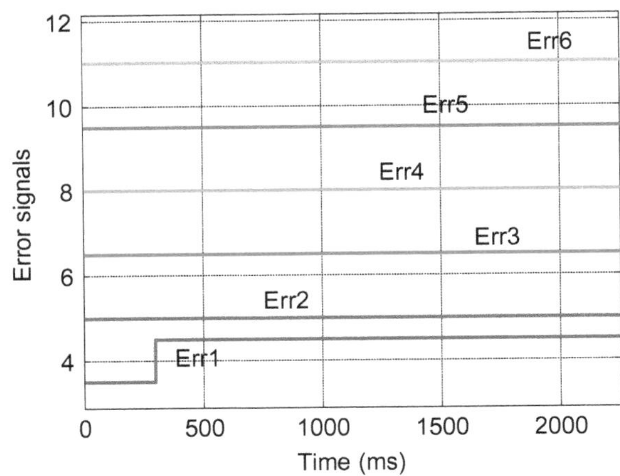

Fig. 15.10 Error signals with error

References

Arof, S., Yaakop, N.M., Jalil, J.A., Mawby, P.A., Arof, H.: Series motor four quadrants drive DC chopper for low cost, electric car: part 1: overall. In: 2014 IEEE International Conference on Power and Energy (PECon), pp. 342–347 (2014)

Arof, S., Muhd Khairulzaman, A.K., Jalil, J.A., Arof, H., Mawby, P.A.: Self tuning fuzzy logic controlling chopper operation of four quadrants drive DC chopper for low-cost electric vehicle. In: 2015 6th International Conference on Intelligent Systems, Modelling and Simulation, pp. 40–45 (2015a)

Arof, S., Zaman, M.K., Jalil, J.A., et al.: Artificial intelligence controlling chopper operation of four quadrants drive DC chopper for low-cost electric vehicle. Int. J. Simul. Syst. Sci. Technol. **16**(4), 1–10 (2015b)

Arof, S., Jalil, J.A., Kamaruddin, N.S., Yaakop, N.M., Mawby, P.A., Hamzah, A.: Series motor four quadrants drive DC Chopper part 2: driving and reverse mode with direct current control. In: 2016 IEEE International Conference on Power and Energy (PECon), pp. 775–780 (2016)

Arof, S., Hassan, H., Rosyidi, M., et al.: Implementation of series motor four quadrants drive DC chopper for dc drive electric car and LRT via simulation model. J. Appl. Environ. Biol. Sci. **7**(3S), 73–82 (2017a)

Arof, S., Noor, N.M., Alias, M.F., et al.: Investigation of chopper operation of series motor four quadrants DC chopper. J. Appl. Environ. Biol. Sci. **7**(3S), 49–56 (2017b)

Arof, S., Diyanah, N.H.N., Noor, N.M. et al.: A new four quadrants drive chopper for separately excited DC motor in low-cost electric vehicle. In: Bakar, M.H.A., Sidik, M.S.M., Öchsner, A. (eds.) Progress in Engineering Technology. Advanced Structures and Materials, 119. Springer, Cham (2019a)

Arof, S., Diyanah, N.H.N., Noor, N.M.N. et al.: Series motor four quadrants drive DC chopper: part 4: generator mode. In: Bakar, M.H.A., Sidik, M.S.M., Öchsner, A. (eds.) Progress in Engineering Technology. Advanced Structures and Materials, 119. Springer, Cham (2019b)

Arof, S., Diyanah, N.H.N., Yaakop, M.A. et al.: Processor in the loop for testing series motor four quadrants drive DC chopper for series motor driven electric car: part 1: Chopper operation modes testing. In: Ismail, A., Bakar, M.H.A., Öchsner, A. (eds.) Advanced Engineering for Processes and Technologies. Advanced Structures Materials, 102. Springer, Cham (2019c)

Arof, S., Diyanah, N.H.N., Mawby, P.A. et al.: Study on implementation of neural network controlling four quadrants direct current chopper: Part1: Using single neural network controller with binary data output. In: Ismail, A., Bakar, M.H.A., Öchsner, A. (eds.) Advanced Engineering for Processes and Technologies. Advanced Structures and Materials, 102. Springer, Cham (2019d)

Arof, S., Diyanah, N.H.N., Noor, N.M. et al.: Genetics algorithm for setting up look up table for parallel mode of new series motor four quadrants DC chopper. In: Bakar, M.H.A., Sidik, M.S.M., Öchsner, A. (eds.) Progress in Engineering Technology. Advanced Structures and Materials, 119. Springer, Cham (2019e)

Arof, S., Diyanah, N.H.N., Mawby, P.A. et al.: Low harmonics plug-in home charging electric vehicle battery charger utilizing multi-level rectifier, zero crossing and buck chopper: part 1: general overview. In: Bakar, M.H.A., Sidik, M.S.M., Öchsner, A. (eds.) Progress in Engineering Technology. Advanced Structures and Materials, 119. Springer, Cham (2019f)

Arof, S., Noor, N.M., Alias, M.F. et al.: Digital proportional integral derivative (PID) controller for closed-loop direct current control of an electric vehicle traction tuned using pole placement. In: Bakar, M.H.A., Zamri, F.A., Öchsner, A. (eds.) Progress in Engineering Technology II. Advanced Structures and Materials, 131. Springer, Cham (2020a)

Arof, S., Noor, N.M., Mohamad, R. et al.: Close loop feedback direct current control in driving mode of a four quadrants drive direct current chopper for electric vehicle traction controlled using fuzzy logic. In: Bakar, M.H.A., Zamri, F.A., Öchsner, A. (eds.) Progress in Engineering Technology II. Advanced Structures and Materials, 131. Springer, Cham (2020b)

Arof, S., Said, M.S., Diyanah, N.H.N. et al.: Series motor four-quadrant direct current chopper: reverse mode, steering position control with double-circle path tracking and control for autonomous reverse parking of direct current drive electric car. In: Bakar, M.H.A., Zamri. F.A., Öchsner, A. (eds.) Progress in Engineering Technology II. Advanced Structures and Materials, 131. Springer, Cham (2020c)

Arof, S., Said. M.S., Diyanah, N.H.N. et al.: Series motor four-quadrant DC chopper: reverse mode, direct current control, triple cascade PIDs, and ascend-descend algorithm with feedback optimization for automatic reverse parking. In: Bakar, M.H.A., Zamri, F.A., Öchsner, A. (eds.) Progress in Engineering Technology II. Advanced Structures Materials, 131. Springer, Cham (2020d).

Arof, S., Sukiman, E.D., Diyanah, N.H.N. et al.: Discrete-time linear system of New series motor four-quadrant drive direct current chopper numerically represented by Taylor series. In: Bakar, M.H.A., Zamri, F.A., Öchsner, A. (eds.) Progress in Engineering Technology II. Advanced Structures and Materials, 131. Springer, Cham (2020e)

Gupta, S., Singhal, R.: Fundamentals and characteristics of an expert system. Int. J. Recent Innov. Trends Comput. Commun. **1**(3), 110–113 (2013)

Lin, F., Chau, K.T., Chan, C.C., Liu, C.: Fault diagnosis of power components in electric vehicles. J. Asian Electr. Veh. **11**, 1659–1666 (2013)

Miljković, D.: Fault detection methods: a literature survey. In: 2011 Proceedings of the 34th International Convention MIPRO, pp. 750–755 (2011)

Oak Ridge National Laboratory: Advanced Brush Technology for DC Motors. https://peemrc.ornl.gov/projects/emdc3.jpg. Accessed 15 May 2009

Sobhani-Tehrani, E., Khorasani, K.: Fault Diagnosis of Nonlinear Systems Using a Hybrid Approach. Springer, London (2009)

Chapter 16
Gradient Descend for Setting Up a Look-Up Table of Series Motor Four Quadrants Drive DC Chopper in Parallel Mode

Saharul Arof, A. Shauqee, M. Rosyidi, N. H. N. Diyanah, Philip Mawby, H. Arof, and Emilia Noorsal

Abstract A series motor has two weaknesses which are overrun when running under load or the drastic speed drop when loaded. The parallel mode allows the series motor to overcome the condition motor speed drop when loaded. For maintaining the torque at its optimum, the field current needs to be controlled. This paper describes the gradient descend method to set up the look-up table of the DC series motor field current for the operation in parallel mode of a new series motor four quadrants DC chopper used for electric car (EC) traction. Matlab/Simulink software is used to test the FQDC chopper and the control algorithm. The result indicates that the gradient descend method is capable of finding the optimum field current to produce the look-up table for parallel mode operation. Matlab/Simulink is also used to establish the system under study, and the results indicate that the proposed gradient descend technique is capable for setting up the look-up table.

S. Arof (✉) · A. Shauqee · M. Rosyidi · N. H. N. Diyanah
Universiti Kuala Lumpur, Malaysian Spanish Institute Kulim Hi-Tech Park, 09000 Kulim, Kedah, Malaysia
e-mail: saharul@unikl.edu.my

A. Shauqee
e-mail: amirshauqee@unikl.edu.my

M. Rosyidi
e-mail: m.rosyidi@unikl.edu.my

N. H. N. Diyanah
e-mail: diyanahhisham94@gmail.com

S. Arof · P. Mawby
University of Warwick School of Engineering, Coventry CV47AL, UK
e-mail: p.a.mawby@warwick.ac.uk

H. Arof
Department of Engineering, Universiti Malaya, Jalan Universiti, 50603 Kuala Lumpur, Malaysia

E. Noorsal
Faculty of Electrical Engineering, Universiti Teknologi MARA (UiTM), Cawangan Pulau Pinang, Permatang Pauh Campus, 13500 Permatang Pauh, Pulau Pinang, Malaysia
e-mail: emilia.noorsal@uitm.edu.my

M. H. Abu Bakar et al. (eds.), *Progress in Engineering Technology III*,
Advanced Structured Materials 148, https://doi.org/10.1007/978-3-030-67750-3_16

Keywords DC drive · EV and HEV · Series motor · Four quadrant chopper · Gradient descend · Parallel mode

16.1 Introduction

Using electric vehicles (EV) and hybrid electric vehicles (HEV) is one of the solutions to reduce global hydrocarbon emission Oak Ridge National Laboratory (2009). In the future, the electric motor propulsion system (electric vehicle) will replace the internal combustion system (mechanical combustion engine). That is not only because of zero-emission, but it also has higher efficiency.

16.2 Literature Review

The success of the Oak Research National Laboratory (ORNL), in 2009, in designing a DC brushed motor with high power output (55 kW), high efficiency (92%) that can operate at low operating voltages (13 V) has started the interest to embark research in DC drive EC. Attempts to improve conventional H-bridge chopper by increasing the operation or to allow the reverse motor action have been continuously carried out. But ever since the foundation of the new DC motor by ORNL, the need for a new series motor four quadrants DC chopper, such as shown in Fig. 16.1, was designed, and the proposed chopper has more operations compared to the earlier version (Arof et al. 2014). Several other studies related to DC drive EV lead to the research on EC battery chargers (Arof et al. 2019a) and different types of DC drive brushed motor such as a separately excited motor for motor traction (Arof et al. 2017a). Detailed investigations on the chopper operation modes led to the establishment of a simulation model to test the chopper operations for the application of electric car and light rail transit (LRT) (Arof et al. 2017b). This simulation model led to further detailed investigations on each of the chopper operations and on the specific pattern of motor voltage, current, torque, speed, of the series motor and FQDC running for dc drive EC application (Arof et al. 2016). For DC series motor traction of EC application, the speed and torque control for the series motor in an attempt to reduce jerk and tire slip have been successfully done and implemented with the direct current control technique (Arof et al. 2019b) For power regeneration, the FQDC offers the generator mode with several methods of starting the regenerated power, and voltage control as studied and discussed in (Arof et al. 2019c). For the FQDC to be applied in the real world, it needs controllers running control algorithms in the embedded system. The controller and its control algorithm are studied and tested using the processor in the loop (PIL) technique (Arof et al. 2015a). Close loop controller using PID tuned using pole placement (Arof et al. 2015b) and Fuzzy logic controller (Arof et al. 2019d) are applied to improve the performance and stability of the motor current in direct current control (DCC) in driving mode. The same DCC method has

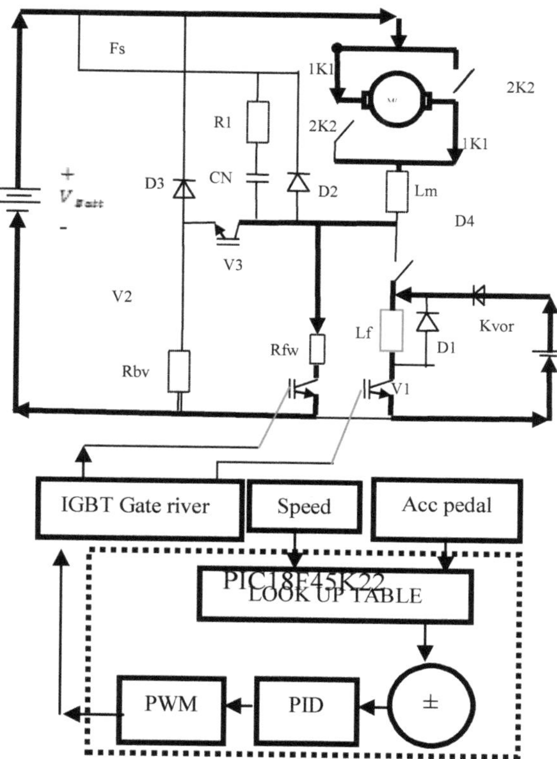

Fig. 16.1 Four quadrants chopper in parallel mode

undergone improvements using cascade PID and ascend descend algorithm and after improvisation made and associated with steering (Arof et al. 2019e) and vehicle movement control (Arof et al. 2019f), it was used for autonomous EV automatic reverse parking. For fault diagnose and online system tuning and optimization, a numerical representation using Taylor series (Arof et al. 2020a) is studied and tried for driving mode. To improve the new FQDC performance, an optimization tool, such as artificial intelligence (AI), is introduced to control all of the chopper operations of the proposed FQDC chopper (Arof et al. 2020b, c). Among the three AI controllers, ANFIS shows the best performance followed by neural network and self-tuning Fuzzy logic controller. Study on the neural network controller to uncover the proper method of tuning has been carried out, such as using a single controller with binary output (Arof et al. 2020d). For each specific FQDC chopper operation mode, the performance can be further improved using an AI optimization tool such as genetics algorithm to set up a particular look-up table for the field current (Arof et al. 2020e).

Gradient descent is a first-order iterative optimization algorithm for finding the local minimum of a function (Luigi et al. 2011). To find a local minimum of a function using the gradient descent approach, one takes steps proportional to the negative of the gradient (or approximate gradient) of the function at the current point (Hanna et al.

2003; Negnevitsky 2005; Mofarreh-Bonab and Ghorashi 2013). If instead, one takes steps proportional to the positive of the gradient, one approaches a local maximum of that function; the procedure is then known as the gradient ascent. Gradient descent was originally proposed by Cauchy in 1847.

16.3 Methodology

16.3.1 DC Series Motor weakness and FQDC in Parallel Mode

The DC series motor has the advantage of a high starting torque. But there is one common weakness of the DC series motor, which is that the motor speed drops drastically when loaded. It is common for any electrical motor to have a decrease in speed when loaded, but for a series motor, it is so apparent that causes disadvantage to this motor. To tackle this issue, the paper proposed a four quadrant choppers that allows the series motor to operate in parallel mode. In this mode, the field current is separately controlled from the armature current, and it can be changed or maintained. When loaded, the speed of the motor will decrease. A decrease in speed will result in a reduced back emf (Eq. 16.2). If the back emf is reduced, the armature current will rise (Eq. 16.1). Since the field current is controlled separately and can be maintained, the increase in armature current will raise the torque (Eq. 16.3). This new torque will overcome the loading effect and overcome a decrease in speed while climbing the steep hill.

During parallel mode, the armature and field current can be represented as equations below. These are general equations describing the voltage and current for the chopper. Equations (16.1)–(16.3) are general equations. B_{emf} is the back emf of the motor, V_{a} and V_{f} are the armature and field voltages, K_{b} is the back emf constant, K_{t} is the torque motor constant, I_{f} is the field current, R_{a} and R_{f} are the motor coil resistances, I_{a} is the armature current, ω is the angular speed, and T_{d} is the motor torque.

$$e_{\mathrm{g}} = K_{\mathrm{bemf}} i_{\mathrm{f}} \omega \tag{16.1}$$

$$T_{\mathrm{d}} = K_{\mathrm{t}} i_{\mathrm{a}} i_{\mathrm{f}} \tag{16.2}$$

$$T_{\mathrm{d}} = J \frac{\mathrm{d}\omega}{\mathrm{d}t} + B\omega + T_{\mathrm{L}} \tag{16.3}$$

$$\frac{\mathrm{d}}{\mathrm{d}t}[I_{\mathrm{f}}] = \left[-\left(\frac{R_{\mathrm{f}}}{L_{\mathrm{f}}} \right) \right] [I_{\mathrm{f}}] \left[\frac{1}{(L_{\mathrm{f}})} \right] [V_{\mathrm{ext}} K_{\mathrm{drv}}] \tag{16.4}$$

$$\frac{\mathrm{d}}{\mathrm{d}t}[I_\mathrm{a}] = \left[-\frac{(R_\mathrm{a})}{L_\mathrm{a}}\right][I_\mathrm{a}]\left[\frac{1}{L_\mathrm{a}} - \frac{1}{L_\mathrm{a}}\right]\left[\begin{matrix} V_\mathrm{dc} \\ E_\mathrm{g} \end{matrix}\right] \tag{16.5}$$

A higher torque can be produced if the armature current is set at a high value. The field current is adjusted to have lower back emf but enough torque to overcome the load.

16.3.2 Control Strategy in Parallel Mode

To get a high torque, the armature current is set to maximum. This can be achieved by setting the PWM value to IGBT V2 to the maximum. The field current is adjusted by FIRING IGBT V1 to the value where it produces low back emf. But this does not always mean the smaller the field current is, the better. To some extent, a small value in field current will not only result in lower back emf but lower in torque too (Arof et al. 2020d). This is due to the fact that a lower field current can result in lower magnetic field and flux. These two are essential elements to produce the motor torque. Figure 16.2 shows the motor torque as a result of changes in field current. As we can see, the maximum field current to achieve the maximum torque is 1A.

To overcome the overfed of field current, a look-up table such as in Fig. 16.8 is used. In the look-up table, the best optimum value of field current is related to the speed of the vehicle, and the accelerator pedal should be set. But in a real application, the value is always an inclusive safety value to avoid overshooting. The look-up table acts as a library where the optimized reference field current value is stored. For so, the genetic algorithm or gradient descend can be used to optimize the required field current.

In this research, GD is used to set up a look-up table for the parallel mode operation. The GD is used to search the optimum point of field current to have lower back emf while at the same time the optimum field current for maximum torque. The whole process is done offline by using compensator Eqs. (16.6) till (16.10) and without the actual system being run. The general control strategy is prescribed as a block diagram in Fig. 16.6. Motor torque and speed are the two crucial signals for the GD

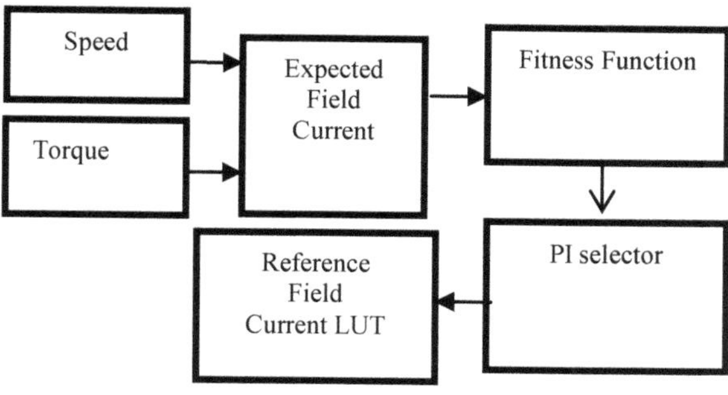

Fig. 16.2 Timing diagram during parallel mode

controller. The output of the controller relies on these two values. The speed of the DC motor value is the input for the GD operator. For the speed, an equation is used to suggest the region of which the best field current where the optimum torque will be found.

The General equation is

$$\text{Range}_{\text{Ifield}} = V\text{dc/Speed} * C \tag{16.6}$$

where C is a constant.

A more precise equation can be gathered from the fitness function equation. When the GD algorithm is activated, it first produces the first expected field current value with respect to the speed, or it can be gathered from the look-up table. This field current, which falls/lies according to the region of field current determined earlier. This value will be tested with a fitness function to examine the strength of each chromosome. The fitness function equation is extracted from Eqs. (16.1) to (16.6), and in steady-state, it can be represented by the below equations.

Fitness Function

$$T_{\text{d}} K_{\text{t}} I_{\text{f}} I_{\text{a}} = \frac{V_{\text{batt}} - K_{\text{v}} I_{\text{f}} \omega}{R_{\text{a}} + Rf w_{\text{a}}} * K_{\text{t}} I_{\text{f}} \tag{16.7}$$

This value will be compared to the expected value. The expected value can be found from the torque-speed curve. The actual output of this equation is the motor torque. The difference between the expected torque and the actual torque values produces the error.

$$E_{rr} K_{\text{t}} I_{\text{f}} I_{\text{a}} = E_{\text{val}} - \text{Act}_{\text{val}} \tag{16.8}$$

The new value of the field current that is going to be used is found using the equation below. The Constant represents learning rate, higher constant value causes a bigger step, and a lower constant value results in smaller steps but precise values.

$$F_{\text{new}} K_{\text{t}} I_{\text{f}} I_{\text{a}} = F_{\text{prev}} - (\text{Constant} * E_{rr}) \tag{16.9}$$

Once the new value has been assigned, the new value is transformed to the previous value.

$$F_{\text{prev}} K_{\text{t}} I_{\text{f}} I_{\text{a}} = F_{\text{new}} \tag{16.10}$$

The new value will undergo a fitness function test again. The process will be repeated depending on the set of number of repetition preset or called number of iteration. If the number of iteration has elapsed, the GD algorithm will stop and wait for the next cycle, it will be reactivated. Changes in speed will reactivate the GD controller. When the GD has finally stopped, it is expected to have already found the best chromosomes, and the system has finally found the best field current value for the motor speed. Figure of the block diagram represents the GD algorithm and is shown in Fig. 16.3.

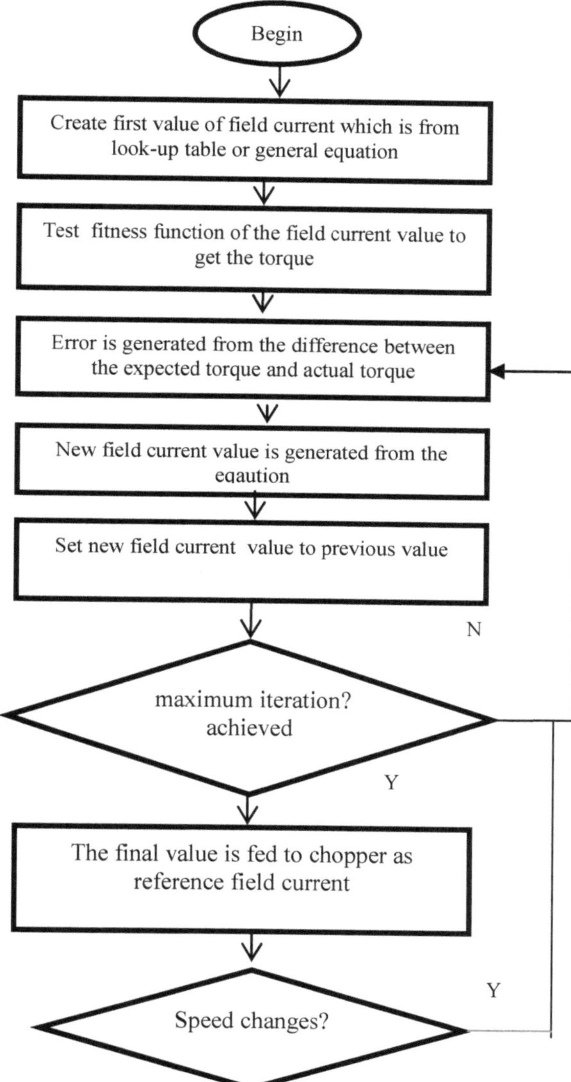

Fig. 16.3 Field weakening

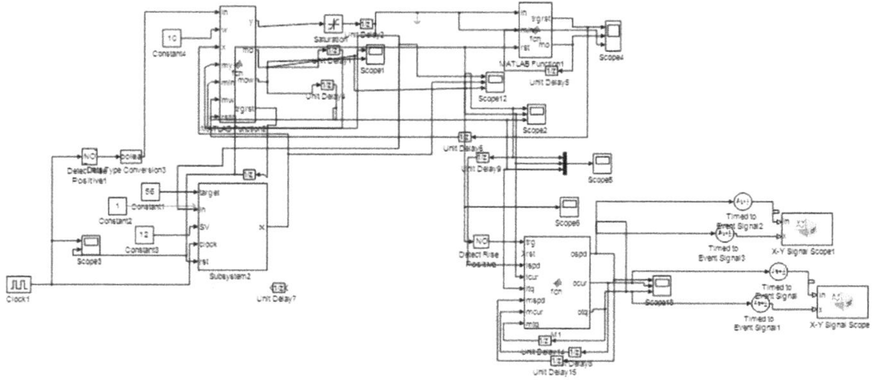

Fig. 16.4 MATLAB/Simulink model of the GD controller

16.3.3 Simulation Model

A simulation model is fast, economical, saves time, and safer for testing any system. A simulation model, as shown in Fig. 16.5 of four quadrants DC chopper using MATLAB/Simulink software, is formed by solving a linear differential equation of DC series motor and is represented in physical based modeling.

The DC EV test consists of a chopper, vehicle dynamics, IGBT firing controller, and LUT in a complete model. The FQDC is tested in parallel mode using a 35 kW motor operating at a maximum power of 22 kW to drive its total weight of 1325 kg and experience inclination about 17.5°. The MATLAB/Simulink simulation model of Gradient Descend control algorithm for setting up a Look Up Table is shown in Fig. 16.4.

For simulation model as shown in Fig. 16.4, In the vehicle dynamics model, the earth profile is set so that the test will experience hill-climbing of 17.5 degrees of inclination illustrated according to Fig. 16.6.

16.4 Results and Discussion

Running the gradient descend control algorithm using MATLAB/Simulink simulation software such as shown in Fig. 16.5 produces results as shown in Fig. 16.7. The Field current is slowly lowered by the GD algorithm and this resulted in a torque increases linearly until at some maximum point before it drops. An algorithm will detect decrease changes in torque, and when this happens, the GD algorithm is stopped, and the last field winding current value being tested before the torque drop is recorded. The same process is repeated using the previous value of field current, and the process continues again and again until it reaches the final iteration.

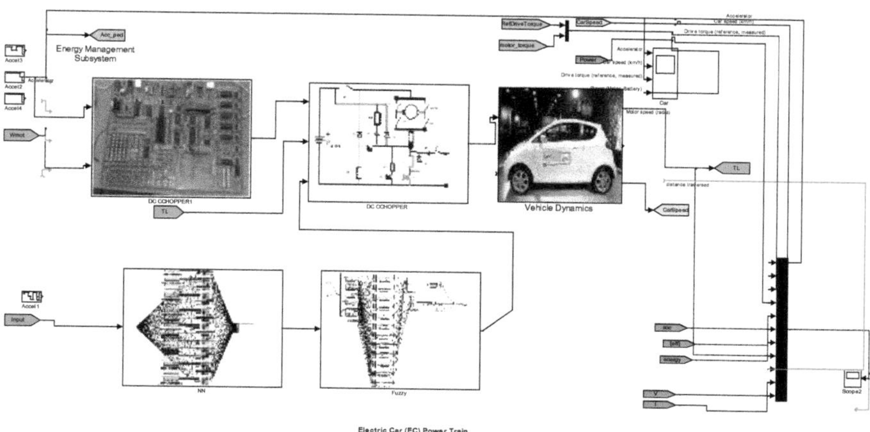

Fig. 16.5 MATLAB/Simulink model

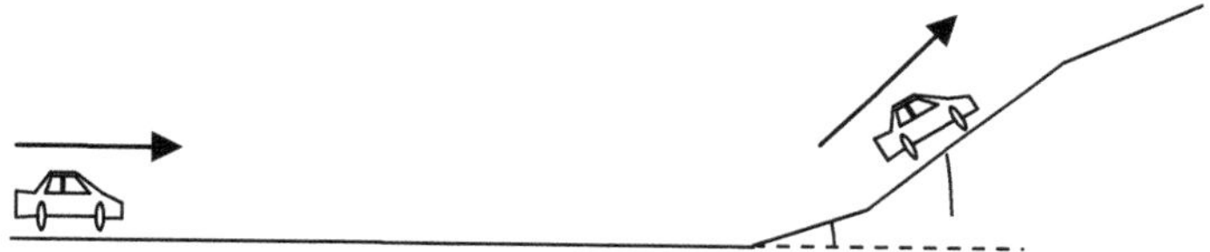

Fig. 16.6 MATLAB/Simulink model

Fig. 16.7 GD performance finding best field current

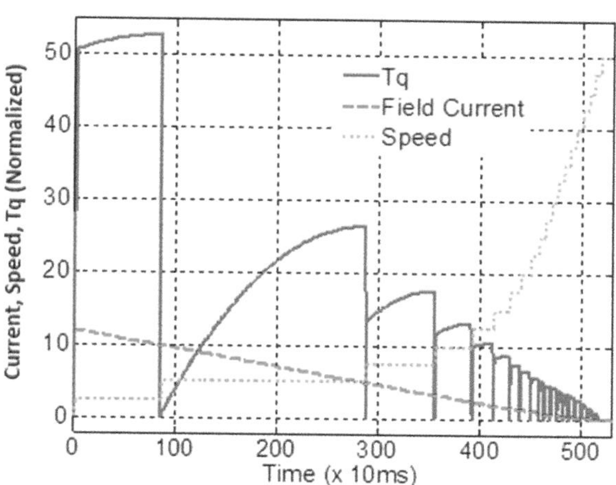

A look-up table, as in Fig. 16.8, is established, and it is extracted from the previously mentioned GD control algorithm process. Interpolation and extrapolation are done whenever necessary to complete the LUT.

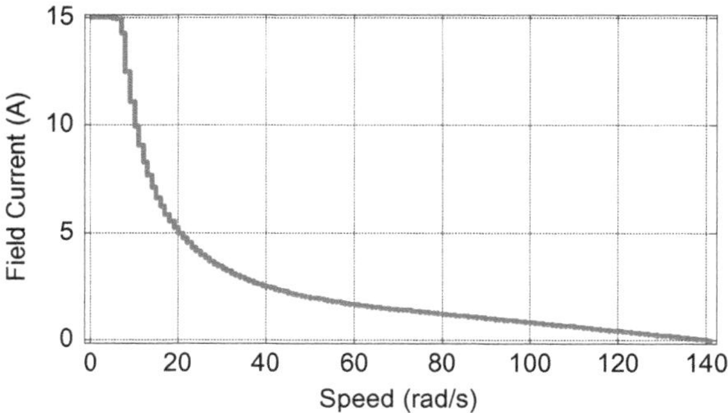

Fig. 16.8 Look-up table of field current versus speed

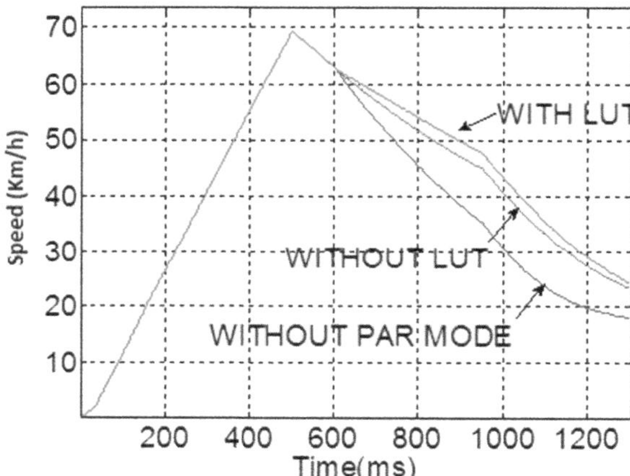

Fig. 16.9 The simulation result of the vehicle speed with the mode of chopper operation plotted

In Fig. 16.9, three simulations were done to test the load effect of the parallel mode using LUT, parallel mode without LUT, and when no parallel mode is taking place. The look-up table is extracted from previously obtained results and amplified. The highest motor speed is achieved when the LUT is used. When parallel mode is done without using LUT, the best field current is not met. The lowest motor speed is when no parallel mode is used.

16.5 Conclusion

The FQDC is able to perform the parallel mode, which causes the vehicle to achieve higher torque to overcome steep hill climbing. The GD successfully achieved the target for finding the best field current to setup a look-up table for maximum torque during the parallel mode of FQDC. DC drive series motor has a high potential to be utilized in EV. This is due to its simple design, low cost, and excellent controllability.

References

Arof, S., Yaakop, N.M., Jalil, J.A., Mawby, P.A., Arof, H.: Series motor four quadrants drive DC chopper for low cost, electric car: part 1: overall. In: 2014 IEEE International Conference on Power and Energy (PECon), pp. 342–347 (2014)

Arof, S., Muhd Khairulzaman, A.K., Jalil, J.A., Arof, H., Mawby, P.A.: Self tuning fuzzy logic controlling chopper operation of four quadrants drive DC chopper for low-cost electric vehicle. In: 2015 6th International Conference on Intelligent Systems, Modelling and Simulation, pp. 40–45 (2015a)

Arof, S., Zaman, M.K., Jalil, J.A., et al.: Artificial intelligence controlling chopper operation of four quadrants drive DC chopper for low-cost electric vehicle. Int. J. Simul. Syst. Sci. Technol. **16**(4), 1–10 (2015b)

Arof, S., Jalil, J.A., Kamaruddin, N.S., Yaakop, N.M., Mawby, P.A., Hamzah, A.: Series motor four quadrants drive DC Chopper part 2: driving and reverse mode with direct current control. In: 2016 IEEE International Conference on Power and Energy (PECon), pp. 775–780 (2016)

Arof, S., Hassan, H., Rosyidi, M., et al.: Implementation of series motor four quadrants drive DC chopper for dc drive electric car and LRT via simulation model. J. Appl. Environ. Biol. Sci. **7**(3S), 73–82 (2017a)

Arof, S., Noor, N.M., Alias, M.F., et al.: Investigation of chopper operation of series motor four quadrants DC chopper. J. Appl. Environ. Biol. Sci. **7**(3S), 49–56 (2017b)

Arof, S., Diyanah, N.H.N., Mawby, P.A. et al.: Low harmonics plug-in home charging electric vehicle battery charger utilizing multi-level rectifier, zero crossing and buck chopper: part 1: general overview. In: Bakar, M.H.A., Sidik, M.S.M., Öchsner, A. (eds.) Progress in Engineering Technology. Advanced Structures and Materials, 119. Springer, Cham (2019a)

Arof, S., Diyanah, N.H.N., Noor, N.M. et al.: A new four quadrants drive chopper for separately excited DC motor in low-cost electric vehicle. In: Bakar, M.H.A., Sidik, M.S.M., Öchsner, A. (eds.) Progress in Engineering Technology. Advanced Structures and Materials, 119. Springer, Cham (2019b)

Arof, S., Diyanah, N.H.N., Noor, N.M.N. et al.: Series motor four quadrants drive DC chopper: part 4: generator mode. In: Bakar, M.H.A., Sidik, M.S.M., Öchsner, A. (eds.) Progress in Engineering Technology. Advanced Structures and Materials, 119. Springer, Cham (2019c)

Arof, S., Diyanah,, N.H.N., Yaakop, M.A. et al.: Processor in the loop for testing series motor four quadrants drive DC chopper for series motor driven electric car: part 1: chopper operation modes testing. In: Ismail, A., Bakar, M.H.A., Öchsner, A. (eds.) Advanced Engineering for Processes and Technologies. Advanced Structurs and Materials, 102. Springer, Cham (2019d)

Arof, S., Diyanah, N.H.N., Mawby, P.A. et al.: Study on implementation of neural network controlling four quadrants direct current chopper: Part1: Using single neural network controller with binary data output. In: Ismail, A., Bakar, M.H.A., Öchsner, A. (eds.) Advanced Engineering for Processes and Technologies. Advanced Structures and Materials, 102. Springer, Cham (2019e)

Arof, S., Diyanah, N.H.N., Noor, N.M. et al.: Genetics algorithm for setting up look up table for parallel mode of new series motor four quadrants DC chopper. In: Bakar, M.H.A., Sidik, M.S.M., Öchsner, A. (eds.) Progress in Engineering Technology. Advanced Structures and Materials, 119. Springer, Cham (2019f)

Arof, S., Noor, N.M., Alias, M.F. et al.: Digital proportional integral derivative (PID) controller for closed-loop direct current control of an electric vehicle traction tuned using pole placement. In: Bakar, M.H.A., Zamri, F.A., Öchsner, A. (eds.) Progress in Engineering Technology II. Advanced Structures and Materials, 131. Springer, Cham (2020a)

Arof, S., Noor, N.M., Mohamad, R. et al.: Close loop feedback direct current control in driving mode of a four quadrants drive direct current chopper for electric vehicle traction controlled using fuzzy logic. In: Bakar, M.H.A., Zamri, F.A., Öchsner, A. (eds.) Progress in Engineering Technology II. Advanced Structures and Materials, 131. Springer, Cham (2020b)

Arof, S., Said, M.S., Diyanah, N.H.N. et al.: Series motor four-quadrant direct current chopper: reverse mode, steering position control with double-circle path tracking and control for autonomous reverse parking of direct current drive electric car. In: Bakar, M.H.A., Zamri, F.A., Öchsner, A. (eds.) Progress in Engineering Technology II. Advanced Structures and Materials, 131. Springer, Cham (2020c)

Arof, S., Said, M.S., Diyanah, N.H.N. et al.: Series motor four-quadrant DC chopper: reverse mode, direct current control, triple cascade PIDs, and ascend-descend algorithm with feedback optimization for automatic reverse parking. In: Bakar, M.H.A., Zamri, F.A., Öchsner, A. (eds.) Progress in Engineering Technology II. Advanced Structures and Materials, 131. Springer, Cham (2020d)

Arof, S., Sukiman, E.D., Diyanah, N.H.N. et al.: Discrete-time linear system of New series motor four-quadrant drive direct current chopper numerically represented by Taylor series. In: Bakar, M.H.A., Zamri, F.A., Öchsner, A. (eds.) Progress in Engineering Technology II. Advanced Structures and Materials, 131. Springer, Cham (2020e)

Hanna, A.I., Yates, I., Mandic, D.P.: Analysis of the class of complex-valued error adaptive normalised nonlinear gradient descent algorithms. IEEE Int. Conf. Acoust. Speech Signal Process. Proc. (ICASSP '03) **2**, 705–708 (2003)

Luigi, M., Matteo, M., Giovanni, P.: Stochastic natural gradient descent by estimation of empirical covariances. Cong. Evol. Comput. 949–259 (2011)

Mofarreh-Bonab, M., Ghorashi, S.A.: A low complexity and high speed gradient descent based secure localization in wireless sensor network. Int. Conf. Comput. Knowl. Eng. 300–303 (2013)

Negnevitsky, M.: Artificial Intelligence: A Guide to Intelligent Systems. Essex, UK (2005)

Oak Ridge National Laboratory: Advanced Brush Technology for DC Motors. https://peemrc.ornl.gov/projects/emdc3.jpg. Accessed May 2009

Chapter 17
DC Drive Electric Car Utilizing Series Motor and Four Quadrants Drive DC Chopper Parameter Determination from General Design Requirements

Saharul Arof, Norramlee Mohamed Noor, N. H. Nur Diyanah, Philip Mawby, Hamzah Arof, and Emilia Noorsal

Abstract The general requirement of a DC drive electric car (EC) is to carry a maximum load of 850–1300 kg. The expected maximum speed of the EC is about 110–120 km/h, and the acceleration is for 60 km/h in less than 10 s. This paper is to study how to determine specific requirements such as motor kilowatt power, gear ratio, maximum motor torque, battery voltage, maximum current, etc. The DC drive an electric car (EC) using a series motor and four quadrants DC chopper (FQDC) to meet the earlier mentioned general requirements. A vehicle dynamic mathematical equation has been used to assist in finding the parameters. A simulation model of this vehicle dynamics equation using MATLAB/Simulink software is developed to study, investigate, and obtain the specific requirements earlier mentioned. Once the specific parameters have been determining, it is tested with the complete electric vehicle model to test the conditions. It concluded that the design requirements parameters

S. Arof (✉) · N. M. Noor · N. H. Nur Diyanah
Universiti Kuala Lumpur, Malaysian Spanish Institute Kulim Hi-Tech Park, 09000 Kulim, Kedah, Malaysia
e-mail: saharul@unikl.edu.my

N. M. Noor
e-mail: noramlee@unikl.edu.my

N. H. Nur Diyanah
e-mail: diyanahhisham94@gmail.com

S. Arof · P. Mawby
University of Warwick School of Engineering, Coventry CV47AL, UK
e-mail: p.a.mawby@warwick.ac.uk

H. Arof
Department of Engineering, Universiti Malaya, Jalan Universiti, 50603 Kuala Lumpur, Malaysia
e-mail: ahamzah@um.edu.my

E. Noorsal
Faculty of Electrical Engineering, Universiti Teknologi MARA, Cawangan Pulau Pinang, Permatang Pauh Campus, 13500 Permatang Pauh, Pulau Pinang, Malaysia
e-mail: emilia.noorsal@uitm.edu.my

M. H. Abu Bakar et al. (eds.), *Progress in Engineering Technology III*,
Advanced Structured Materials 148, https://doi.org/10.1007/978-3-030-67750-3_17

and the FQDC could represent using a vehicle dynamics mathematical model, and it can perform all requirements for the DC drive EC.

Keywords DC drive · FQDC · Design requirement · EV · Mathematical model chopper · NEDC · Series motor · Four quadrants chopper

17.1 Introduction

Using electric vehicles is one of the solutions to reduce global hydrocarbon emissions. Unfortunately, the price of EVs is expensive and unattainable to many people, especially those living in developing countries. This fact has led to the study on the possibility for DC drive EV, which is known to be economical.

17.2 Review Stage

Oak Research National Laboratory (ORNL) (2009), United States in 2009, had succeeded in designing a DC brushed motor with high power output (55 kW), high efficiency (92%) that can operate at low operating voltages (13 V), and this has started the interest to embark research in DC drive EC. A new series motor four quadrants DC chopper such as shown in Fig. 17.1, was designed, and the proposed chopper has multiple operations (Arof et al. 2014). Several other studies related to DC drive EV led to research on EC battery chargers (Arof et al. 2019a) and different types of DC drive motors such as separately excited (Arof et al. 2019b). Detailed investigations on the chopper operation modes led to the establishment of a simulation model to test the chopper operations for the application of electric car and light rail transit (LRT) been done (Arof et al. 2017a). This led to further investigations on each of the chopper operations in detail on the specific pattern of voltage, current, torque, speed, of the FQDC running DC series motor been carried out (Arof et al. 2017b). For the DC series motor traction EC application, the speed, and torque control has been successfully done and implemented with direct current control (Arof et al. 2016). For power regeneration, the FQDC offers the generator mode, and several techniques of regenerating the power are studied and discussed in (Arof et al. 2019c). In order for the FQDC has to be applied in the real world, it needs controllers running control algorithms in the embedded system. The controller and its control algorithm are studied and tested using a processor in the loop technique (Arof et al. 2019d). The new FQDC can improve the performance of optimization tool such as artificial intelligence (AI). It is introduced to control all the chopper operations of the proposed FQDC chopper (Arof et al. 2015a, b, 2019e). Each operation of FQDC modes performance can be improved using an AI optimization tool such as the genetics algorithm to set up a specific look-up table (Arof et al. 2019f). The pole placement (Arof et al. 2020a) method is used to tune the close loop PID controller to improve controller

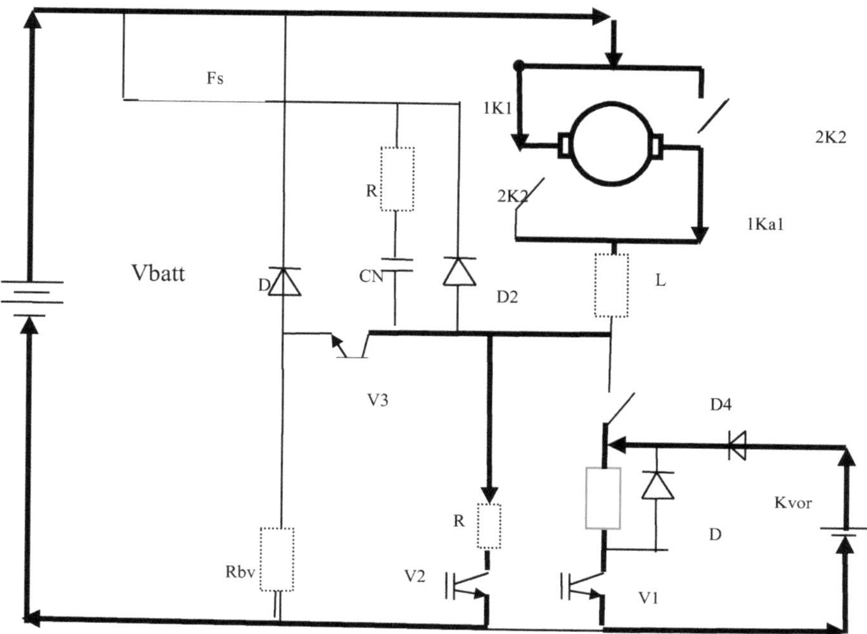

Fig. 17.1 Proposed four quadrants DC chopper (FQDC)

performance while the fuzzy logic controller is used to control the motor current (Arof et al. 2020b) and applied to improve the control performance and system stability. For fault diagnosis and online system tuning and optimization, a numerical representation using Taylor's series (Arof et al. 2020c) has been studied and tried for the driving mode. This method has improved performance using PID cascade and algorithm fluctuations. After improvization made and associated with steering (Arof et al. 2020d) and vehicle movement control (Arof et al. 2020e) was used for autonomous EV automatic reverse parking. As the conducted research studies and investigations are expected to realize in a real prototype of an electric vehicle (EV), all design requirements are needed to fulfill the actual need of the electric vehicle (Gao and Ehsani 2010).

17.2.1 General Requirements of Low Cost, Fuel Economics EV Design

The requirements of the DC drive EV powered by FQDC as in Fig. 17.1, are based on the 35–50 kW of DC series motor and a maximum weight with a load of 800–1300 kg (Westbrook 2001). The maximum speed is expected to be 100–120 km/h (Husain 2003; Bansal 2005). In this paper, the design requirements of a DC drive

EV using a series motor and FQDC are presented. This paper is to study how to determine specific requirements of series motor FQDC DC DRIVE EV such as the motor kilowatt power, gear ratio, maximum motor torque, battery voltage, maximum current, etc. The design requirements and the expected parameters to be determined are needed to allow the DC drive EV using a series motor and FQDC that have been designing to pass the New European Driving Cycle (NEDC) test.

17.2.2 *Vehicle Dynamics*

The propulsion unit of the vehicle needs a tractive force F_{TR} to propel the vehicle. The tractive force must fulfill the requirements of vehicle dynamics to overcome forces such as the rolling resistance, gravitational, and aerodynamically which are summed together as the road load force F_{RL} as shown in Fig. 17.2 (Arof et al. 2020a, b, c, d).

17.2.2.1 Grading Resistance—Gravitational Force

The gravitational force F_{g} depends on the slope of the roadway, as shown in Eq. (17.1). It is positive when climbing and negative when descending a downgrade roadway.

$$F_{\mathrm{g}} = \mathrm{mg} \sin \alpha \tag{17.1}$$

α is the grade angle, m is the total mass of the vehicle, g is the gravity constant.

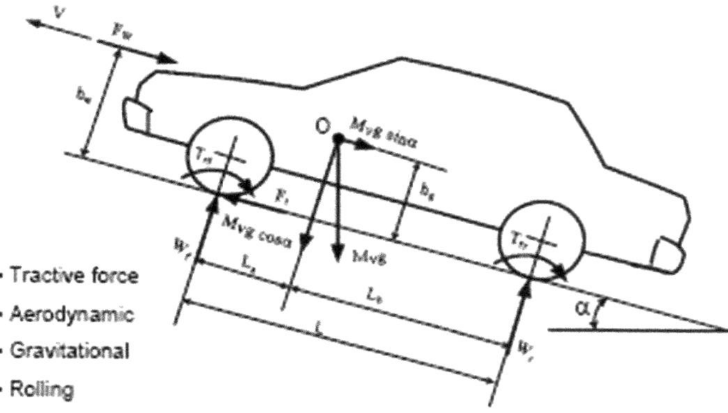

Fig. 17.2 Vehicle dynamic of a car

17.2.2.2 Rolling Resistance

The hysteresis of the tire material causes it at the contact surfaces with the roadway. The centroid of the vertical forces on the wheel moves forward when the tire rolls. Therefore, from beneath the axle toward the direction of motion by the vehicle, as shown in Fig. 17.3.

Tractive Force

The tractive force was used to overcome the F_{roll} force along with the gravity force and the aerodynamic drag force. The rolling resistance has been minimized by keeping the tires as inflated as possible by reducing the hysteresis. The ratio of retarding forces due to rolling resistance and the vertical load on the wheel known as the coefficient of rolling resistance C_0. The rolling resistance force is given by

$$E_{\mathrm{roll}} = \left\{ \begin{array}{ll} \mathrm{sgn}[v_{xt}]\mathrm{mg}(C_0 + C_1 v_{xT}^2) \text{ if} & v_{xt} \neq 0 \\ F_{\mathrm{TR}} - F_{gxT} & \text{if } V_{xt} = 0 \text{ and } \left|F_{\mathrm{TR}} - F_{gxT}\right| \leq C_0\mathrm{mg} \\ \mathrm{sgn}(F_{\mathrm{TR}} - F_{gxT})(C_0\mathrm{mg}) \text{ if } & V_{xt} = 0 \text{ and } \left|F_{\mathrm{TR}} - F_{gxT}\right| > C_0\mathrm{mg} \end{array} \right\} \tag{17.2}$$

where V_{xt} is the vehicle speed, F_{TR} is the total tractive force, C_0 and C_1 are rolling coefficients Table 17.1.

Typical rolling coefficients are $0.004 < C_0 < 0.02$ (unit less) and $C_1 \ll C_0$ ($\mathrm{S}^2/\mathrm{m}^2$), $C_0\mathrm{mg}$ is the maximum rolling resistance at standstill. The

$$\mathrm{sgn}[V_{xt}] = \left\{ \begin{array}{ll} 1 & v_{xT} > 0 \\ -1 & v_{xT} < 0 \end{array} \right\}, \text{ Approximation, } C_0 = 0.01,$$

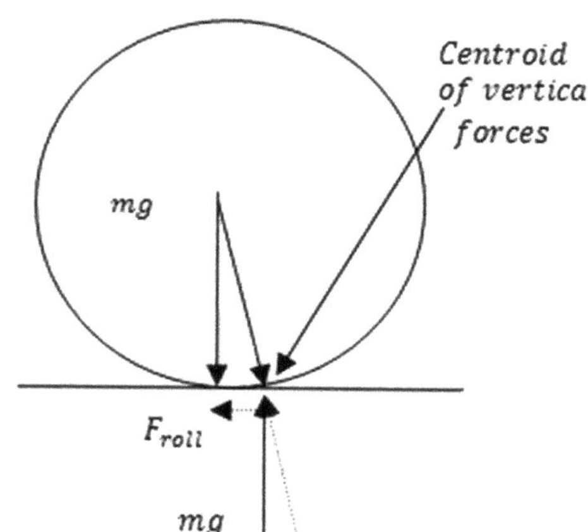

Fig. 17.3 Rolling resistance

Table 17.1 The parameter of road condition

Road condition	Rolling coefficient C_0
Car tire on concrete or asphalt	0.013
Rolled gravel	0.02
Unpaved road	0.05
Field	0.1–0.35
Truck Tires on concrete or asphalt	0.006–0.01
Wheels on rails	0.001–0.002

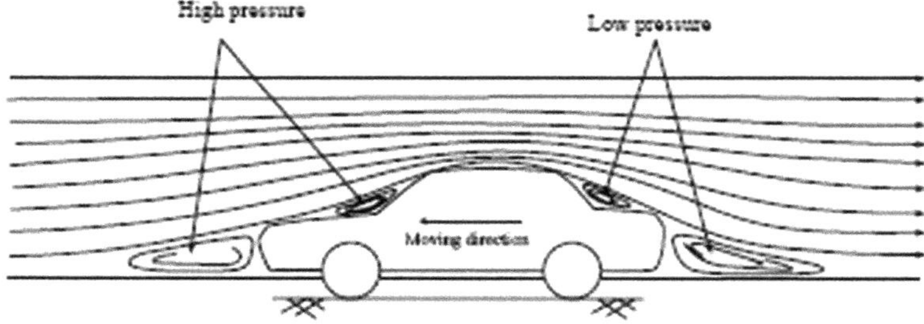

Fig. 17.4 Aerodynamic drag fore

$$C_1 = C_0 \frac{V_{xt}}{100} \tag{17.3}$$

Aerodynamics Force

The aerodynamic drag force is the result of the viscous resistance of the air against the motion, as shown in Fig. 17.4. The aerodynamic drag force $\boldsymbol{F}_{\mathbf{AD}}$ is

$$F_{\mathrm{AD}} = \mathrm{sgn}[V_{xT}]\left\{0.5\rho C_{\mathrm{D}} A_{\mathrm{F}} (V_{xT} + V\omega)^2\right\} \tag{17.4}$$

where ρ is the air density, C_{D} is an aerodynamic drag coefficient, A_{F} is an equivalent frontal area of the vehicle V_{ω} is the head-wind velocity.

17.2.2.3 Propulsion Power

The desired power rating for the electric motor has calculated based on the system constraints of starting acceleration, vehicle rated, and maximum velocity, and vehicle gradability. The torque at the vehicle wheels obtained from the power relation.

$$P = T_{\mathrm{TR}} \omega_{\mathrm{wh}} = F_{\mathrm{TR}} . v_{xT} \, \mathrm{watts} \tag{17.5}$$

where T_{TR} is the tractive torque in Nm, ω_{wh} is the wheel angular velocity in rads/sec, F_{TR} is in N, and v_{xT} is in m/s., assuming no slip, the angular velocity, and the vehicle speed are related by

$$v_{xT} = \omega_{wh} r_{wh} \tag{17.6}$$

where r_{wh} is the radius of the wheel in a meter.

Tractive force versus steady-state velocity characteristics obtained from the equation of motion. When the steady-state velocity is reached $dv/dt = 0$; and $\sum F = 0$.

Therefore

$$\begin{aligned} F_{TR} - F_{AD} - F_{ROLL} - F_{gxT} &= 0 \\ \Rightarrow F_{TR} &= mg\left[\sin\beta + C_0 \mathrm{sgn}(V)\right] \\ &+ \mathrm{sgn}(V)\left[mgC_1 + \frac{\rho}{2} C_D A_F\right] V^2 \end{aligned} \tag{17.7}$$

17.2.2.4 Maximum Gradeability

The vehicle has expected to move forward very slowly when climbing a steep slope. Hence, the following assumptions for maximum gradeability made according to the vehicle move very slowly v is 0, F_{AD} F_r are negligible. The vehicle is not accelerating, dv/dt is 0, and F_{TR} is the maximum tractive force delivered by the motor at or near zero speed.

With the assumptions, at near stall conditions,

$$\sum F = 0 \rightarrow F_{TR} - F_{gxT} = 0 \rightarrow F_{TR} = mg \sin\beta \tag{17.8}$$

If required mass $(m) = 1300$ kg, (full load)

$$F_{TR} = 1300 \times 9.81 \sin 18 = 3940\,\mathrm{N}$$

the maximum percent grade is max % grade = 100 tan β,

$$\text{max\% grade} = \frac{100 F_{TR}}{\sqrt{(mg)^2 - F_{TR}{}^2}} = \frac{100 \times 3940}{\sqrt{(1300 \times 9.81)^2 - (3940)^2}} = 32.48\%$$

17.2.3 Ideal Gearbox: Steady-State Model

The EV transmission equation has established by assuming an ideal gearbox as shown in Fig. 17.5 with P_{losses} is 0, and the efficiency is 100%, perfectly rigid gears, and no gear backlash.

17.2.3.1 Gear Ratio

For a tire wheel with radius r, the tangential and the angular velocity are related by:

$$\omega r = v,\ \omega = \frac{v}{r} \tag{17.9}$$

The tangential velocity at the gear teeth contact point is the same for the two gears with different radius.

$$r_{\text{in}}\omega_{\text{in}} = v = r_{\text{out}}\omega_{\text{out}} \tag{17.10}$$

The gear ratio has defined in terms of speed transformation between the input shaft and the output shaft.

$$\text{GR} = \frac{\omega_{\text{in}}}{\omega_{\text{out}}} = \frac{r_{\text{out}}}{r_{\text{in}}} \tag{17.11}$$

Assuming 100% efficiency of the gear train:

$$P_{\text{out}} = P_{\text{in}},\Rightarrow T_{\text{out}}\omega_{\text{out}} = T_{\text{in}}\omega_{\text{in}} \tag{17.12}$$

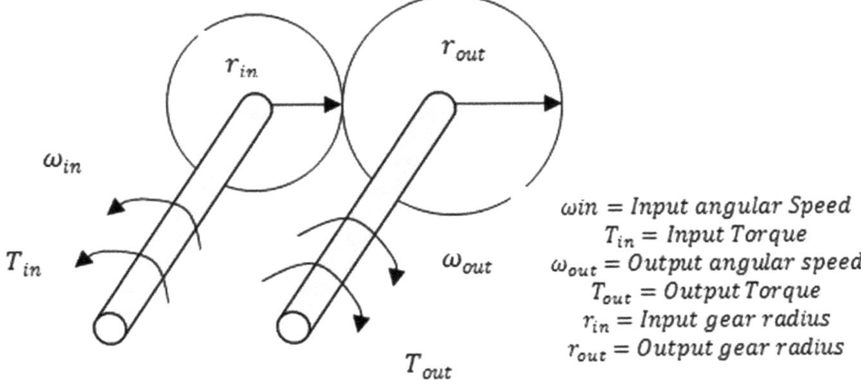

Fig. 17.5 Ideal gearbox: steady-state model

The gear ratio in terms of the torque at the two shafts is

$$GR = \frac{\omega_{\text{in}}}{\omega_{\text{out}}} = \frac{T_{\text{out}}}{T_{\text{in}}} \tag{17.13}$$

At the point of gear mesh, the supplied and delivered forces are the same. The torque at the shaft is the force at the mesh divided by the radius of the disk. In the two-gear combination, the torque ratio between the two gears is proportional to the ratio of gear disk radius. The torque of the inner disk in terms of its radius and force at the gear mesh is

$$T_{\text{in}} = F r_{\text{in}} \Rightarrow F = \frac{T_{\text{in}}}{r_{\text{in}}} \tag{17.14}$$

Similarly, for the other disk with radius r_{out}, the force at the gear mesh is

$$F_{\text{wheel}} = \frac{T_{\text{out}}}{r_{\text{out}}},\ T_{\text{out}} = F_{\text{wheel}} \times r_{\text{out}} \tag{17.15}$$

Therefore, the gear ratio is

$$\text{GR} = \frac{T_{\text{out}}}{T_{\text{in}}} = \frac{r_{\text{in}}}{r_{\text{out}}} \tag{17.16}$$

17.2.4 Initial Acceleration

The initial acceleration is specified as 0 to v_f in t_fs. v_f is the vehicle rated speed obtained $v_f = \omega_{f\text{wh}}.r_{\text{wh}}$. The acceleration of the vehicle in terms of these variables has given by Eq. (17.17):

$$a = \frac{\text{d}v}{\text{d}t} = \frac{F_{\text{TR}} - F_{\text{RL}}}{m} \tag{17.17}$$

17.3 Methodology

MATLAB/Simulink is the medium to simulate the DC drive EV for vehicle design. The previous vehicle dynamics mathematical equations are grouped to form a physical-based model, as shown in Fig. 17.6 (Arof et al. 2017a). The model is used to determine specific requirements parameters of DC drive EC. Some general parameters of the model are preset as follows:

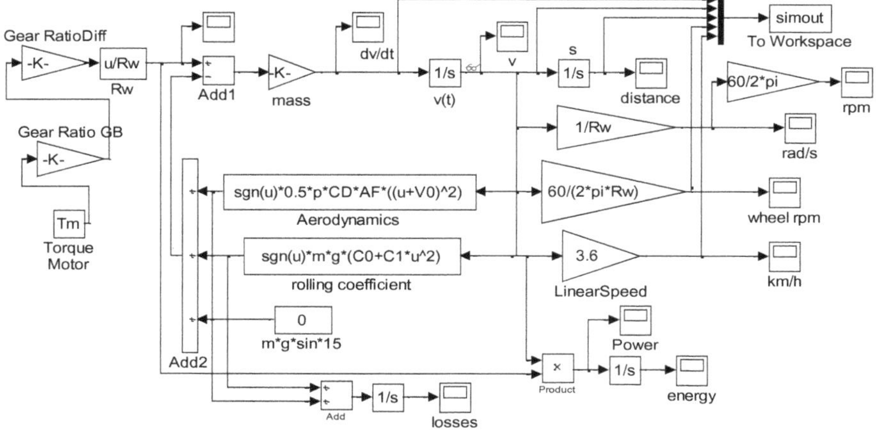

Fig. 17.6 Simulation of the vehicle

$m = 1350$ kg, (full load) $C_D = 0.2$, $A_F = 2\text{m}^2$, $C_O = 0.009$, $C_1 = 0$, $\rho = 1.1614$ kg/m^3, and $g = 9.81$ m/s^2, r_{wh} = radius of wheel 0.28 m = 11 in, gear ratio = 4.2

A complete simulation model of the series motor FQDC DC drive EV using the MATLAB/Simulink software that includes the vehicle dynamics model, four quadrants DC chopper, DC series motor and controllers is as shown in Fig. 17.7. The model is used to determine the EV acceleration from a standstill. It required torque for the acceleration, motor traction power, the EV maximum speed and maximum EV speed when climbing a steep hill from a standstill. The suitable EV battery voltage supply and the equivalent expected maximum battery current associated with it.

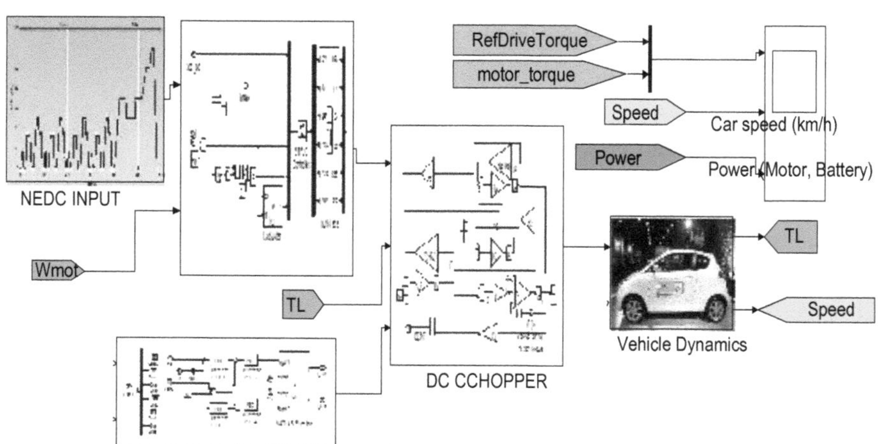

Fig. 17.7 Mathematical model of FQDC for EV

17.4 Result and Discussion

The first requirement is tested for acceleration from standstill to 60 km/h in 10 s, as shown in Fig. 17.8. From the figure, the constant acceleration has been obtained from 0 to 10 s. The speed of 100 km/h is also shown in the figure at 30 s and the maximum speed has expected to be slightly higher than 100 km/h.

The previous acceleration and speed have been tested. The required torque to achieve the acceleration and speed determined using the complete DC EC simulation model is 240 Nm. The required 240 Nm torque is as shown in Fig. 17.9. In this figure, the constant acceleration earlier mentioned, is obtained during the constant torque from 0 to 10 s.

From the torque, we determined the required motor traction power. To produce the required acceleration and torque the series motor and FQDC EC is requiring 41 kW/ motor power, as shown in Fig. 17.10.

The required current to produce 40 kW of series motor traction power is shown in Fig. 17.11. In the figure, there are several options of required voltage and battery current that can be selected. For instance, if the 150 V of battery voltage has set, therefore the expected maximum battery current to maintain the 40 kW of motor

Fig. 17.8 Acceleration test

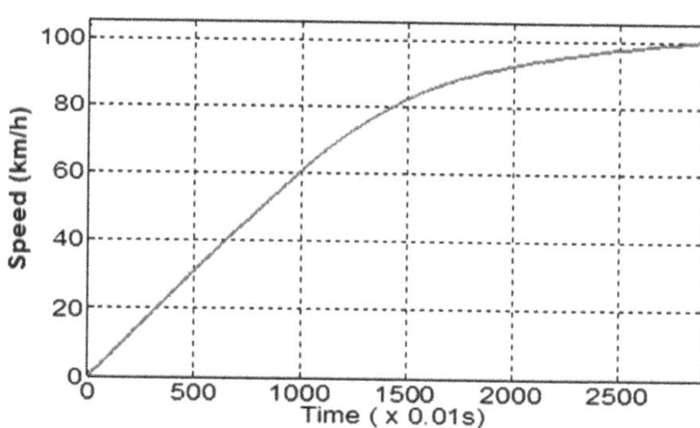

Fig. 17.9 Torque

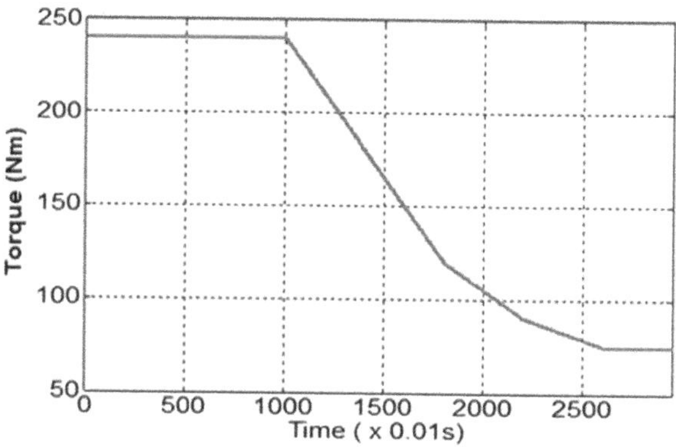

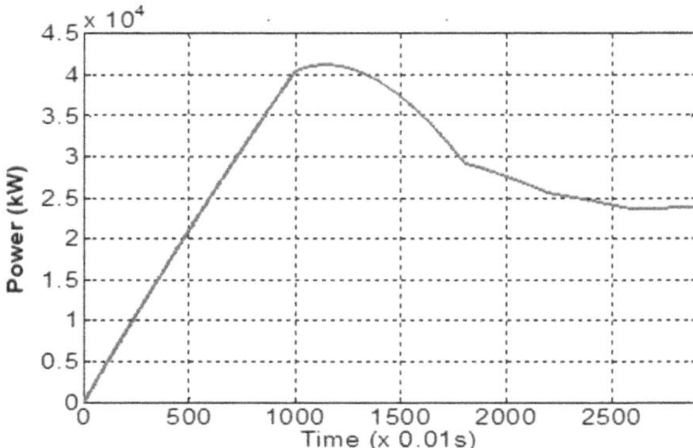

Fig. 17.10 Motor traction power

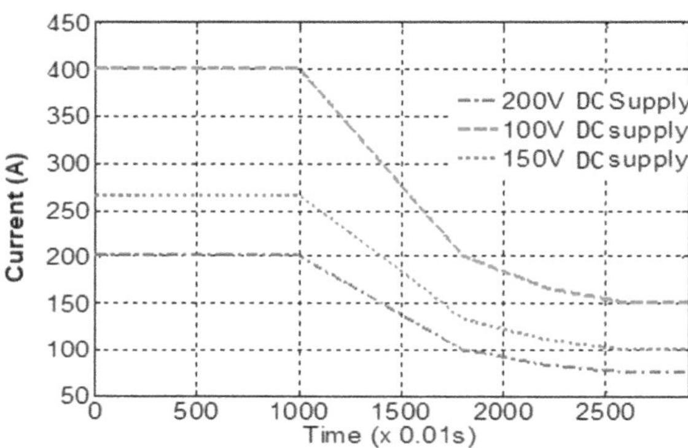

Fig. 17.11 The required current

power is required to be 275 A. While for 100 V of the battery voltage, the expected battery of maximum needed current is 400 A, and for 200 V of the battery voltage, 200 A of battery currents required.

The vehicle has been tested to climb a steep hill at a 21° angle from standstill carrying a maximum load of 1300 kg. The vehicle speed is as shown in Fig. 17.12. The speed is about 1.03 km/h, which is very slow but enough to meet the design requirements.

17.5 Conclusions

The vehicle dynamics mathematical model can determine the requirements of the specific parameter of DC drive EC such as the traction motor. The complete simulation model of EC, which consists of the vehicle dynamics model, FQDC, controllers and series motor can be used to find, test, and verify the design requirement parameters for EC application. These parameters have been used for testing with the New

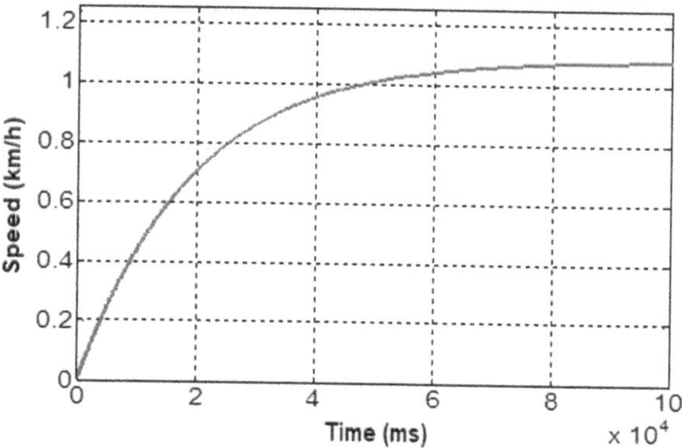

Fig. 17.12 Vehicle speed

European Driving Cycle (NEDC) Test. The proposed four quadrants DC chopper for DC drive series motor has a high potential to be utilized in EV. It is due to a simple design, low cost and excellent controllability.

References

Arof, S., Yaakop, N.M., Jalil, J.A., Mawby, P.A., Arof, H.: Series motor four quadrants drive DC chopper for low cost, electric car: part 1: overall. In: 2014 IEEE International Conference on Power and Energy (PECon), pp. 342–347 (2014)

Arof, S., Muhd Khairulzaman, A.K., Jalil, J.A., Arof, H., Mawby, P.A.: Self tuning fuzzy logic controlling chopper operation of four quadrants drive DC chopper for low-cost electric vehicle. In: 2015 6th International Conference on Intelligent Systems, Modelling and Simulation, pp. 40–45 (2015a)

Arof, S., Zaman, M.K., Jalil, J.A., et al.: Artificial intelligence controlling chopper operation of four quadrants drive DC chopper for low-cost electric vehicle. Int. J. Simul. Syst. Sci. Technol. **16**(4), 1–10 (2015b)

Arof, S., Jalil, J.A., Kamaruddin, N.S., Yaakop, N.M., Mawby, P.A., Hamzah, A.: Series motor four quadrants drive DC chopper part2: driving and reverse mode with direct current control. In: 2016 IEEE International Conference on Power and Energy (PECon), pp. 775–780 (2016)

Arof, S., Hassan, H., Rosyidi, M., et al.: Implementation of series motor four quadrants drive DC chopper for dc drive electric car and LRT via simulation model. J. Appl. Environ. Biol. Sci. **7**(3S), 73–82 (2017a)

Arof, S., Noor, N.M., Alias, M.F., et al.: Investigation of chopper operation of series motor four quadrants DC chopper. J. Appl. Environ. Biol. Sci. **7**(3S), 49–56 (2017b)

Arof, S., Diyanah, N.H.N., Mawby, P.A. et al.: Low harmonics plug-in home charging electric vehicle battery charger utilizing multi-level rectifier, zero crossing and buck chopper: part 1: general overview. In: Bakar, M.H.A., Sidik, M.S.M., Öchsner, A. (eds.) Progress in Engineering Technology. Advanced Structures and Materials, 119. Springer, Cham. (2019a)

Arof, S., Diyanah, N.H.N., Noor, N.M. et al.: A new four quadrants drive chopper for separately excited DC motor in low-cost electric vehicle. In: Bakar, M.H.A., Sidik, M.S.M., Öchsner, A. (eds.) Progress in Engineering Technology. Advanced Structures and Materials, 119. Springer, Cham (2019b)

Arof, S., Diyanah, N.H.N., Noor, N.M.N. et al.: Series motor four quadrants drive DC chopper: Part 4: generator mode. In: Bakar, M.H.A., Sidik, M.S.M., Öchsner, A. (eds.) Progress in Engineering Technology. Advanced Structures and Materials, 119. Springer, Cham (2019c)

Arof, S., Diyanah, N.H.N., Yaakop, M.A. et al.: Processor in the loop for testing series motor four quadrants drive DC chopper for series motor driven electric car: part 1: chopper operation modes testing. In: Ismail, A., Bakar, M.H.A., Öchsner, A. (eds.) Advanced Engineering for Processes and Technologies. Advanced Structures and Materrials, 102. Springer, Cham (2019d)

Arof, S., Diyanah, N.H.N., Mawby, P.A. et al.: Study on implementation of neural network controlling four quadrants direct current chopper: part1: using single neural network controller with binary data output. In: Ismail, A., Bakar, M.H.A., Öchsner, A. (eds.) Advanced Engineering for Processes and Technologies. Advanced Structures and Materials, 102. Springer, Cham (2019e)

Arof, S., Diyanah, N.H.N., Noor, N.M. et al.: Genetics algorithm for setting up look up table for parallel mode of new series motor four quadrants DC chopper. In: Bakar, M.H.A., Sidik, M.S.M., Öchsner, A. (eds.) Progress in Engineering Technology. Advanced Structures and Materials, 119. Springer, Cham (2019f)

Arof, S., Noor, N.M., Alias, M.F. et al.: Digital proportional integral derivative (PID) controller for closed-loop direct current control of an electric vehicle traction tuned using pole placement. In: Bakar, M.H.A., Zamri, F.A., Öchsner, A. (eds.) Progress in Engineering Technology II. Advanced Structures and Materials, 131. Springer, Cham (2020a)

Arof, S., Noor, N.M., Mohamad, R. et al.: Close loop feedback direct current control in driving mode of a four quadrants drive direct current chopper for electric vehicle traction controlled using fuzzy logic. In: Bakar, M.H.A., Zamri, F.A., Öchsner, A. (eds.) Progress in Engineering Technology II. Advanced Structures and Materials, 131. Springer, Cham (2020b)

Arof, S., Sukiman, E.D., Diyanah, N.H.N. et al.: Discrete-time linear system of New series motor four-quadrant drive direct current chopper numerically represented by Taylor series. In: Bakar, M.H.A., Zamri, F.A., Öchsner, A. (eds.) Progress in Engineering Technology II. Advanced Structures and Materials, 131. Springer, Cham (2020c)

Arof, S., Said, M.S., Diyanah, N.H.N. et al.: Series motor four-quadrant direct current chopper: reverse mode, steering position control with double-circle path tracking and control for autonomous reverse parking of direct current drive electric car. In: Bakar, M.H.A., Zamri, F.A., Öchsner, A. (eds.) Progress in Engineering Technology II. Advanced Structures and Materials, 131. Springer, Cham (2020d)

Arof, S., Said, M.S., Diyanah, N.H.N. et al. Series motor four-quadrant DC chopper: reverse mode, direct current control, triple cascade PIDs, and ascend-descend algorithm with feedback optimization for automatic reverse parking. In: Bakar, M.H.A., Zamri, F.A., Öchsner, A. (eds.) Progress in Engineering Technology II. Advanced Structures and Materials, 131. Springer, Cham (2020e)

Bansal, R.C.: Electric Vehicle, Handbook of Automotive Power Electronics and Motor Drives. CRC Press, India (2005)

Gao, Y., Ehsani, M.: Design and control methodology of plug-in hybrid electric vehicle. IEEE Trans. Ind. Electron. **57**, 633–640 (2010)

Husain, I.: Electric and Hybrid Electric Vehicles, Design Fundamentals. CRC Press, United States (2003)

Oak Ridge National Laboratory: Advanced Brush Technology for DC Motors. https://peemrc.ornl.gov/projects/emdc3.jpg. Accessed May 2009

Westbrook, M.H.: The Electric and Hybrid Electric Vehicle. Society of Automotive Engineers, United States (2001)

Chapter 18
Low Harmonics Plug-in Home Charging Electric Vehicle Battery Charger Utilizing Multi-level Rectifier, Zero Crossing, and Buck Chopper: BMS Battery Charging Control Algorithm

Saharul Arof, M. S. Sazali, Norramlee Mohamed Noor, J. Nur Amirah, Philip Mawby, Emilia Noorsal, Aslina Abu Bakar, and Yusnita Mohd Ali

Abstract This paper focuses on a series motor direct current (DC) drive electric car (EC) battery charger control algorithm for controlling and synchronizing Low Harmonics Plug-in Home Charging Electric Vehicle Battery Charger that consists of zero crossing, multi-level rectifier (MLR), and buck chopper (BC). The MATLAB/Simulink software was used to test the control algorithm and proposed battery charger. The simulation results showed that the proposed battery charger control algorithm successfully performed the expected operation.

S. Arof (✉) · M. S. Sazali · N. M. Noor · J. Nur Amirah
Malaysian Spanish Institute Kulim Hi-Tech Park, Universiti Kuala Lumpur, Kulim 09000, Kedah, Malaysia
e-mail: saharul@unikl.edu.my

M. S. Sazali
e-mail: msazali@unikl.edu.my

N. M. Noor
e-mail: noramlee@unikl.edu.my

J. Nur Amirah
e-mail: noramirah0706@gmail.com

P. Mawby
University of Warwick School of Engineering, Coventry CV47AL, UK
e-mail: p.a.mawby@warwick.ac.uk

E. Noorsal · A. A. Bakar · Y. M. Ali
Faculty of Electrical Engineering, Universiti Teknologi MARA (UiTM) Cawangan Pulau Pinang, Bukit Mertajam 13500, Pinang, Malaysia
e-mail: emilia.noorsal@uitm.edu.my

A. A. Bakar
e-mail: aslina060@uitm.edu.my

Y. M. Ali
e-mail: yusnita082@uitm.edu.my

M. H. Abu Bakar et al. (eds.), *Progress in Engineering Technology III*,
Advanced Structured Materials 148, https://doi.org/10.1007/978-3-030-67750-3_18

Keywords DC drive · Zero crossing · Buck chopper · MLR · Charging · Battery charger · EC

18.1 Introduction

The emission of hydrocarbons not only pollutes the environment but also contributes to global warming, which melts the icebergs and increases the sea level. Using efficient electric vehicles (EVs) for transportation is one of the solutions to reduce global hydrocarbon emission.

18.2 Literature Review

In 2009, the Oak Research National Laboratory (ORNL) (Oak Ridge National Laboratory 2009) succeeded in designing a DC brushed motor with high power output (55 kW) and high efficiency (92%) that can operate at low operating voltages (13 V). This started the interest to embark on a DC drive EC research. Attempts to improve the conventional H-bridge chopper by increasing more operations or allowing motor reverse action has been continuously made. However, since the discovery of the new DC motor by ORNL, a new series motor four-quadrant DC chopper (FQDC) has been designed, which has more operations compared to the conventional version (Arof et al. 2014). Other studies related to DC drive EV have led to the research on different types of DC drive brushed motor, such as the separately excited DC motor that can be used for traction motor of DC EV (Arof et al. 2019). Moreover, detailed investigations on the chopper operation modes have led to the establishment of a simulation model to test the chopper operations for the application of EC and light rail transit (LRT) (Arof et al. 2017). This simulation model prompts further detailed investigation on each chopper operation and specific pattern of motor voltage, current, torque, speed of the series motor, and FQDC running for DC drive EC application (Arof et al. 2017). For the DC series traction motor of EC application, an attempt to reduce jerk and tire slip of the speed and torque control for the series motor has been successfully made and implemented with the DC control technique (Arof et al. 2016).

For power regeneration, (Arof et al. 2019) studied and discussed the FQDC, which offers a generator mode with several techniques of starting the regenerated power and voltage control. In order for the FQDC to be applied in the real world, controllers are needed to run a control algorithm in the embedded system. The controller and its control algorithm have been studied and tested by using the processor-in-the-loop (PIL) technique (Diyanah 2019). To improve the new FQDC performance, optimization tools, such as the artificial intelligence (AI) are introduced to control all chopper operations of the proposed FQDC chopper (Arof et al. 2015; Arof et al. 2015). Among the three AI controllers, the adaptive network-based fuzzy inference system (ANFIS) shows the best performance followed by the neural network and self-tuning fuzzy logic controllers. A study on the neural network controller to uncover a proper method of tuning has been carried out, i.e., by using a single controller with binary output (Arof et al. 2019). For each specific FQDC chopper operation mode, the performance can be further improved by using AI optimization tools, such as the genetic algorithm to set up a specific lookup table for field current in parallel mode operation (Arof et al. 2019). Close loop controller using PID tuned using pole placement (Arof et al. 2020) and Fuzzy logic (Arof et al. 2020) are applied to improve the performance and stability of armature current in direct current control (DCC) in the driving mode. This method has undergone improvements using cascade PID and ascend descend algorithm and after improvisations made and associated with steering (Arof et al. 2020) and vehicle movement control (Said MS and Diyanah 2020), it was used for autonomous EV automatic reverse parking. For fault diagnose and system online tuning and optimization, a numerical representation using Taylor series was applied (Arof et al. 2020).

Several studies related to DC drive EV have led to a research on EC battery charger, such as in Fig. 18.1, which offers low harmonics (Arof et al. 2019). This battery charger uses zero cross, multi-level rectifier (MLR), and buck chopper (BC). To make the battery charger efficient, it requires a battery management system (BMS) to make it a safe, reliable, and cost-efficient solution. The BMS controls the operational conditions of the battery to prolong its life and guarantee its safety. The BMS, in

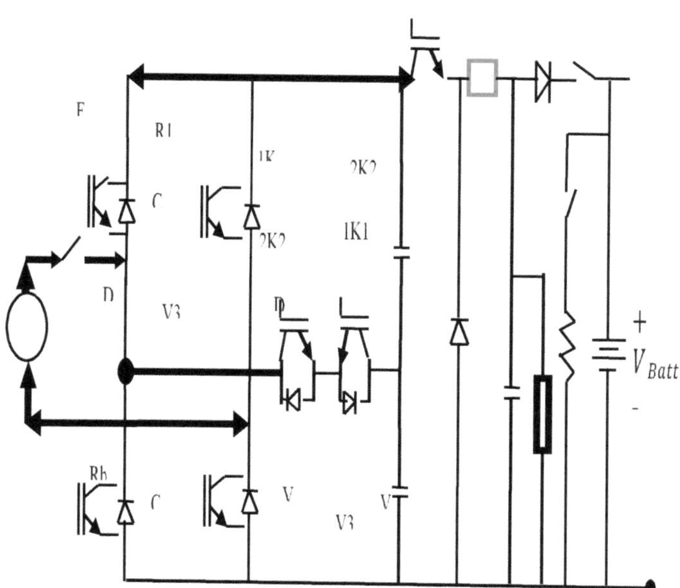

Fig. 18.1 Battery charger

specific, also has to control the charging process of the battery, which will be further discussed in this paper.

18.3 Methodology

There are two modes of charging a battery when a vehicle is running, which are through FQDC chopper during generator (Arof et al. 2019), regenerative braking, and resistive braking modes and the other one is when the vehicle is in standstill via a battery charger circuit. The battery charger (as in Fig. 18.2), which is for standstill charging consists of a multi-level inverter/rectifier (MLR) and a buck chopper (BC). For charging when the EV is not running, the BMS controller passes over the signal, i.e., battery state of charge (SOC) to the data distribution controller. The data is then passed to the battery charging controller. Next, the battery charging controller will pass an instruction to the zero cross, MLR, and a BC controller. The controllers need this data to determine the generator voltage and current to charge the batteries.

The charging process via the two modes is explained in the block diagram in Fig. 18.2 and via the flow chart in Fig. 18.3. The BMS controller provides battery SOC to the data distribution controller and segregates it to the zero cross, MLR, and BC controllers. Once the SOC value is known, the charging controller will control the charging voltage according to the SOC value.

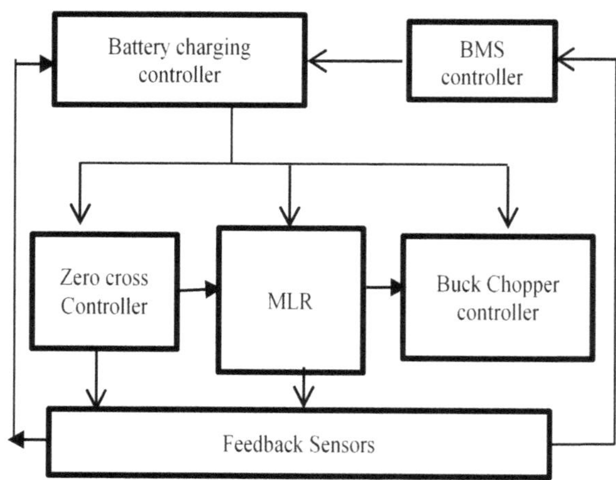

Fig. 18.2 Block diagram of BMS controller data path

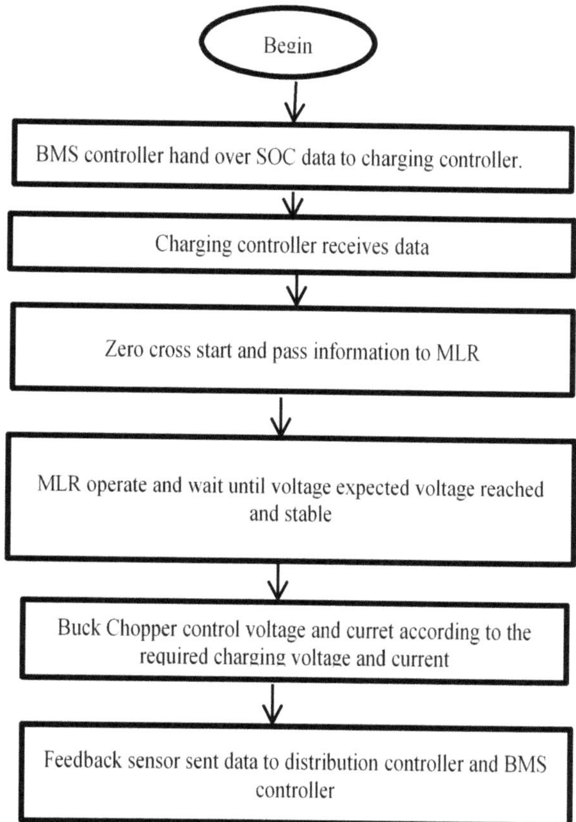

Fig. 18.3 Flow chart of BMS controller operation

18.3.1 Charging Controller Control Algorithm

The charging controller algorithm includes controlling the battery charging current and voltage. While charging, the most important element to be controlled is the current. However, if the voltage is not controlled, it could produce high instant charging current at the beginning of the charging process, which is unwanted. High rising of current causes heat to be produced and this can shorten the life of a battery or destroy it. At the start of the battery charging process, zero cross is initiated. Once operated, the zero cross will provide a start signal to the MLR. Without this signal, the MLR will not be fired at the right sequence, which is not according to the frequency and the positive voltage source of the AC grid. It can cause the MLR to fail. A successive MLR will produce DC voltage at the output and the value will be observed by the BC controller. Once the expected voltage is reached and it is stable, then only the BC will start. Its control algorithm reduces the voltage before it becomes stable; this is called the steady state. Next, the batteries will be ready for charging. The process is explained in Fig. 18.4.

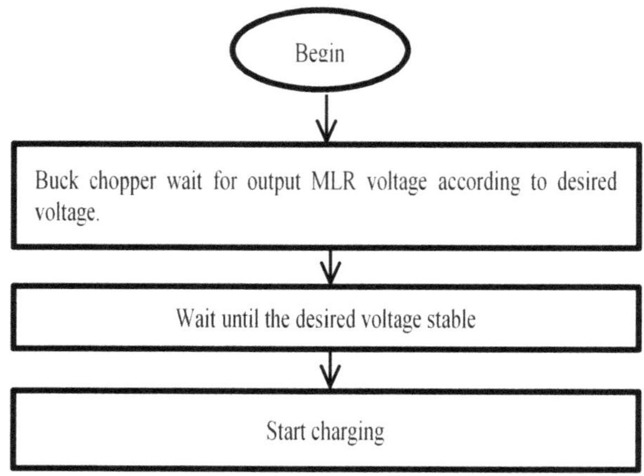

Fig. 18.4 Block diagram of BC control voltage charging

18.3.2 Buck Chopper Control Charging Current

The strategy of the BMS control is to have a low rising current at the beginning of the battery charging process. Firstly, the BC needs to chop the input voltage so that the output of the BC is low, i.e., nearly approaching the terminal battery voltage. This type of feedback control is called voltage control. The input reference is the battery terminal voltage plus some offset voltages and the control feedback is the BC output voltage. Once the expected voltage is achieved, the battery will be connected to the BC. As the voltage difference between the BC output and the battery terminal is low, the charging current will start with a low current at the beginning. The voltage will slowly increase to increase the charging current and finally, the control will change to current control and the control feedback of the battery current control will be passed over to the current sensor. The operation is explained in block diagram in Fig. 18.10 and the results of the BMS control charging current are shown in Fig. 18.5.

18.3.3 Zooming Effect by Using Amplifier and Offset Circuit

To get a better voltage of BC, a combination of an amplifier and offset circuits is required to have the zooming effect for better voltage control. This circuit is used once the expected BC voltage output is nearly approaching the desired voltage value. The BC controller must switch from reading the usual voltage sensor to this zooming effect circuit.

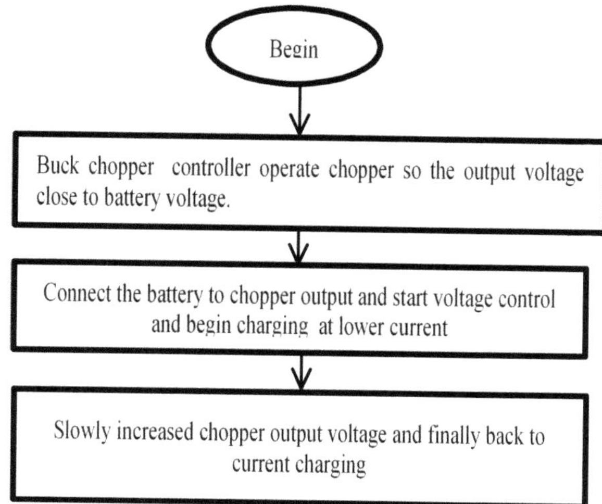

Fig. 18.5 Buck chopper controller control charging current operation

18.3.4 Control Battery Charging Temperature

During the battery charging process, the temperature increases because the current that flows to the battery produces heat. If the same amount of charging current can remain, the battery temperature will continue to increase up to 60 °C and above. To overcome this problem, the charging controller is equipped with a battery cooling apparatus, such as a cooling fan to lower the battery temperature. This is one of the countermeasures to handle overheating. Another alternative is that the charging controller will lower the expected charging current by lowering the charging voltage to decrease the charging current.

$$\begin{aligned}\text{Expected current} = {} & \text{expected current} - (\text{offset voltage constant} \\ & * (\text{current temperature} - \text{base temperature}) \\ & /(\text{maximum deviation temperature}))\end{aligned}$$

18.3.5 Simulation Model of Battery Charger and Charging Controller

A Simulink model that consisted of zero crossing, MLR and BC, and a charging controller was developed to test the battery charger and control algorithm (Fig. 18.6). The offset and amplifier simulation circuit model are shown in Fig. 18.7.

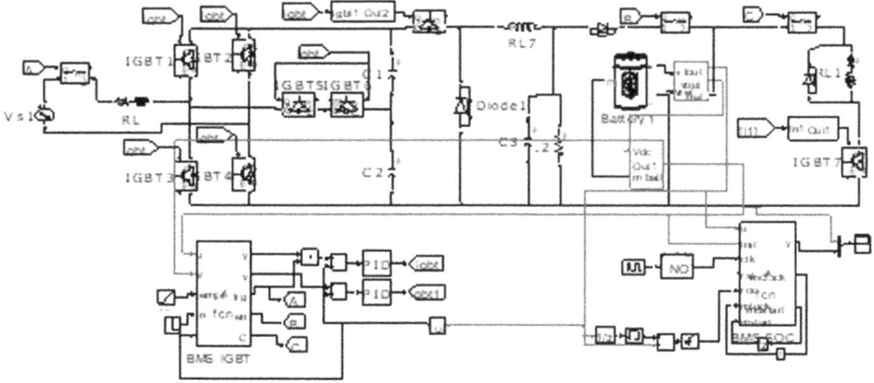

Fig. 18.6 MATLAB/Simulink model

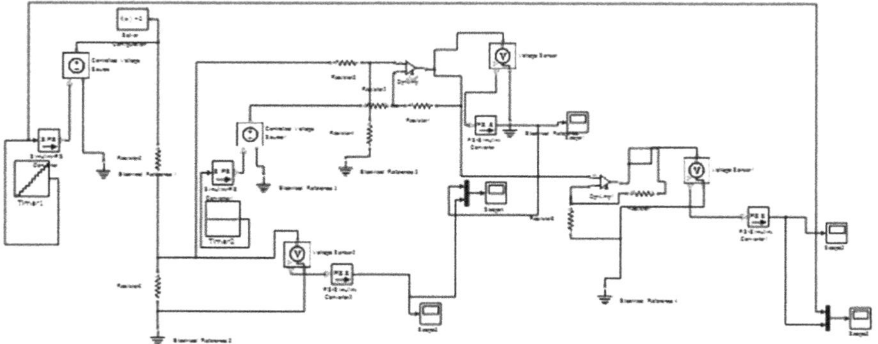

Fig. 18.7 Offset and amplifier circuit

18.4 Results and Discussion

Figure 18.8 shows the battery charging process. Once activated, the battery charging starts by sending a start signal to the zero crossing (ZC) circuit. Immediately, when the ZC circuit receives the signal, it will produce ZC signals and ZC start the signal to the MLR controller to start the MLR operation. Without the ZC signal, MLR will not produce the IGBT gate firing signals, the MLR will operate like a normal full bridge rectifier. When the MLR starts and the MLR output voltage has reached the expected voltage and is stable, the BC controller will start to operate the BC. The BC controller will regulate the voltage input (MLR output) to the desired BC voltage output. Firstly, the BC must chop the output voltage of the MLR to the desired voltage, then, the battery charging process will start.

Based on Fig. 18.9, the BC controller will first wait for the MLR voltage to reach the expected voltage and becomes stable before it operates.

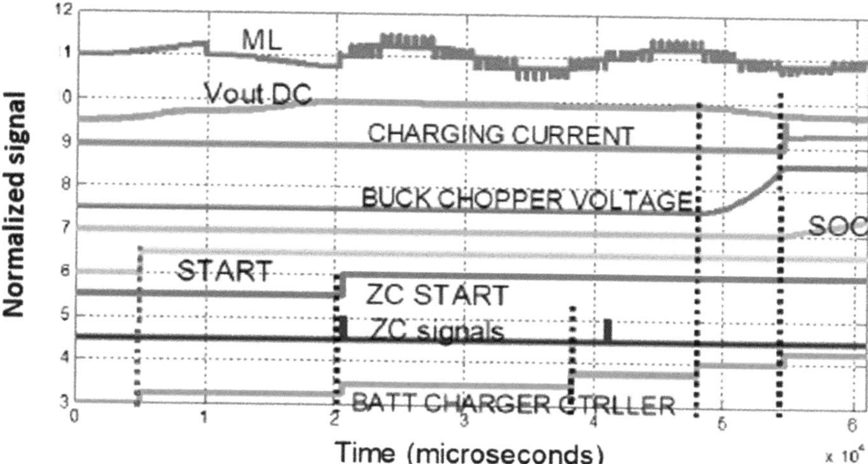

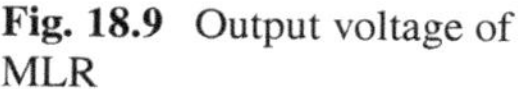
Fig. 18.8 Overall result

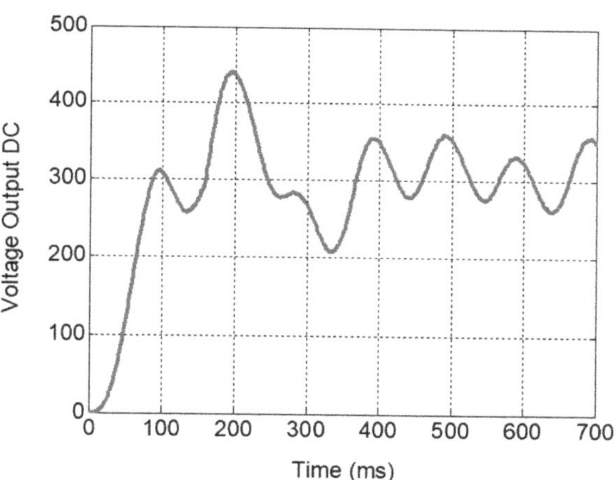

Fig. 18.9 Output voltage of MLR

Figure 18.10 shows that the BC controller lowers the voltage before it starts the battery charging process.

As shown in Fig. 18.11, the BC controller controls the BC output voltage to nearly approaching the battery voltage before the charging process starts to disallow high starting current.

Increase in SOC value during charging will not significantly increase in battery voltage. The increase in voltage is small compared to the increase in SOC. For this reason, a circuit called zooming circuit can be used. The circuit is to amplify the battery voltage so that the increase in battery voltage becomes obvious. Figure 18.12

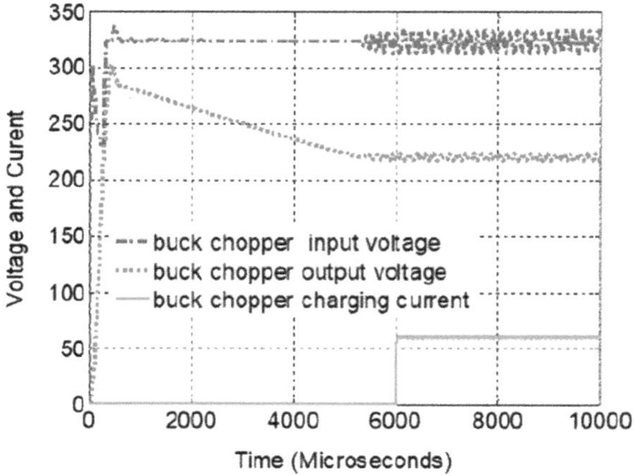

Fig. 18.10 BMS control charging voltage

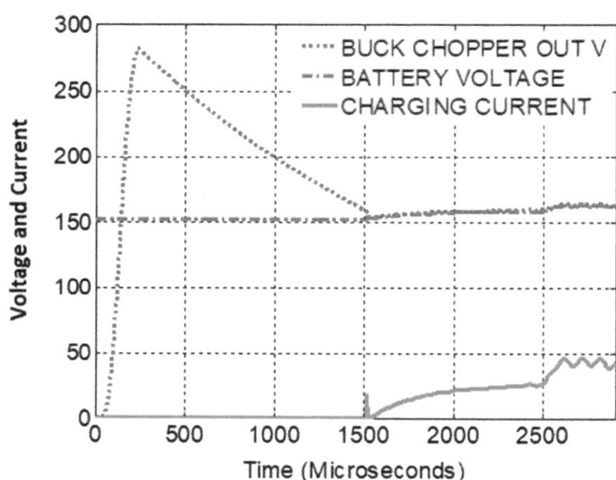

Fig. 18.11 BMS control charging current

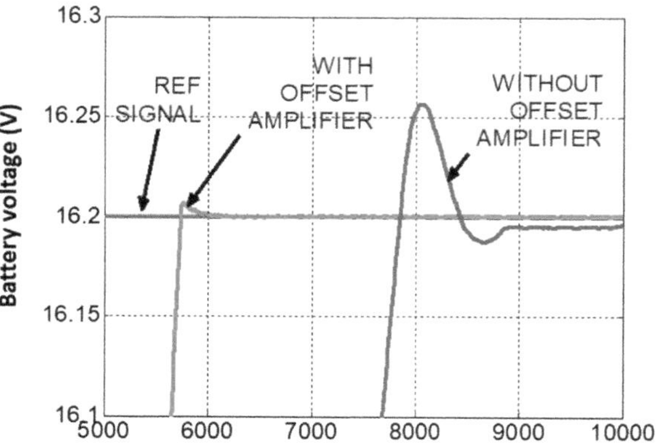

Fig. 18.12 Comparison of circuits with and without zooming effect

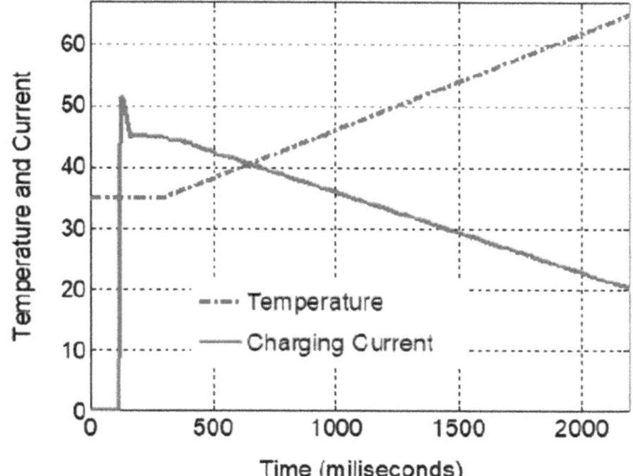

Fig. 18.13 Controlling charging current to control temperature

shows the zooming circuit effects. Based on the figure, the circuit with the zooming effect produces better results in the close loop feedback of battery voltage control.

Based on Fig. 18.13, the charging controller will lower the expected charging current by lowering the charging voltage in the case of increases of battery temperature to the state that can cause defect to the batteries. The battery temperature increment will result in battery charging controller decrease the charging current.

18.5 Conclusion

The proposed battery charging controller is able to handle the battery charging process for DC drive EV on static charging with an AC power source. The DC EV incorporated with the proposed battery charger is also able to be an alternative for economical EV.

References

Arof, S., Diyanah, N.H.N., Mawbym P.A. et al.: Study on implementation of neural network controlling four quadrants direct current chopper: Part1: using single neural network controller with binary data output. In: Ismail, A., Bakar, M.H.A., Öchsner, A. (eds.) Advanced Engineering for Processes and Technologies. Adv Struc Mat 102. Springer, Cham (2019)

Arof, S., Diyanah, N.H.N., Noor, N.M., et al.: A new four quadrants drive chopper for separately excited DC motor in low-cost electric vehicle. In: Bakar, M.H.A., Sidik, M.S.M., Öchsner, A. (eds.) Progress in Engineering Technology. Adv Struc Mat 119. Springer, Cham (2019)

Arof, S., Diyanah, N.H.N., Noor, N.M. et al.: Genetics algorithm for setting up look up table for parallel mode of new series motor four quadrants DC chopper. In: Bakar, M.H.A., Sidik, M.S.M.,

Öchsner, A. (eds.) Progress in Engineering Technology. Adv Struc Mat 119. Springer, Cham (2019)

Arof, S., Diyanah, N.H.N., Yaakop, M.A. et al.: Processor in the loop for testing series motor four quadrants drive DC chopper for series motor driven electric car: Part 1: Chopper operation modes testing. In: Ismail, A., Bakar, M.H.A., Öchsner, A. (eds.) Advanced Engineering for Processes and Technologies. Adv Struc Mat 102. Springer, Cham (2019)

Arof, S., Muhd Khairulzaman, A.K., Jalil, J.A., Arof, H., Mawby, P.A.: Self tuning fuzzy logic controlling chopper operation of four quadrants drive DC chopper for low-cost electric vehicle. In: 2015 6th International Conference on Intelligent Systems, Modelling and Simulation, pp. 40–45 (2015)

Arof, S., Noor, N.M., Alias, M.F., et al.: Digital proportional integral derivative (PID) controller for closed-loop direct current control of an electric vehicle traction tuned using pole placement. In: Bakar, M.H.A., Zamri, F.A., Öchsner, A. (eds.) Progress in Engineering Technology II. Adv Struc Mat 131. Springer, Cham (2020)

Arof, S., Noor, N.M., Mohamad, R., et al.: Close loop feedback direct current control in driving mode of a four quadrants drive direct current chopper for electric vehicle traction controlled using fuzzy logic. In: Bakar, M.H.A., Zamri, F.A., Öchsner, A. (eds.) Progress in Engineering Technology II. Adv Struc Mat 131. Springer, Cham (2020)

Arof, S., Said, M.S., Diyanah, N.H.N., et al.: Series motor four-quadrant DC chopper: reverse mode, direct current control, triple cascade PIDs, and ascend-descend algorithm with feedback optimization for automatic reverse parking. https://doi.org/10.1007/978-3-030-46036-5_13

Arof, S., Said, M.S., Diyanah, N.H.N., et al.: Series motor four-quadrant direct current chopper: reverse mode, steering position control with double-circle path tracking and control for autonomous reverse parking of direct current drive electric car. In: Bakar, M.H.A., Zamri, F.A., Öchsner, A. (eds.) Progress in Engineering Technology II. Adv Struc Mat 131. Springer, Cham (2020)

Arof, S., Sukiman, E.D., Diyanah, N.H.N., et al.: Discrete-time linear system of New series motor four-quadrant drive direct current chopper numerically represented by Taylor series. In: Bakar, M.H.A., Zamri, F.A., Öchsner, A. (eds.) Progress in Engineering Technology II. Adv Struc Mat 131. Springer, Cham (2020)

Arof, S., Yaakop, N.M., Jalil, J.A., Mawby, P.A., Arof, H.: Series motor four quadrants drive DC chopper for low cost, electric car: Part 1: Overall. In: 2014 IEEE International Conference on Power and Energy (PECon), pp. 342–347 (2014)

Arof, S., Zaman, M.K., Jalil, J.A., et al.: Artificial intelligence controlling chopper operation of four quadrants drive DC chopper for low-cost electric vehicle. Int. J. Simul. Syst. Sci. Technol. **16**(4), 1–10 (2015)

Arof, S., Jalil, J.A., Kamaruddin, N.S., Yaakop, N.M., Mawby, P.A., Hamzah, A.: Series motor four quadrants drive DC Chopper Part2: driving and reverse mode with direct current control. In: 2016 IEEE International Conference on Power and Energy (PECon), pp. 775–780 (2016)

Arof, S., Hassan, H., Rosyidi, M., et al.: Implementation of series motor four quadrants drive DC chopper for dc drive electric car and LRT via simulation model. J. Appl. Environ. Biol. Sci. **7**(3S), 73–82 (2017a)

Arof, S., Noor, N.M., Alias, M.F., et al.: Investigation of chopper operation of series motor four quadrants DC chopper. J. Appl. Environ. Biol. Sci. **7**(3S), 49–56 (2017b)

Arof, S., Diyanah, N.H.N., Noor, N.M.N., et al.: Series motor four quadrants drive DC chopper: Part 4: Generator mode. In: Bakar, M.H.A., Sidik, M.S.M., Öchsner, A. (eds.) Progress in Engineering Technology. Adv Struc Mat 119. Springer, Cham (2019)

Arof, S., Diyanah, N.H.N., Mawby, P.A., et al.: Low harmonics plug-in home charging electric vehicle battery charger utilizing multi-level rectifier, zero crossing and buck chopper: Part 1: general overview. In: Bakar, M.H.A., Sidik, M.S.M., Öchsner, A. (eds.) Progress in Engineering Technology. Adv Struc Mat 119. Springer, Cham (2019)

Oak Ridge National Laboratory (2009) Advanced brush technology for DC Motors. http://peemrc.ornl.gov/projects/emdc3.jpg. Accessed 15 May 2009

Chapter 19
Low Harmonics Plug-in Home Charging Electric Vehicle Battery Charger Utilizing Multilevel Rectifier, Zero Crossing, and Buck Chopper

State of Charge Estimator Using Current Integration Algorithm for Embedded System

Saharul Arof, Mohamad Rosyidi Ahmad, Nurazlin Mat Yaakop, Philip Mawby, H. Arof, Emilia Noorsal, Aslina Abu Bakar, and Yusnita Mohd Ali

Abstract SOC is an indicator that represents the available charge stored in the battery under an energy management system (EMS) of an electric vehicle(EV). The SOC is necessary not only for optimal management of the energy in the EV but also to protect the battery from going to the deep discharge or overcharge. During

S. Arof (✉) · M. R. Ahmad · N. M. Yaakop
Universiti Kuala Lumpur, Malaysian Spanish Institute, Kulim Hi-Tech Park, 09000 Kulim, Kedah, Malaysia
e-mail: saharul@unikl.edu.my

M. R. Ahmad
e-mail: mrosyidi@unikl.edu.my

N. M. Yaakop
e-mail: nurazlin@unikl.edu.my

S. Arof · P. Mawby
School of Engineering, University of Warwick, Coventry CV47AL, UK
e-mail: p.a.mawby@warwick.ac.uk

H. Arof · A. A. Bakar
Engineering Department, Universiti Malaya, Jalan Universiti, 50603 Kuala Lumpur, Malaysia
e-mail: ahamzah@um.edu.my

A. A. Bakar
e-mail: aslina060@uitm.edu.my

E. Noorsal · Y. M. Ali
Faculty of Electrical Engineering, Universiti Teknologi MARA (UiTM), Cawangan Pulau Pinang, Permatang Pauh Campus, 13500 Permatang Pauh, Pulau Pinang, Malaysia
e-mail: emilia.noorsal@uitm.edu.my

Y. M. Ali
e-mail: yusnita082@uitm.edu.my

M. H. Abu Bakar et al. (eds.), *Progress in Engineering Technology III*,
Advanced Structured Materials 148, https://doi.org/10.1007/978-3-030-67750-3_19

charging the battery voltage cannot be determined as it is equivalent to the terminal charging voltage. This causes the difficulty to determine whether the battery has fully charged using the voltage reading method. To determine the exact battery voltage while charging, it requires to detach the charging voltage connection before making the voltage reading, and this causes a difficulty and slows down the charging process. This paper focuses on the current integration method for estimating the battery's state of charge which is inclusively associated with a control algorithm for controlling the process to determine the battery state of charge (SOC). The MATLAB/Simulink software is used to test the control algorithm. The simulation results with MATLAB/Simulink software shows that the proposed control algorithm is able to perform the expected operation.

Keywords DC drive · Four quadrants chopper · Series motor · BMS and EMS controller · Charging · Generator · Battery charger · EC

19.1 Introduction

The emission of hydrocarbons not only pollutes the environment but also contributes to global warming which melts the icebergs and increases the sea level. Using efficient electric vehicles (EV) and hybrid electric vehicles (HEV) for transportation is one of the solutions to reduce global hydrocarbon emissions (Oak Ridge National Laboratory 2009).

19.2 Review Stage

Oak Research National Laboratory (ORNL) (Oak Ridge National Laboratory 2009), the United States in 2009, had succeeded in designing a DC brushed motor with high power output (55 kW), high efficiency (92%) that can operate in low operating voltages (13 V). This has started the interest to embark research in DC Drive EC. Attempts to improve conventional H-bridge chopper by an increased number of operations or to allow the motor reverse action have been continuously carried out. But ever since the development of the new DC motor by ORNL, a new series motor four quadrants DC chopper was designed and the proposed chopper has more operations compared to the conventional version (Arof et al. 2014). Other studies related to DC drive EV led to research on different types of DC drive brushed motors such as the separately excited DC motor that can be used for motor traction for DC EV (Arof et al. 2019). Detailed investigations on the chopper operation modes led to the establishment of a simulation model to test the chopper operations for the application of electric car and light rail transit (LRT) (Arof et al. 2017). This simulation model led to further detailed investigations on each of the chopper operations and on the specific pattern of motor voltage, current, torque, speed, of the series motor and FQDC running for

DC drive EC application (Arof et al. 2017). For the DC series motor traction of EC application, the speed and torque control for the series motor in an attempt to reduce jerk and tire slip has been successfully done and implemented with the direct current control technique (Arof et al. 2016). Close-loop controller using PID tuned using pole placement (Arof et al. 2020) and Fuzzy logic (Arof et al. 2020) are applied to improve the performance and stability. This method has undergone improvements using cascade PID and ascend descend algorithm and after improvisation made and associated with steering (Arof et al. 2020) and vehicle movement control (Said MS and Diyanah 2020), it was used for autonomous EV automatic reverse parking. For fault diagnose and online system tuning and optimization, a numerical representation using Taylor's series (Arof et al. 2020) is studied and tried for the driving mode. For power regeneration, the FQDC offers the generator mode with several techniques of starting the regenerated power and voltage control and this is studied and discussed in (Arof et al. 2019). In order for the FQDC to be applied in the real world it needs controllers running control algorithm in the embedded system. The controller and its control algorithm are studied and tested using the processor in the loop (PIL) technique (Diyanah 2019). To improve the new FQDC performance, an optimization tool such as artificial intelligence(AI) is introduced to control all of the chopper operations of the proposed FQDC chopper (Arof et al. 2015a, b). Among the three AI controllers, ANFIS shows the best performance followed by neural network and the self-tuning Fuzzy logic controller. A study on neural network controller to uncover the proper method of tuning has been carried out such as using a single controller with binary output (Arof et al. 2019). On each specific FQDC chopper operation mode the performance can be further improved using an AI optimization tool such as genetics algorithm to set up a specific lookup table for the field current in parallel mode operation (Arof et al. 2019). Several other studies related to DC drive EV led to research on EC battery chargers such as shown in Fig. 19.1, that offers low harmonics (Arof et al. 2019). This battery charger is using zero cross, multilevel rectifier, and buck chopper.

To make the battery charger efficient, it requires a battery management system (BMS) to ensure that the battery is safe, reliable, and a cost-efficient solution. The BMS should contain accurate algorithms to measure and estimate the functional

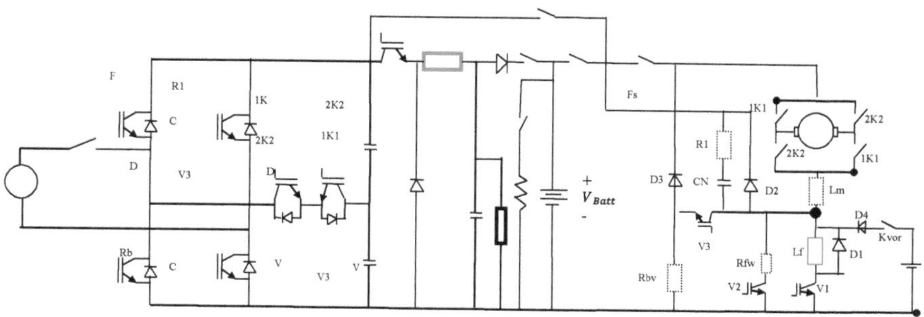

Fig. 19.1 Integrated FQDC chopper and battery charger

status of the battery and has state-of-the-art mechanisms to protect the battery from hazardous and inefficient operating conditions. This is to prolong the battery life and guarantee its safety. Estimation of the SOC is an indicator that represents the available charge stored in the battery compared to the full capacity charge of the battery. An accurate estimation of the SOC is necessary not only for optimal management of the energy in the EVs but also to protect the battery from going to the deep discharge or overcharge conditions. SOC algorithms and approaches have been proposed to estimate the SOC from the battery's available measurements. Coulomb counting or Ah counting is one of the most conventional methods. Measuring the open circuit voltage is another approach to calculate the SOC based on the static relationship between the OCV and the SOC. The state of charge estimator for series motor four quadrants drive DC chopper (FQDC) for DC drive electric vehicles (EVs) battery charger as in Fig. 19.1 application is using the current integration method in which the algorithm can be run in an embedded system such as a PIC microcontroller as in Fig. 19.2. This approach will be discussed in this paper.

There are two modes of charging batteries which are when the vehicle is running which is through the FQDC chopper during generator, regenerative braking, and resistive braking modes and the other one when the vehicle is at standstill via a

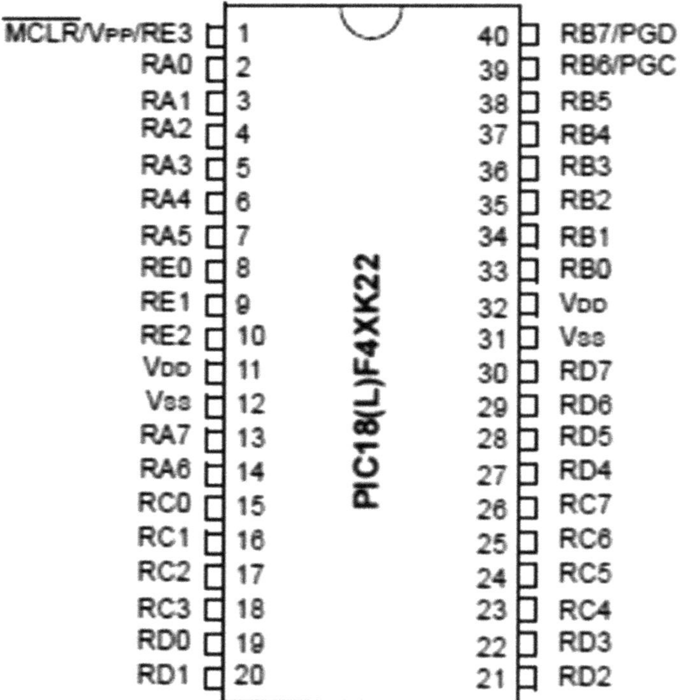

Fig. 19.2 PIC microcontroller

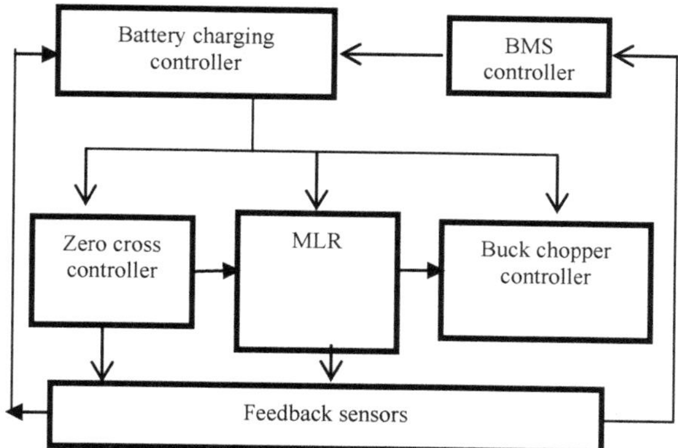

Fig. 19.3 Block diagram of BMS controller data path

battery charger circuit. The battery charger as in Fig. 19.1 is for standstill charging and is consists of zero crossing, multilevel rectifier and buck chopper. The charging operation is simplified in a block diagram as shown in Fig. 19.3.

The charging process is explained via the flow chart in Fig. 19.4.

19.3 Methodology

A. Battery SOC Estimator from Known SOC Battery

The Battery SOC is not hard to be determined and updated if the initial value of SOC is known. But in the case of this is not known, it requires a combination of hardware and software to determine the SOC. The algorithm for updating the SOC with known SOC is first discussed in this paper, and it requires a current sensor as it operates using curent integration method. Equations (1) and (2) are transformed into an algorithm code to find the SOC value.

$$\begin{aligned}\text{SOC} = ((\text{SOC}) \pm (&\text{battery current charging/discharging} \\ &* \text{charging/discharging constant} * \text{Error})).\end{aligned} \tag{1}$$

$$\text{Error} = \text{SOC actual} - \text{SOC estimate} \tag{2}$$

The charging/discharging constant can be determined by trial and error or by experiment.

The algorithm when the SOC value is known is presented in a block diagram as shown in Fig. 19.5. The algorithm is tested for charging, discharging, and charging–discharging.

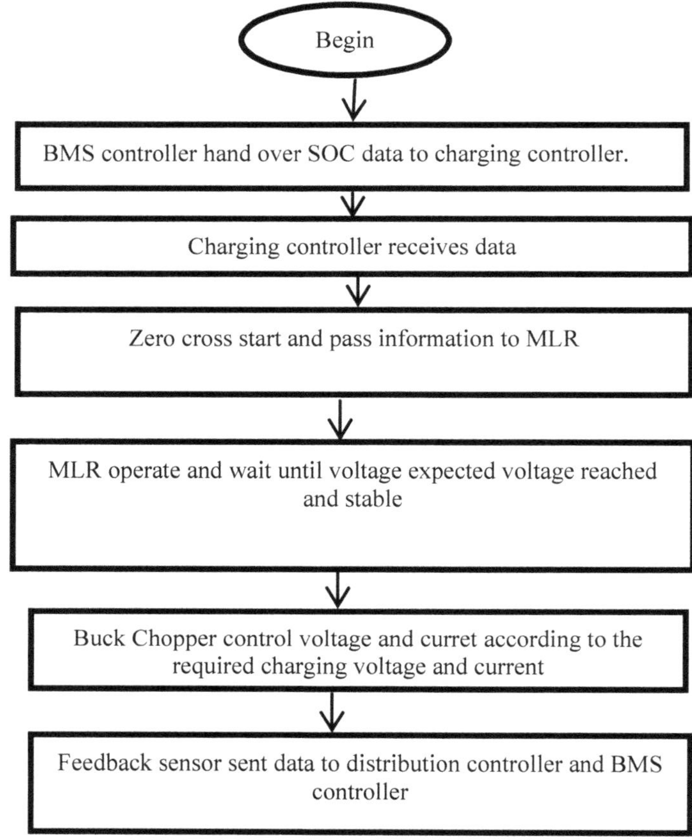

Fig. 19.4 Flow chart of BMS controller operation

B. Battery SOC Estimator from an Unknown SOC Battery

The algorithm when the SOC value is not known is presented in the block diagram as shown in Fig. 19.6. The algorithm is tested for charging, discharging, and charging–discharging. A lookup table from the datasheet is used to predict the current SOC taken from the LUT by comparing the current battery voltage and the expected SOC stated in the table is needed as the first step to preestimate the SOC. Interpolation and extrapolation of the lookup table is necessary to find the exact SOC from the range given in the table. Once this value is obtained on the next process of charging and discharging and by referring to the lookup table it will correct this value until it becomes reliable/accurate. The algorithm code which is transformed from Eqs. (1) and (2) is used to find the SOC value. When this is achieved a process flow of SOC estimator as in Fig. 19.5 will be used rather than in Fig. 19.6.

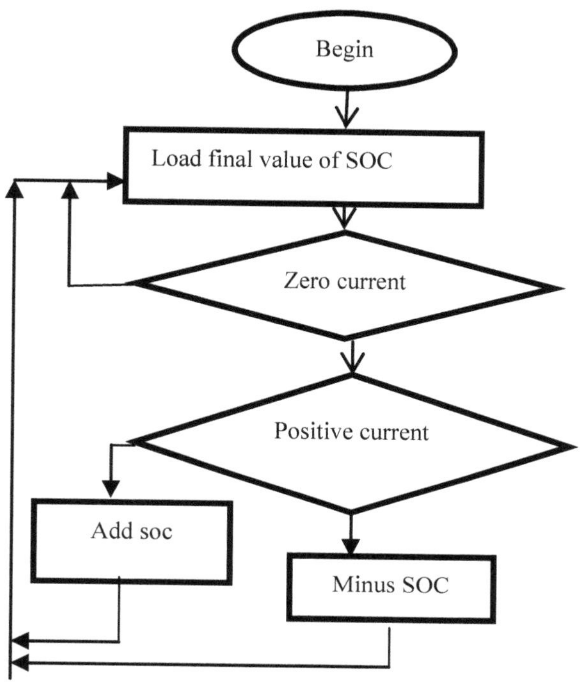

Fig. 19.5 SOC algorithm

To determine the SOC using LUT of batteries open circuit voltage versus state of charge (SOC) such as shown in Fig. 19.7, the voltage sensors are required to be high precision because, for some batteries, the SOC rated from 20% to 80% of SOC, as shown in Fig.19.7(a), shows very small voltage differences. A zooming circuit using a substractor circuit and an amplifier circuit is needed for accuracy and better performance but this is not discussed in this paper.

The other method to determine the battery SOC has to undergo several steps as shown in Fig. 19.8. The battery voltage will be first determined using a battery voltage sensor for reference. Then the battery is loaded (connected to a resistor) and the battery voltage and current is recorded. If it is about the same with some minor tolerance, the SOC is determined with the equation. Some values need to be determined by experiments or trial and error, while others require the datasheet of the battery.

$$\text{SOC\%} = 100 - (((\text{max volt} - \text{current reading volt}) / (\text{maximum voltage} - \text{minimum voltage})) * 100) \quad (3)$$

The expected current reading is given in the following equation:

$$\text{CR} = (\text{V}/\text{Rbatt} + \text{R}) \quad (4)$$

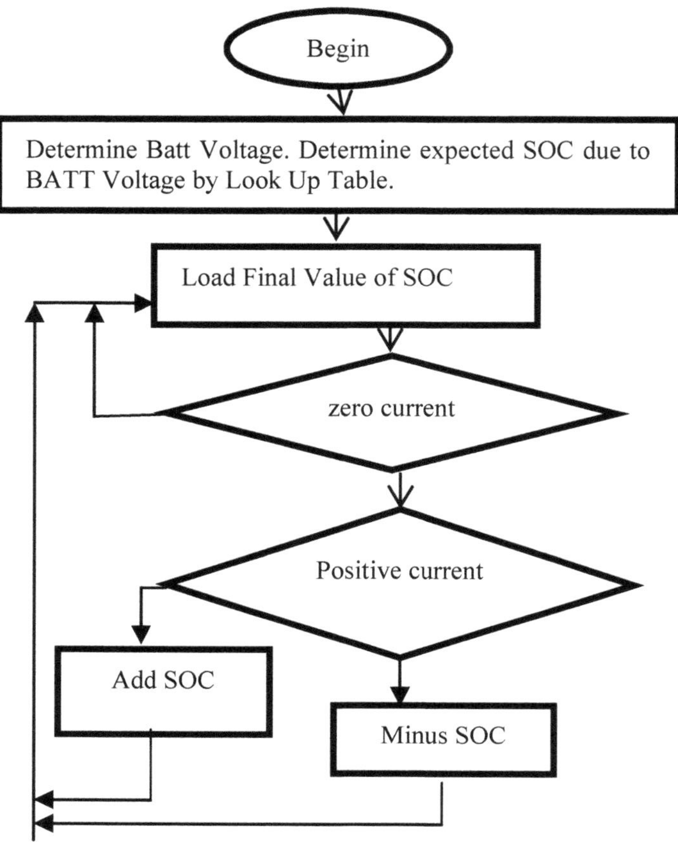

Fig. 19.6 SOC estimator algorithm

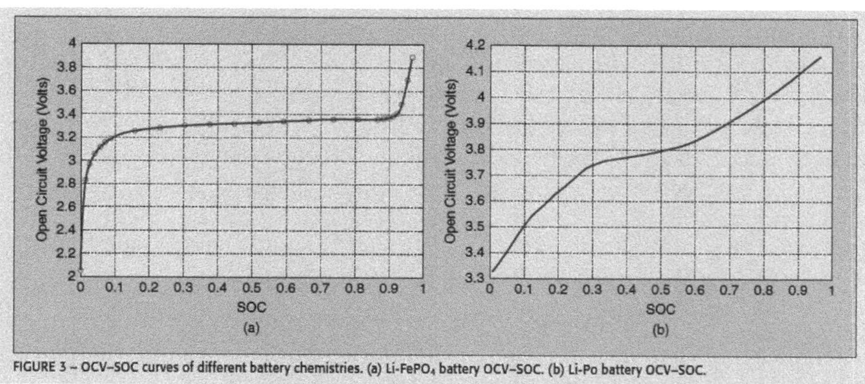

FIGURE 3 – OCV–SOC curves of different battery chemistries. (a) Li-$FePO_4$ battery OCV–SOC. (b) Li-Po battery OCV–SOC.

Fig. 19.7 Battery voltage and SOC graph

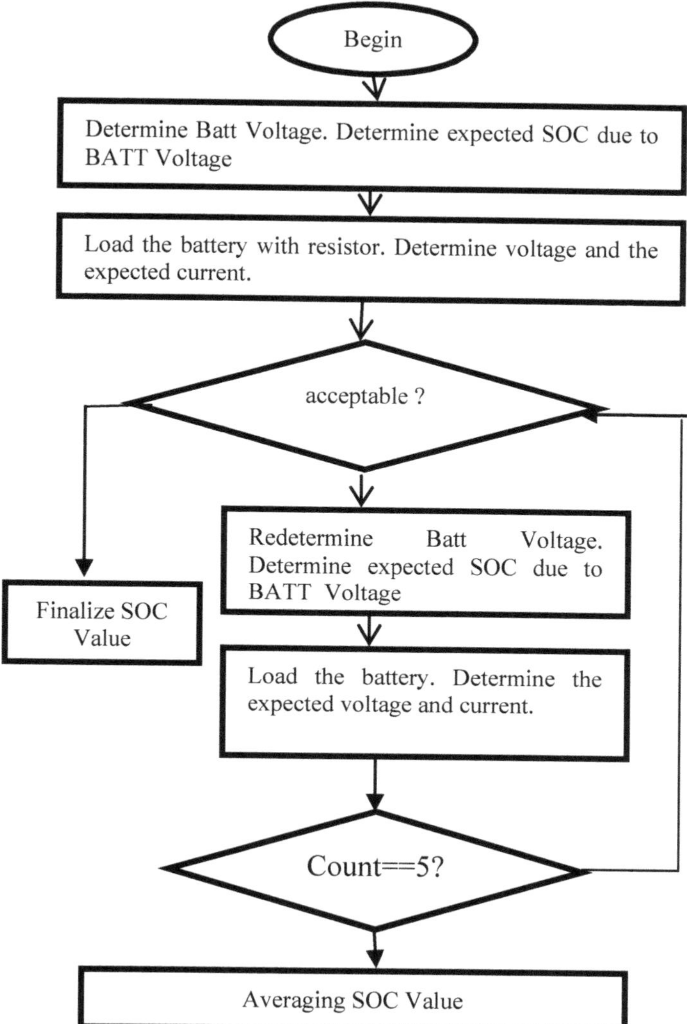

Fig. 19.8 SOC estimator algorithm

The Error Current reading

$$ECR = (V/R + Rbatt) - CR; \tag{5}$$

The Error Ratio equation is

$$ER = ECR/CR * 100 \tag{6}$$

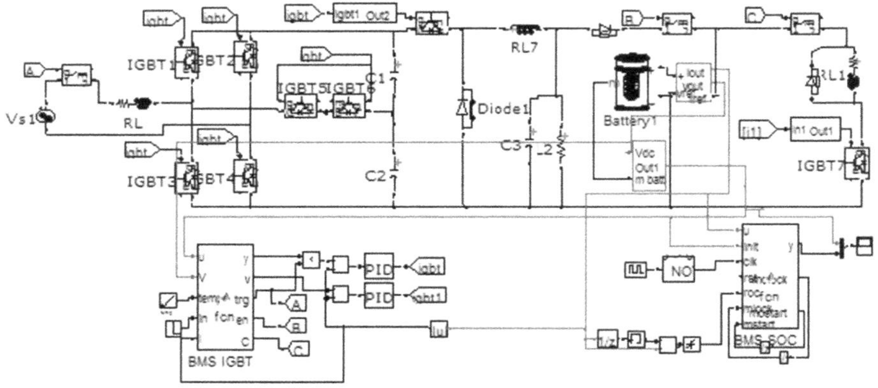

Fig. 19.9 MATLAB/Simulink BMS and charger model

If the error current reading more than 5% is not as expected as the voltage reading the SOC can be further rectified as equation

$$\begin{aligned}\text{SOC\%} = {} & 100 - (((\text{max volt} - \text{current reading volt})/(\text{maximum voltage} \\ & -\text{minimum voltage}) * 100)) - (((\text{abs}(\text{msocc})/100) \\ & * (100 - (((\text{max volt} - \text{V})/\text{maximum voltage} \\ & -\text{minimum voltage}) * 100))) * \text{cf});\end{aligned} \tag{7}$$

where Cf is the correction factor gain.

If the battery voltage is different from the previous voltage of the first voltage reading the process will be repeated five times and the SOC value will be averaged.

C. **Simulation Model of Battery Charger and BMS Controller Tested Using MATLAB/Simulink model**.

A Simulink model that consists of the battery charger and BMS controller is developed for testing the battery charger and BMS control algorithm. This is shown in Fig. 19.9.

19.4 Results and Discussion

Figures 19.10, 19.11, 19.12 and 19.13 show the results of the current integrator control algorithm to determine SOC when the initial SOC value is known. The estimator result is stair-like while the actual is more linear. Figures 19.13–19.14, show the control algorithm together with the current integrator control algorithm to determine SOC value. At first, the SOC is predicted using the lookup table. This

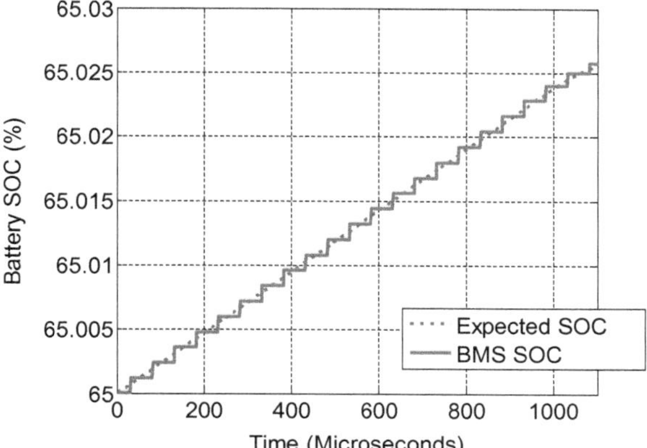

Fig. 19.10 SOC value while charging

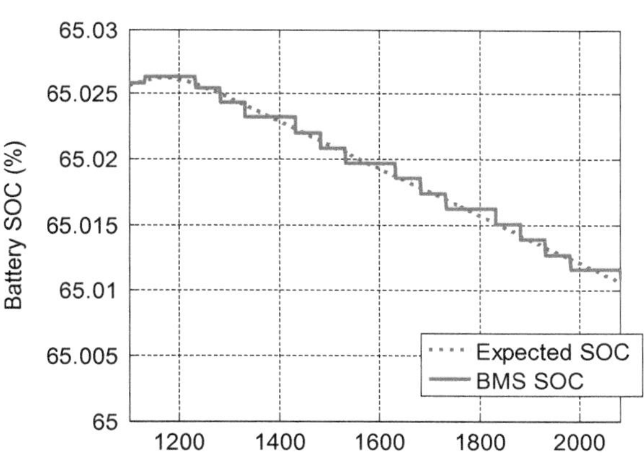

Fig. 19.11 SOC value when discharging

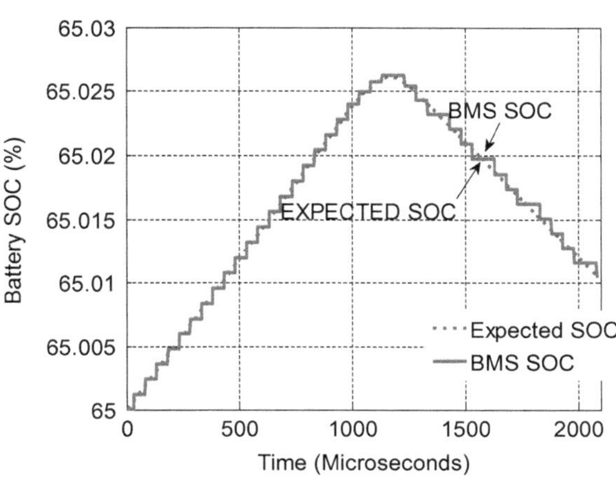

Fig. 19.12 SOC value while charging and discharging

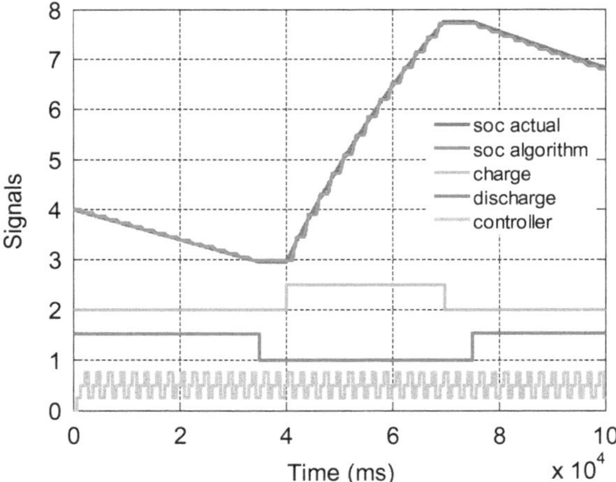

Fig. 19.13 SOC value while charging and discharging

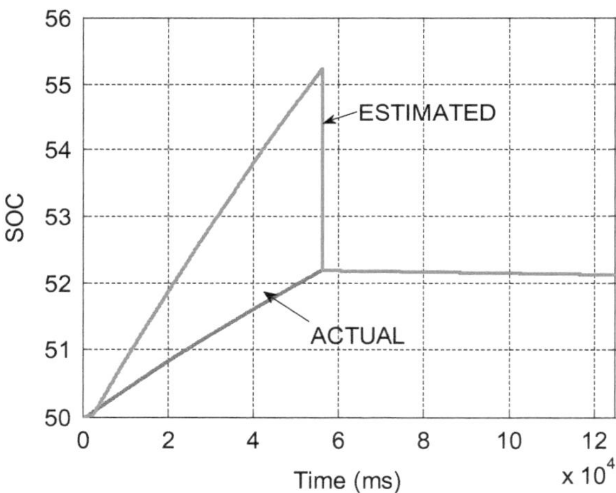

Fig. 19.14 SOC algorithm determines SOC while charging and discharging

value is slightly above the actual SOC but at the end after several times of correction it has almost about the same value (Fig. 19.15).

19.5 Conclusion

Three types of control algorithms have successfully been tested using the simulation model with the MATLAB/Simulink software. The results are the proposed current integrator and its associated control algorithms are able to determine the battery SOC monitoring. As the final conclusion the battery charger and its control algorithm

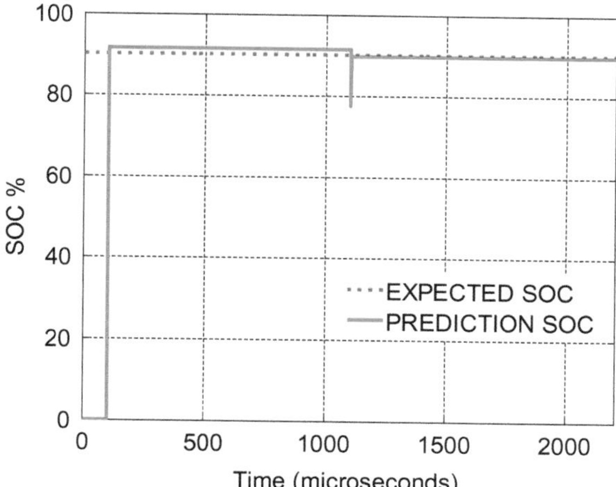

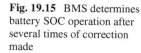

Fig. 19.15 BMS determines battery SOC operation after several times of correction made

inclusive of the current integrator SOC estimator is suitable for economical DC drive EV.

References

Arof, S., Diyanah, N.H.N., Mawby, P.A. et al.: Study on implementation of neural network controlling four quadrants direct current chopper: Part1: Using single neural network controller with binary data output. In: Ismail, A., Bakar, M.H.A., Öchsner, A. (eds.) Advanced Engineering for Processes and Technologies. Adv Struc Mat 102. Springer, Cham (2019)

Arof, S., Diyanah, N.H.N., Noor, N.M., et al.: A new four quadrants drive chopper for separately excited DC motor in low-cost electric vehicle. In: Bakar, M.H.A., Sidik, M.S.M., Öchsner, A. (eds.) Progress in Engineering Technology. Adv Struc Mat 119. Springer, Cham (2019)

Arof, S., Diyanah, N.H.N., Noor, N.M. et al.: Genetics algorithm for setting up look up table for parallel mode of new series motor four quadrants DC chopper. In: Bakar, M.H.A., Sidik, M.S.M., Öchsner, A. (eds.) Progress in Engineering Technology. Adv Struc Mat 119. Springer, Cham (2019)

Arof, S., Diyanah, N.H.N, Yaakop, M.A., et al.: Processor in the loop for testing series motor four quadrants drive DC chopper for series motor driven electric car: Part 1: Chopper operation modes testing. In: Ismail, A., Bakar, M.H.A., Öchsner, A. (eds.) Advanced Engineering for Processes and Technologies. Adv Struc Mat 102. Springer, Cham (2019)

Arof, S., Muhd Khairulzaman, A.K., Jalil, J.A., Arof, H., Mawby, P.A.: Self tuning fuzzy logic controlling chopper operation of four quadrants drive DC chopper for low-cost electric vehicle. In: 2015 6th International Conference on Intelligent Systems, Modelling and Simulation, pp. 40–45 (2015)

Arof, S., Noor, N.M., Alias, M.F., et al.: Digital proportional integral derivative (PID) controller for closed-loop direct current control of an electric vehicle traction tuned using pole placement. In: Bakar, M.H.A., Zamri, F.A., Öchsner, A. (eds.) Progress in Engineering Technology II. Adv Struc Mat 131. Springer, Cham (2020)

Arof, S., Noor, N.M., Mohamad, R. et al.: Close loop feedback direct current control in driving mode of a four quadrants drive direct current chopper for electric vehicle traction controlled using fuzzy logic. In: Bakar, M.H.A., Zamri, F.A., Öchsner, A. (eds.) Progress in Engineering Technology II. Adv Struc Mat 131. Springer, Cham (2020)

Arof, S., Said, M.S., Diyanah, N.H.N., et al.: Series motor four-quadrant DC chopper: reverse mode, direct current control, triple cascade PIDs, and ascend-descend algorithm with feedback optimization for automatic reverse parking. In: Bakar, M.H.A., Zamri, F.A., Öchsner. A. (eds.) Progress in Engineering Technology II. Adv Struc Mat 131. Springer, Cham (2020)

Arof, S., Said, M.S., Diyanah, N.H.N., et al.: Series motor four-quadrant direct current chopper: reverse mode, steering position control with double-circle path tracking and control for autonomous reverse parking of direct current drive electric car. In: Bakar, M.H.A., Zamri, F.A., Öchsner, A. (eds.) Progress in Engineering Technology II. Adv Struc Mat 131. Springer, Cham (2020)

Arof, S., Sukiman, E.D., Diyanah, N.H.N., et al.: Discrete-time linear system of New series motor four-quadrant drive direct current chopper numerically represented by Taylor series. In: Bakar, M.H.A., Zamri, F.A., Öchsner, A. (eds.) Progress in Engineering Technology II. Adv Struc Mat 131. Springer, Cham (2020)

Arof, S., Yaakop, N.M., Jalil, J.A., Mawby, P.A., Arof, H.: Series motor four quadrants drive DC chopper for low cost, electric car: Part 1: Overall. In: 2014 IEEE International Conference on Power and Energy (PECon), pp. 342–347 (2014)

Arof, S., Zaman, M.K., Jalil, J.A., et al.: Artificial intelligence controlling chopper operation of four quadrants drive DC chopper for low-cost electric vehicle. Int. J. Simul. Syst. Sci. Technol. **16**(4), 1–10 (2015)

Arof, S., Jalil, J.A., Kamaruddin, N.S., Yaakop, N.M., Mawby, P.A., Hamzah, A.: Series motor four quadrants drive DC Chopper Part2: driving and reverse mode with direct current control. In: 2016 IEEE International Conference on Power and Energy (PECon), 775–780 (2016)

Arof, S., Hassan, H., Rosyidi, M., et al.: Implementation of series motor four quadrants drive DC chopper for dc drive electric car and LRT via simulation model. J. Appl. Environ. Biol. Sci **7**(3S), 73–82 (2017a)

Arof, S., Noor, N.M., Alias, M.F., et al.: Investigation of chopper operation of series motor four quadrants DC chopper. J Appl. Environ. Biol. Sci. **7**(3S), 49–56 (2017b)

Arof, S., Diyanah, N.H.N., Mawby, P.A. et al.: Low harmonics plug-in home charging electric vehicle battery charger utilizing multi-level rectifier, zero crossing and buck chopper: Part 1: General overview. In: Bakar, M.H.A., Sidik, M.S.M., Öchsner, A. (eds.) Progress in Engineering Technology. Adv Struc Mat 119. Springer, Cham (2019)

Arof, S., Diyanah, N.H.N., Noor, N.M.N., et al.: Series motor four quadrants drive DC chopper: Part 4: Generator mode. In: Bakar, M.H.A., Sidik, M.S.M., Öchsner, A. (eds.) Progress in Engineering Technology. Adv Struc Mat 119. Springer, Cham

Oak Ridge National Laboratory (2009) Advanced brush technology for DC Motors. http://peemrc.ornl.gov/projects/emdc3.jpg. Accessed 15 May 2009

Chapter 20
Series Motor Four Quadrants Drive DC Chopper

Reverse Mode with Automatic Reverse Parking of DC Drive Electric Car with Constant Brake Motor Control Combine to the Propulsion Motor Torque

Saharul Arof, M. S. Sazali, N. H. N. Diyanah, Philip Mawby, H. Arof, and Emilia Noorsal

Abstract This paper is focused on the motor torque control of a DC motor attached to a hydraulic brake pedal to provide a friction-braking effect (the combination makes an electrohydraulic brake). This combination provides a braking effect to control the vehicle speed, and makes the final stop to ensure a correct position of EV during deceleration and stop of DC drive electric car for automatic reverse parking. The DC drive electric car uses a series motor powered by a four-quadrant drive DC chopper. The integration of the electrohydraulic brake with propulsion motor control is used to provide a braking action to decelerate and finally stop the vehicle. The control technique was simulated by using MATLAB/Simulink and results indicated that the technique had successfully met the objective of torque, current, speed and position

S. Arof (✉) · M. S. Sazali
Universiti Kuala Lumpur, Malaysian Spanish Institute, Kulim Hi-Tech Park, 09000 Kulim, Kedah, Malaysia
e-mail: saharul@unikl.edu.my

M. S. Sazali
e-mail: msazali@unikl.edu.my

N. H. N. Diyanah · P. Mawby
School of Engineering, University of Warwick, Coventry CV47AL, UK
e-mail: diyanahhisham94@gmail.com

P. Mawby
e-mail: p.a.mawby@warwick.ac.uk

H. Arof
Engineering Department, Universiti Malaya, Jalan Universiti, 50603 Kuala Lumpur, Malaysia
e-mail: ahamzah@um.edu.my

E. Noorsal
Faculty of Electrical Engineering, Universiti Teknologi MARA (UiTM) Cawangan Pulau Pinang, Pinang 13500, Malaysia
e-mail: emilia.noorsal@uitm.edu.my

M. H. Abu Bakar et al. (eds.), *Progress in Engineering Technology III*,
Advanced Structured Materials 148, https://doi.org/10.1007/978-3-030-67750-3_20

control for automatic reverse parking, and thus was suitable for implementation with a DC drive electric car by using the series motor and four-quadrant DC chopper.

Keywords DC drive · Reverse parking · Four-quadrant DC chopper · Series motor · Torque · Position and speed · Torque control · EV

20.1 Introduction

In Yemen climate change has caused famine when the ability to grow its own food and access to own clean water decreased. Moreover, the use of electric vehicles (EV) and hybrid electric vehicles (HEV) in transportation is one of the solutions to reduce global hydrocarbon emission that causes climate change.

20.2 Review Stage

Unfortunately, the price of EV and HEV is expensive, making it impossible for many people, especially for those living in poor countries. This led to a study on the possibility of DC drive EV (Oak Ridge National Laboratory 2009), which is known to be economical. The interest to embark research on DC drive EC started in 2009 when the Oak Research National Laboratory (ORNL) (Oak Ridge National Laboratory 2009), United States, successfully designed a DC brushed motor with high power output (55 kW), high efficiency (92%) and it can operate under low operational voltage (13 V). A new series motor four-quadrant DC chopper, as shown in Fig. 20.1, was designed and the proposed chopper had multiple operations (Arof et al. 2014). Several studies on DC drive EV have led to research on EC battery charger (Arof et al. 2019) and different types of DC drive motor (Arof et al.2019). The detailed investigation done on chopper operational modes led to the establishment of a simulation model to test the chopper operations for the application of electric car and light rail transit (LRT) (Arof et al. 2017). This provoked further detailed investigation on each chopper operation, such as specific pattern of voltage, current, torque and speed of the FQDC running DC series motor (Arof et al. 2017). For DC series motor traction EC application, the speed and torque control was successfully done and implemented with direct current control (Arof et al. 2016). For power regeneration, the FQDC offer generator mode with several techniques of regenerated power. It was studied and discussed in (Arof et al. 2019). For FQDC the application in the real world, it needs controllers which are run by a control algorithm in the embedded system. The controller and its control algorithm are studied and tested by using a processor in the loop technique (Diyanah 2019). To improve the new FQDC performance, optimisation tools, such as artificial intelligence (AI) with self-tuning fuzzy (Arof et al. 2015), neural network and ANFIS, were introduced to control all proposed FQDC chopper operations (Arof et al. 2015a; b; Arof et al. 2019). Each

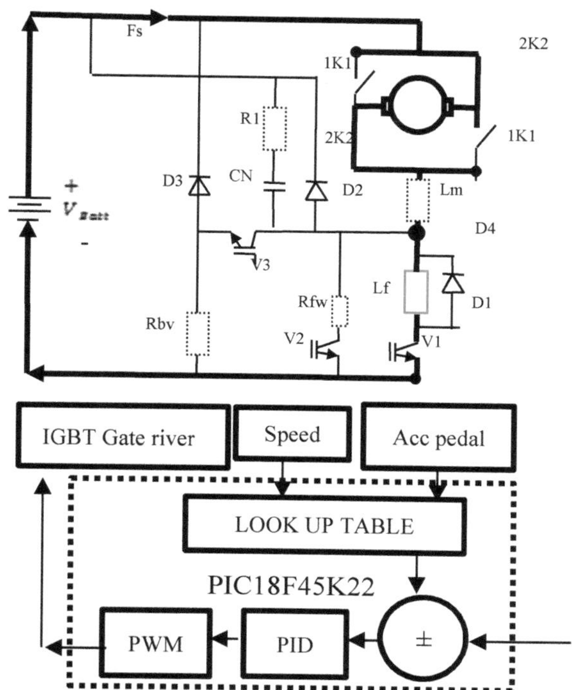

Fig. 20.1 Current paths in reverse mode

FQDC mode operational performance can be improved by using AI optimisation tools such as genetic algorithm to set up a specific lookup table (Arof et al. 2019). Pole placement (Arof et al. 2020) method used to tune close-loop PID controller to improve the controller performance while Fuzzy logic controller is used to control the motor current (Arof et al. 2020) is applied to improve the control performance and system stability. For fault diagnose and online system tuning and optimisation a numerical representation using the Taylor series (Arof et al. 2020) is studied and tried for driving mode. The cascade PID with ascend descend algorithm and after improvisation made associated to steering (Arof et al. 2020) and vehicle movement control (Said MS and Diyanah 2020) was used for autonomous EV for automatic reverse parking. This paper is the completion and continuation of the previous study on the automatic reverse parking and the focus is on automatic braking system.

A. *Series Motor Four-Quadrant DC Drive Chopper*

1. **Automatic Parking Review with DC Motor Assist Brake**

Automatic parking is important when a vehicle driver deals with a constrained environment, whereby much attention and skills are required. The automatic parking requires coordinated control of the steering angle with the surrounding to avoid collision (Yuan et al. 2018; Anwar 2004; Castillo et al. 2016; Simonik et al. 2018; Ho et al. 2017).

The mathematical model for the DC motor for power steering

$$I_m = \frac{V_{\text{batt}} - I_m(R_m) - B_{\text{emf}}}{L_m} \tag{1}$$

$$B_{\text{emf}} = K_v I_m \omega \tag{2}$$

$$T_d = K_t I_m \tag{3}$$

$$T_d = J\frac{d\omega}{dt} + B_w + T_L \tag{4}$$

The parking assistance system was developed by using image assistance by Mercedes, Volvo and BMW. A disadvantage age of the current technology is the captured image signal that is influenced by environmental brightness. In 2004 a commercial version of the automatic parking assistance was introduced by Toyota Motor Corporation in the Toyota Prius. Lexus 2007 LS also has an advanced parking guidance system.

20.3 Methodology

In automatic parking, there are two required movements, which are the vehicle movement (in reverse mode) and the steering angle. These movements are not discussed in this paper. The electrohydraulic brake shown in Fig. 20.2 is important for automatic reverse parking because the braking effect helps to decelerate the vehicle. The system has an electric motor and when operated can replace the action of manual brake by using the feet. The speed of car in the reverse mode automatic parking is determined by the propulsion motor and brake system. The brake system is used to decelerate, and stop at a fixed parking before the final hand brake action is applied. The brake force needs to be applied to the disc brake while controlling the motor speed so as to create the required deceleration effect. The brake torque required to decelerate the EV is guarded by controlling the motor current. High current causes a high motor torque, resulting in a high force which is transferred to the brake pad to cause a high braking power.

In normal condition, if the EV is allowed to accelerate and then decelerate by removing the traction power and without braking action interference, then Fig. 20.3 will exhibit the vehicle speed pattern.

If the same driving pattern is repeated but the mechanical brake is applied during deceleration the speed pattern will be shown as in Fig. 20.4.

For automatic reverse parking the speed of the vehicle is required as shown in Fig. 20.5, whereby the acceleration, constant speed and deceleration are shown. The

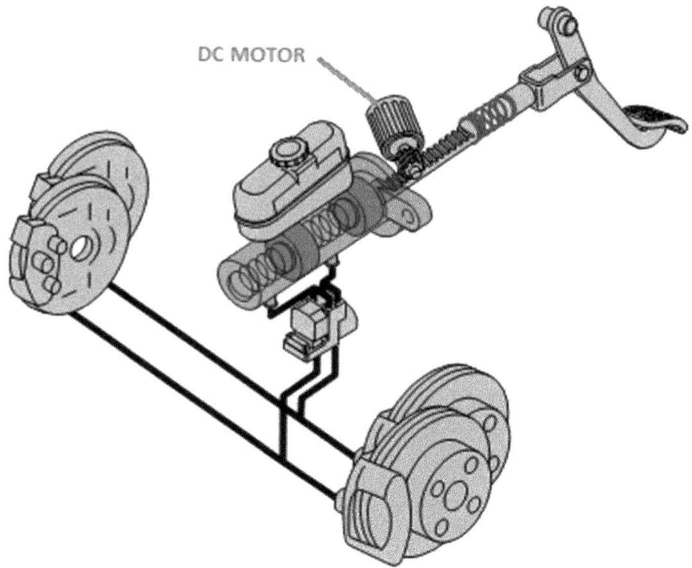

Fig. 20.2 Electrical braking system (EBS)

Fig. 20.3 Normal speed pattern

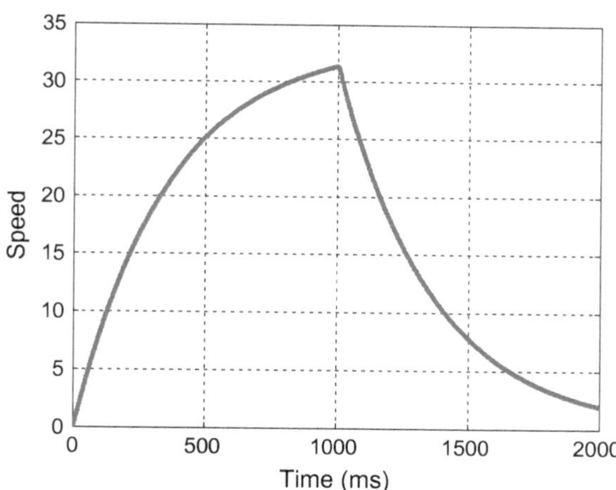

acceleration effect made is to avoid tire slip. The deceleration is to avoid overshoot and a clean stop.

If the speed shown in Fig. 20.5 is required and the action is while running the propulsion motor, the mechanical brake is then applied to get the acceleration and deceleration effect as shown in the Fig. 20.5, this results in the propulsion motor torque will exhibit a torque pattern as shown in Fig. 20.6. This motor torque pattern

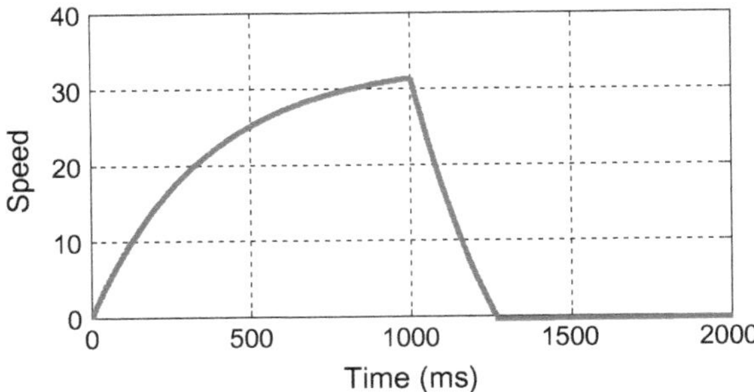

Fig. 20.4 speed pattern with braking action

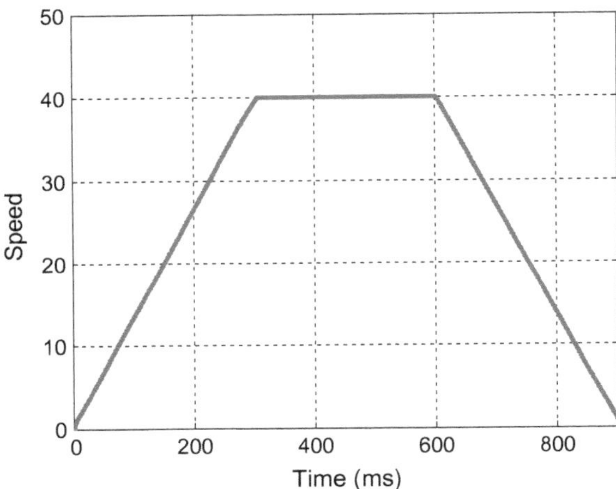

Fig. 20.5 Required speed pattern for automatic reverse parking

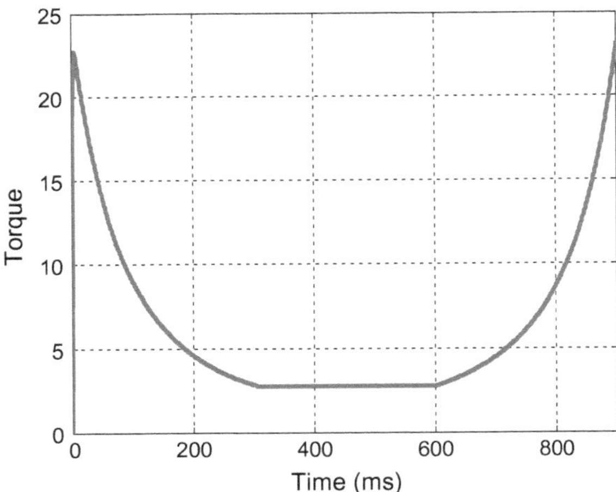

Fig. 20.6 Propulsion motor torque with mechanical brake being applied while electrical motor is running concurrently

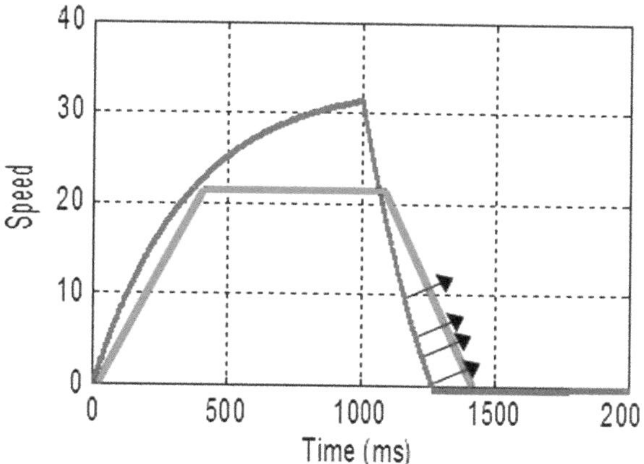

Fig. 20.7 Shifting the vehicle movement by using propulsion motor torque to overcome constant effect of mechanical brake

will draw high motor current which causes overheating and probably destroys the propulsion motor.

Figure 20.6. Propulsion motor torque with applied mechanical brake.

While the method with mechanical brake seems possible to be done, not only it causes overheat to the propulsion motor, the truth is the mechanical braking with friction pad braking is difficult to control to get the linear braking effect.

Braking with Vehicle Movement

Therefore, a better technique is required to guarantee propulsion motor safety while simultaneously producing the same linear acceleration and deceleration, as shown in Fig. 20.5. The acceleration effect can be done using direct current control or direct torque control using propulsion motor only. The idea to get the same deceleration effect is to combine a constant mechanical braking effect (the mechanical brake is applied first) and then varying the propulsion motor torque to oppose the braking effect as shown in Fig. 20.7, to get the same deceleration effect during vehicle deceleration, as shown in Figs. 20.5 and 20.8. This is to prevent the propulsion motor from being overheated.

The propulsion motor can be controlled and it can produce linear acceleration during acceleration (red colour in Fig. 20.8). However, the propulsion does not slow down during deceleration, as shown in Fig. 20.8 (red colour). If a constant mechanical brake torque is the only force applied to the vehicle and without the propulsion motor force, the vehicle speed will decelerate as shown in Fig. 20.8 (blue colour). This speed is unwanted because it is not linear and similar to Fig. 20.5. To solve the problem and make it linear, a propulsion motor which is running with the brake is applied. The brake force is reduced and becomes just enough for the propulsion motor to counter the braking torque so that the speed and deceleration will shift from blue to red as required (Fig. 20.7).

This method will not cause the propulsion motor to overheat because only a smaller motor torque force is required to counter a smaller braking effect. Details

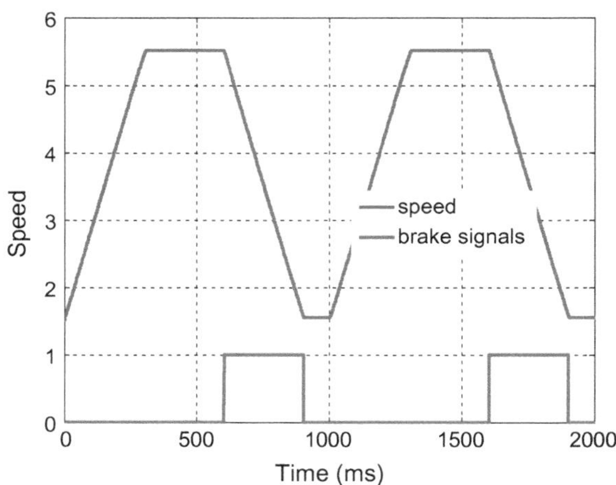

Fig. 20.8 Brake signals during deceleration

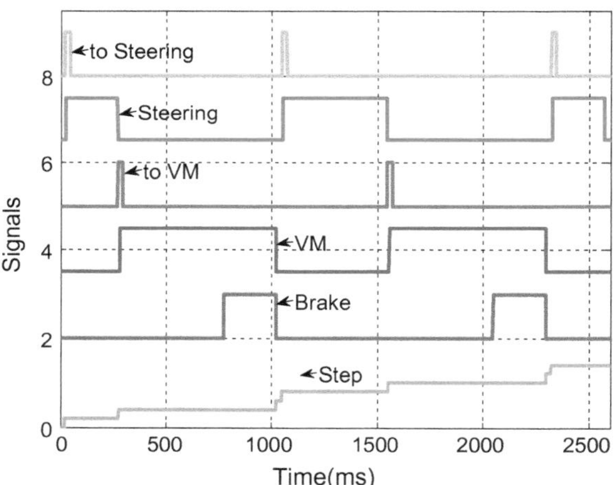

Fig. 20.9 Sequences of braking signals of overall EV, with vehicle movement, steering and brake

on the explanation of how the motor torque is controlled during acceleration and deceleration were not covered in this paper.

The overall automatic reverse parking operation with vehicle movement by using a propulsion motor, steering movement and braking movement is shown in Fig. 20.9. At the upper left of the figure, the first signal is to steer. The steering reacts by turning left or right according to the requirement. Then a signal is passed to vehicle movement. The vehicle will move via the propulsion motor. Then, at the end of the vehicle movement a brake signal is activated and the process is repeated before the final stop which means that the vehicle is successfully parked.

Brake Motor Current Control

The braking effect is guarded by controlling the flow of current to the motor. The higher the current the greater the motor torque, and the greater the braking torque. This motor torque is transferred to the hydraulic brake and disc, causing the brake disc to grip and produce the braking effect.

PID controller is used to control the brake motor current.

$$K_p K_p + \frac{K_i}{s} + = \frac{K_p s + K_i}{s} = K_p\left(\frac{\left(s + \frac{K_i}{K_p}\right)}{s}\right) \quad (5)$$

Tuning PID by Using MATLAB/Simulink Tools

For tuning, PID MATLAB/Simulink offers tools, such as the use of the SISO tool which can tune the PID controller. Equations (1)–(4) can be used to form a physical-based model which represents the motor torque and motor current. The MATLAB/Simulink tool can linearise the model and find the PID gain for the system. The MATLAB/Simulink drag and drop PID has a function button that can be clicked to automatically tune the PID.

MATLAB/Simulink system identification can be used to find the system transfer function, system order and test system stability by using root locus or bode plot. By using MATLAB/Simulink SISO tool the transfer function for the brake motor current can be loaded and the PID controller gain can be obtained by using the PID autotune function. The result after PID controller tuning for brake torque control by controlling the motor current is as seen in Fig. 20.10.

Simulation Model with MATLAB/Simulink

A simulation model is developed to test the control technique.

For monitoring the car trajectory a simulation model is established by using the mathematical equation, as shown in Figs. 20.11 and 20.12.

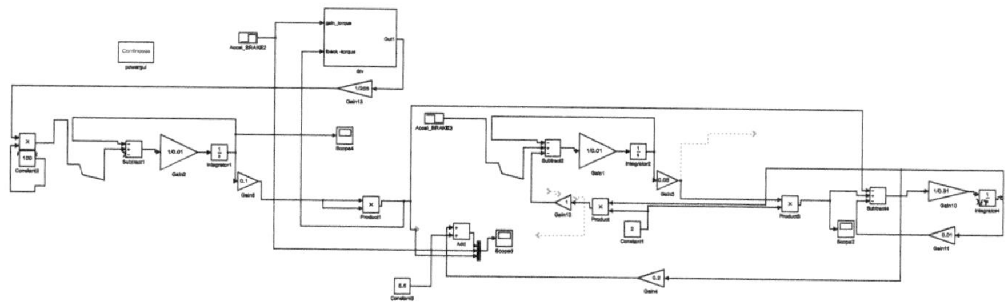

Fig. 20.10 Brake simulation model brake

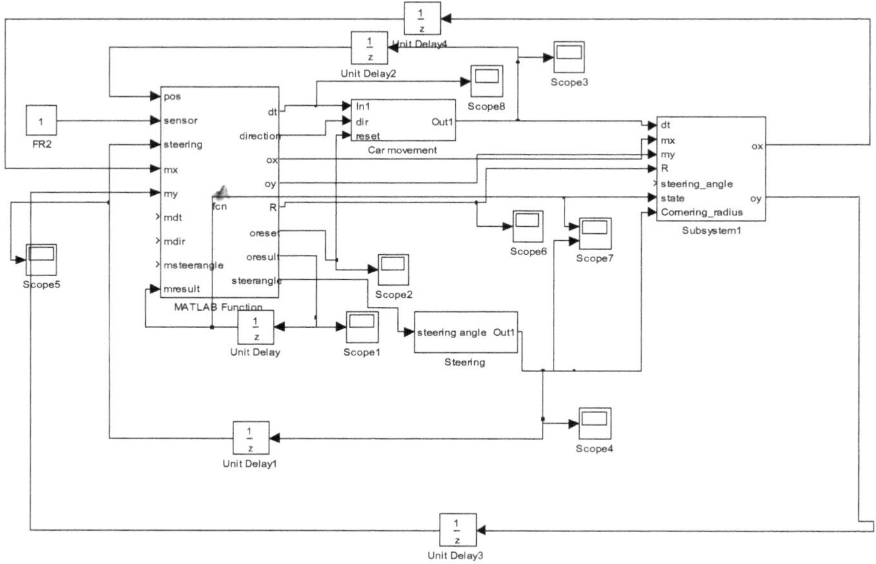

Fig. 20.11 Simulation model of auto parking

Fig. 20.12 Motor brake current

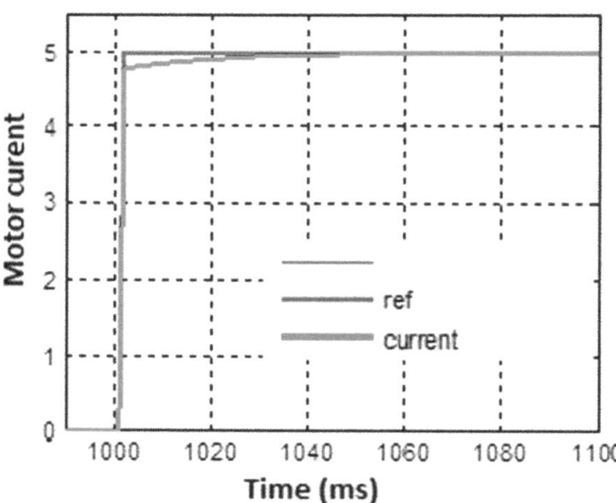

Figure 20.13 shows the application of a motor current to create the mechanical brakes for braking and decelerating the EV. Figure 20.14 shows the propulsion motor current to overcome the applied mechanical brakes for braking and decelerating the EV.

Figure 20.14 shows the vehicle trajectory resulted from the combination of vehicle movement and steering movement for automatic parallel parking.

Figure 20.15 shows the vehicle trajectory which resulted from the combination of vehicle and steering movement for automatic reverse parking.

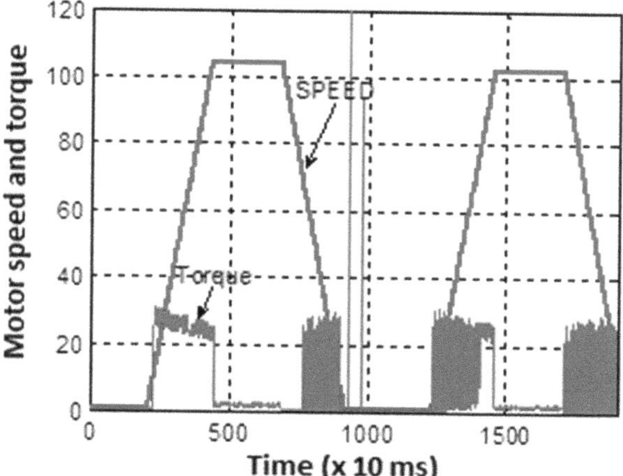

Fig. 20.13 Motor speed and torque

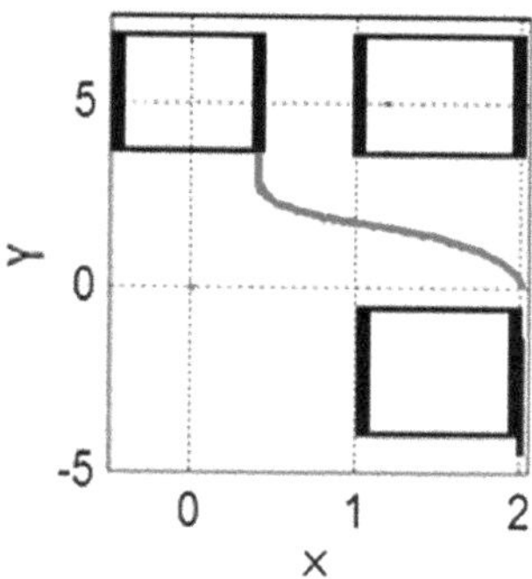

Fig. 20.14 Paralle park measured from back tire right side

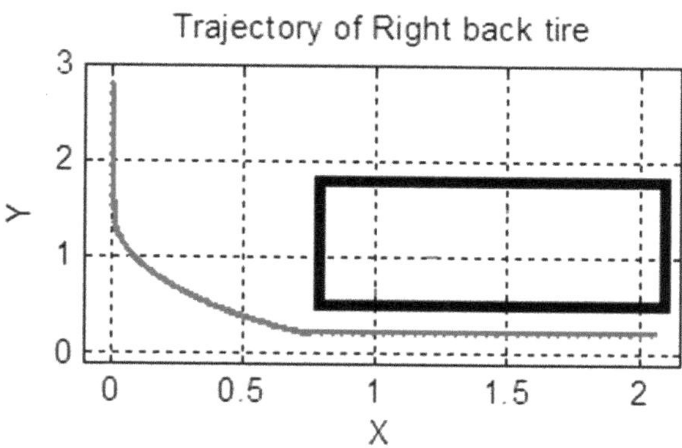

Fig. 20.15 Trajectory of right back tire with reverse parking

20.4 Conclusion

The control technique proposed for automatic parking for parallel and reverse parking was successfully performed and simulated. The speed, torque and position of steering and vehicle movement were successfully controlled for the application of automatic parking for reverse and parallel parking. The DC drive electric car powered by four-quadrant drive DC chopper for series motor is suitable for the application of DC drive electric car and can have an automatic car parking as an extra feature.

References

Anwar, S.: An anti-lock braking control system for a hybrid electromagnetic/ electrohydraulic brake-by-wire system. In: Proceedings of the 2004 American Control Conference, vol. 3, pp. 2699–2704 (2004)

Arof, S., Diyanah, N.H.N., Mawby, P.A. et al.: Study on implementation of neural network controlling four quadrants direct current chopper: Part1: using single neural network controller with binary data output. In: Ismail, A., Bakar, M.H.A., Öchsner, A. (eds.) Advanced Engineering for Processes and Technologies. Adv Struc Mat 102. Springer, Cham (2019)

Arof, S., Diyanah, N.H.N., Noor, N.M., et al.: A new four quadrants drive chopper for separately excited DC motor in low-cost electric vehicle. In: Bakar, M.H.A., Sidik, M.S.M., Öchsner, A. (eds.) Progress in Engineering Technology. Adv Struc Mat 119. Springer, Cham (2019)

Arof, S., Diyanah, N.H.N., Noor, N.M. et al.: Genetics algorithm for setting up look up table for parallel mode of new series motor four quadrants DC chopper. In: Bakar, M.H.A., Sidik, M.S.M., Öchsner, A. (eds.) Progress in Engineering Technology. Adv Struc Mat 119. Springer, Cham (2020)

Arof, S., Diyanah, N.H.N., Yaakop, M.A., et al.: Processor in the loop for testing series motor four quadrants drive DC chopper for series motor driven electric car: Part 1: Chopper operation modes testing. In: Ismail, A., Bakar, M.H.A, Öchsner, A. (eds.) Advanced Engineering for Processes and Technologies. Adv Struc Mat 102. Springer, Cham (2019)

Arof, S., Muhd Khairulzaman, A.K., Jalil, J.A., Arof, H., Mawby, P.A.: Self tuning fuzzy logic controlling chopper operation of four quadrants drive DC chopper for low-cost electric vehicle. In: 2015 6th International Conference on Intelligent Systems, Modelling and Simulation, pp. 40–45 (2015)

Arof, S., Noor, N.M., Alias, M.F. et al.: Digital proportional integral derivative (PID) controller for closed-loop direct current control of an electric vehicle traction tuned using pole placement. In: Bakar MHA, Zamri FA, Öchsner A (eds) Progress in Engineering Technology II. Adv Struc Mat 131. Springer, Cham (2020)

Arof, S., Noor. N.M., Alias, M.F., et al.: Investigation of chopper operation of series motor four quadrants DC chopper. J. Appl. Environ. Biol. Sci. **7**(3S): 49–56 (2017)

Arof, S., Noor, N.M., Mohamad, R. et al.: Close loop feedback direct current control in driving mode of a four quadrants drive direct current chopper for electric vehicle traction controlled using fuzzy logic. In: Baka,r M.H.A., Zamri, F.A., Öchsner, A. (eds.) Progress in Engineering Technology II. Adv Struc Mat 131. Springer, Cham (2020)

Arof, S., Said, M.S., Diyanah, N.H.N., et al.: Series motor four-quadrant DC chopper: reverse mode, direct current control, triple cascade PIDs, and ascend-descend algorithm with feedback optimization for automatic reverse parking. In: Bakar, M.H.A., Zamri, F.A., Öchsner, A. (eds.) Progress in Engineering Technology II. Adv Struc Mat 131. Springer, Cham (2020)

Arof, S., Said, M.S., Diyanah, N.H.N. et al.: Series motor four-quadrant direct current chopper: reverse mode, steering position control with double-circle path tracking and control for autonomous reverse parking of direct current drive electric car. In: Bakar, M.H.A., Zamri, F.A., Öchsner, A. (eds.) Progress in Engineering Technology II. Adv Struc Mat 131. Springer, Cham (2020)

Arof, S., Sukiman, E.D., Diyanah, N.H.N. et al.: Discrete-time linear system of New series motor four-quadrant drive direct current chopper numerically represented by Taylor series. In: Bakar, M.H.A., Zamri, F.A., Öchsner, A. (eds.) Progress in Engineering Technology II. Adv Struc Mat 131. Springer, Cham (2020)

Arof, S., Yaakop, N.M., Jalil, J.A., Mawby, P.A., Arof, H.: Series motor four quadrants drive DC chopper for low cost, electric car: Part 1: Overall. In: 2014 IEEE International Conference on Power and Energy (PECon), pp. 342–347 (2014)

Arof, S., Zaman, M.K., Jalil, J.A., et al.: Artificial intelligence controlling chopper operation of four quadrants drive DC chopper for low-cost electric vehicle. Int. J. Simul. Syst. Sci. Technol. **16**(4), 1–10 (2015)

Arof, S., Jalil, J.A., Kamaruddin, N.S., Yaakop, N.M., Mawby, P.A., Hamzah, A.: Series motor four quadrants drive DC chopper Part2: driving and reverse mode with direct current control. In: 2016 IEEE International Conference on Power and Energy (PECon), pp. 775–780 (2016)

Arof, S., Hassan, H., Rosyidi, M., et al.: Implementation of series motor four quadrants drive DC chopper for dc drive electric car and LRT via simulation model. J. Appl. Environ. Biol. Sci. **7**(3S), 73–82 (2017)

Arof, S., Diyanah, N.H.N., Mawby, P.A. et al.: Low harmonics plug-in home charging electric vehicle battery charger utilizing multi-level rectifier, zero crossing and buck chopper: Part 1: General overview. In: Bakar, M.H.A., Sidik, M.S.M., Öchsner, A. (eds.) Progress in Engineering Technology. Adv Struc Mat 119. Springer, Cham (2019)

Arof, S., Diyanah, N.H.N., Noor, N.M.N., et al.: Series motor four quadrants drive DC chopper: Part 4: generator mode. In: Bakar, M.H.A., Sidik, M.S.M., Öchsner, A. (eds.) Progress in Engineering Technology. Adv Struc Mat 119. Springer, Cham (2019)

Castillo, J., Cabrera, J., Guerra, A.J., et al.: A novel electrohydraulic brake system with tire–road friction estimation and continuous brake pressure control. IEEE Trans. Ind. Electron. **63**(3), 1863–1875 (2016)

Ho, L.M., Satzger, C., Castro, R.: Fault-tolerant control of an electrohydraulic brake using virtual pressure sensor. In: International Conference on Robotics and Automation Sciences(ICRAS), Hong Kong, pp. 76–82 (2017)

Oak Ridge National Laboratory (2009) Advanced brush technology for DC Motors. http://peemrc.ornl.gov/projects/emdc3.jpg. Accessed 15 May 2009

Simonik, P., Mrovec, T., Przeczek, S. et al.: Brake by wire for remotely controlled vehicle. In: International Conference on Electrical Systems for Aircraft, Railway, Ship Propulsion and Road Vehicles & International Transportation Electrification Conference (ESARS-ITEC), Nottingham, pp. 1–4 (2018)

Yuan, Y., Zhang, J., Li, Y., et al.: A novel regenerative electrohydraulic brake system: development and hardware-in-loop tests. IEEE Trans. Veh. Technol. 11440–11452 (2018)

Chapter 21
Oxy Cutting for Mild Steel Response to Cutting Gap Using the Taguchi Method—An Experimental Study

Zahirul Irfan Zainudin, Mohd Riduan Ibrahim, Tajul Adli Abdul Razak, and Pranesh Krishnan

Abstract Oxy-fuel cutting uses oxygen and fuel gases to cut and weld metals. A cutting torch is used to heat a metal to its kindling temperature. Computer Numerical controlled oxy cuts thick materials and is adapted for multi-axis cutting. Furthermore, the technique allows prospects for intricate welding layers that are considered to be difficult. The Taguchi method is functionally used to determine significant factors to ensure the best possible conditions. The orthogonal array L9 that is simple is used in the method. Flame cutting requires a careful balance of the parameters, which are the nozzle gap, feed rate, oxygen pressure, and the layer. The process characteristic was drawn from the experimental data to obtain the optimum parameters.

Keywords Oxy-fuel cutting · Computer numerical control oxy · Taguchi method · Orthogonal array L9 · Nozzle gap · Feed rate · Oxygen pressure · Optimum parameter

21.1 Introduction

Oxy-acetylene is a fusion of ethyne (acetylene) and oxygen, as it is utilized in welding or cutting metals to produce a scorching flame. Common gases are natural gas, propane, hydrogen, MAPP gas, liquefied petroleum, propylene, and acetylene, the

Z. I. Zainudin · T. A. Abdul Razak
Mechanical Section, Malaysian Spanish Institute, Universiti Kuala Lumpur, Kulim 09000, Kedah, Malaysia
e-mail: tajuladli@unikl.edu.my

M. R. Ibrahim (✉)
Manufacturing Section, Malaysian Spanish Institute, Universiti Kuala Lumpur, Kulim 09000, Kedah, Malaysia
e-mail: mohdriduan@unikl.edu.my

P. Krishnan
Intelligent Automotive Systems Research Cluster, Electrical Electronic and Automation Section, Malaysian Spanish Institute Universiti Kuala Lumpur, Kulim Hi-Tech Park, 09000 Kulim, Kedah, Malaysia
e-mail: pranesh@unikl.edu.my

M. H. Abu Bakar et al. (eds.), *Progress in Engineering Technology III*,
Advanced Structured Materials 148, https://doi.org/10.1007/978-3-030-67750-3_21

most common of which is acetylene. In many cases, one gas is not beneficial over another, although, in some situations, a particular gas may be desirable. For instance, the heating capacity of gases are different, that makes it more accessible to utilize with particular metals.

Oxy-acetylene cutting uses oxygen and ethyne to heat up a metal to red-hot and later consumes pure oxygen to burn away the warmed metal. Oxy-acetylene cutting is the most commonly used method of cutting ferrous metals by the application of heat. The principle of oxy-acetylene cutting is simple. An oxy-acetylene flames heat the metal to its ignition temperature. Then a pure oxygen jet is directed toward the hot metal, and there is a chemical reaction called oxidation. Oxidation is a well-known chemical reaction. It is called combustion or burning when it occurs quickly; it is called rusting when it happens slowly. When a metal is being cut by the oxy-acetylene torch method, the oxidation of the metal is extremely rapid in short, the metal burns. The heat released by iron or steel burning melts the iron oxide formed by the chemical reaction, and the pure iron or steel is also heated. The fused material runs off as slag, exposing the oxygen jet to more iron or steel.

Most of the material is in the form of oxides removed from the kerf. The remainder of the material removed from the kerf is pure metal, which is blown or washed out of the kerf by the force of the oxygen jet. Since oxidation of the metal is a vital part of the oxy-acetylene cutting process, this process is not eligible for metals that do not oxidize readily, such as copper, brass, stainless steel, and so on. The oxy-acetylene cutting process easily cuts low carbon steels.

The Taguchi method, as an experimental method, is used to achieve the required process quality. This method is designed to optimize process parameters and improve the quality. There are three designs in this method which are system design, parameter design, and tolerance design. There is Minitab software that has been using in this method.

The factor information of experiments was chosen, as shown in Table 21.1. Then, the series of experiments and the number of output factors which were performed have been tabulated in Table 21.2. The graphs are created with the corresponding output values. The investigation is conducted out in a Taguchi L9 orthogonal array (Mustafa et al. 2014; Reddy et al. 2017), and the results are presented in Table 21.2.

Table 21.3 presents the MRR value for the L9 orthogonal array (Pang et al. 2014). Figure 21.1 depicts the graphs of the main effects of the means of MRR using Minitab17. The inferences showed the increment in speed, stand-off-distance, and oxygen pressure. The difference in stand-off-distance is compared to the rest. The

Table 21.1 Table for factor information

Factor	Levels	Levels		
		−1 Level	0 Level	+1 Level
Speed (mm/min)	3	450	475	500
Stand-off-distance (mm)	3	4	63.5	8
Oxygen pressure (kgf/cm^2)	3	3.5	4.5	5.5

Table 21.2 Experiments conducted for investigation

S. No.	Speed (mm/min)	Stand-off-diatance (mm)	Oxygen Pressure (kg/cm^2)	KERF (mm)	M.R.R (gm/sec)
1	450	4	3.5	2.6	3.439
2	450	6	4.5	2.96	4.635
3	450	8	5.5	3.33	5.924
4	475	4	4.5	2.45	4.018
5	475	6	5.5	2.94	5.424
6	475	8	3.5	3.45	5.223
7	500	4	5.5	2.47	4.806
8	500	6	3.5	3.25	4.507
9	500	8	4.5	3.55	5.019

Table 21.3 Response table for means of MRR

Level	Speed (A) (mm/min)	Standoff distance (B) (mm)	Oxygen pressure (C) (kg/cm^2)
1	4.703	3.993	4.330
2	4.853	5.080	4.493
3	5.050	5.533	5.303
Delta	0.347	1.540	0.973
Rank	1	1	2

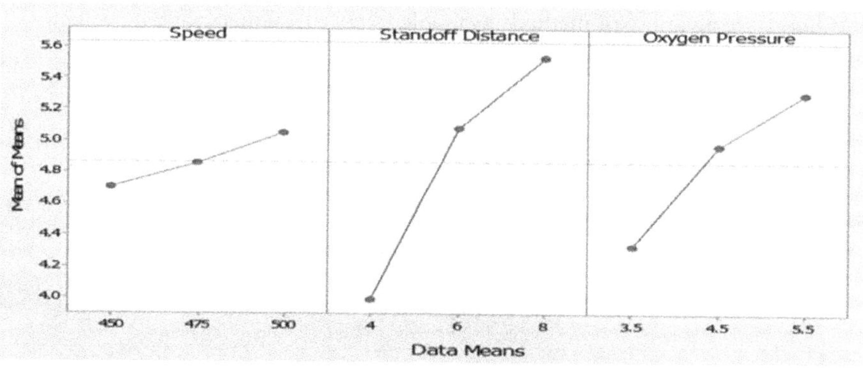

Fig. 21.1 Main effects plot for means of MRR

response table of means of MRR is produced, and the ranking for the factors affecting the MRR is given in Table 21.3 centered on the delta values.

Table 21.3 details the effect of limits on MRR. The standoff distance affects the MRR to the upper limit and then oxygen pressure and then speed. The optimum mixture of factors is A3B3C3, i.e., speed: 500 mm/min, standoff distance: 8 mm, oxygen pressure: 5.5 kg/cm (Ramakrishna et al. 2018).

21.2 Methodology and Experimental Setup

21.2.1 Fabrication of the Product

CNC or computer numerical controls is the operating system or brain of a CNC system. A controller completes the critical connection between a computer system and a CNC machine's mechanical components (Minquiz et al. 2014). The primary task of the controller is to receive conditioned signals from a computer or indexer and to interpret those signals through motor output into mechanical motion. Several components make up a controller, and each component works in unison to produce the desired motor movement, which are the power supply, electronic handwheel, USB controller, micro step driver, and stepper motor (Bologa et al. 2016; Deng et al. 2018).

21.2.2 Fabrication Process

The platform machine of oxy-acetylene welding as shown in Fig. 21.2 need to be build from several types of metals, and each require various metal cutting techniques. Thus, there are a variety of methods available to cut metal materials.

- For smaller cutting jobs, hand tools, including hand shears, cutting machines, and hacksaws, may be used for cutting the metal into the desired shape. This cutting method is best suited for projects using more pliable metals.

Fig. 21.2 Platform machine of oxy-acetylene welding

- A grinder requires the finished part to be extremely smooth, a grinding machine maybe your tool of choice. Using a rotating blade or wheel made of abrasive material, a grinder uses friction to wear down the surface of the metal until it is smooth, similar to sanding wood.
- Welding is the process of combining two pieces of metal. The welding process not only binds the two pieces together as in brazing and soldering but causes the metal structures of the two pieces to join together and become one by using extreme heat and sometimes by adding other metals or gasses.
- After a workpiece is laid out and adequately mounted, the drilling process can begin. Drilling is the process where a round hole is created within a workpiece or enlarged by rotating an end-cutting tool, a drill.
- Tapping is the process of cutting a thread inside a hole so that a cap screw or bolt can be threaded into the hole. Also, it is used to make a thread on nuts. Tapping can be done on the lathe by power feed or by hand.
- Wiring control panel is a complex process, and it needs several carefully planned and performer details. Highly reliable panel wiring and control systems are made to connect to the machine to include automation in manufacturing processes.

21.2.2.1 MACH3 Software

Mach3 is a popular Windows operating CNC machine tool control software that runs on desktop computers and laptops. Mach3 uses the parallel (printer) port of the computer by default and sends all signals via this port. It allows productivity and works to be done without the software being worried. Thus, a user needs to spend less time programming and as a consequence, more cutting time is available. Mach3 has been used successfully for lathes, mills, routers, lasers, plasma cutters, engravers, and gear cutting control.

21.2.2.2 Specimen Test

As the experiments are carried out using the L9 Taguchi method (Wakjira et al. 2019), all the specimens are fabricated based on parameters and levels that have been set up in the orthogonal array as shown in Fig. 21.3.

21.2.2.3 Profile Projector

A profile projector (see Fig. 21.4) is is a specialized measuring system that is versatile and indispensable in the manufacturing sector, where two-dimensional measurements of small parts are required in the departments of R&D, production, or quality control. A profile projector brings together beautifully optics, mechanics, electronics,

Fig. 21.3 Specimens fabricated based on L9 Taguchi method

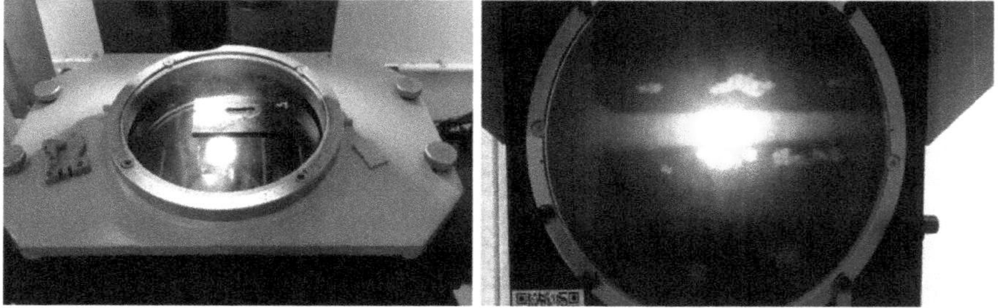

Fig. 21.4 Specimens at the profile projector

and software to provide two-dimensional measurements for small parts. Alternatively, it can identify and mark points, and the system automatically calculates the distance, angle, arc length, etc., of the shadow and consequently calculates the same for the part.

21.2.2.4 Taguchi Method Using Minitab

Step 1: Go to Minitab18, then Stat> DOE> Taguchi> Create Taguchi Design to generate a Taguchi design (orthogonal array). Each column in the orthogonal array represents a specific factor, and each row represents a run. After that, the Taguchi design box will appear. Choose 3-Level Design and Number of factors: 4, which means four parameters will be used. Select Designs and use L9 for Runs.

Step 2: Next, key in the data required in the L9 Runs. Then, select DOE> Taguchi > Define Custom Taguchi Design. The Define Custom Taguchi Design box will come out. Select all factors and click OK. After that, put in RESULT at C5.

Step 3: Go to DOE> Taguchi > Analyze Taguchi Design. Select all the factors. Open the Graphs and tick Signal to Noise ratios and Means. Open the analysis and tick the same to Graphs for Display response tables. Do the same thing for Storage. Next, select the best Signal to Noise Ratio based on an experiment. Lastly, click, OK, and the graphs will appear.

Step 4: Then, select DOE > Taguchi > Predict Taguchi Results. Tick Mean and Signal to Noise ratio only. Next, select Levels and choose Select levels from a list. Key in the levels as optimum at Means graphs and click OK. The Taguchi analysis RESULT will come out.

21.3 Results and Discussion

The goal of this experimental work is to investigate the effects of cutting gap parameters on the specimen and to establish a correlation between them. For this, nozzle gap, feed rate, oxygen pressure, and the layer were chosen as process parameters as shown in Table 21.4.

Taguchi's methodology involves the use of specially constructed "Orthogonal Array" (OA) tables that require a very small number of consistent and straightforward design experimental runs as shown in Table 21.5 for overall layout and Table 21.6 for this experimental layout.

The value for the L9 orthogonal array is tabulated, and the graphs of main effects for the means are developed using Minitab18 as shown in Fig. 21.5. The change in the layer is more significant and optimum compared to the other three.

Table 21.7 details the effect of parameters in means. The layer effects to the maximum and then nozzle gap and then oxygen pressure and then feed rate. The best combination of parameters is A1B2C3D2, i.e., nozzle gap: 1 mm, feed rate: 10 mm/min, oxygen pressure: 6 bar, layer: 2.

The value for the L9 orthogonal array is tabulated, and the graphs of main effects for the SN ratios are developed using Minitab18 as shown in Fig. 21.6. The SN

Table 21.4 Machining parameters

Parameters	Level 1	Level 2	Level 3	Units
Nozzle gap (A)	1	2	3	mm
Feed rate (B)	5	10	15	mm/min
Oxygen pressure (C)	4	5	6	bar
Layer (D)	1	2	3	pcs

Table 21.5 Layout using an L9 orthogonal array

Test No.	A	B	C	D
1	1	1	1	1
2	1	2	2	2
3	1	3	3	3
4	2	1	2	3
5	2	2	3	1
6	2	3	1	2
7	3	1	3	2
8	3	2	1	3
9	3	3	2	1

Table 21.6 Experimental layout using an L9 orthogonal array

Test No.	A	B	C	D
1	1	5	4	1
2	1	10	5	2
3	1	15	6	3
4	2	5	5	3
5	2	10	6	1
6	2	15	4	2
7	3	5	6	2
8	3	10	4	3
9	3	15	5	1

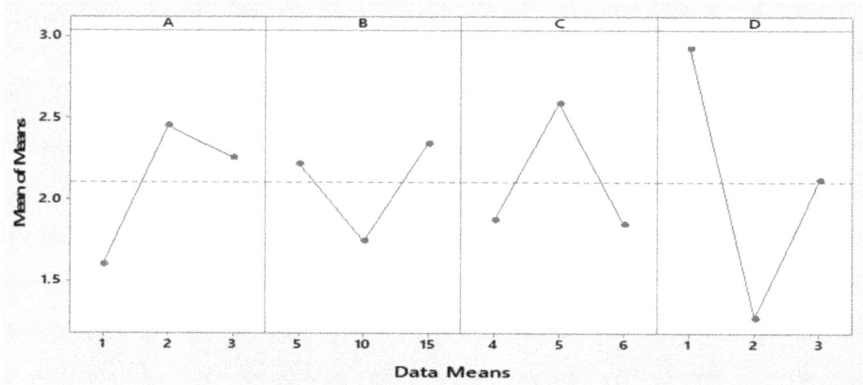

Fig. 21.5 Main effects plot for means

Table 21.7 Response for means

Level	A	B	C	D
1	1.605	2.224	1.878	2.926
2	2.452	1.747	2.589	1.274
3	2.261	2.347	1.851	2.117
Delta	0.847	0.600	0.738	1.652
Rank	2	4	3	1

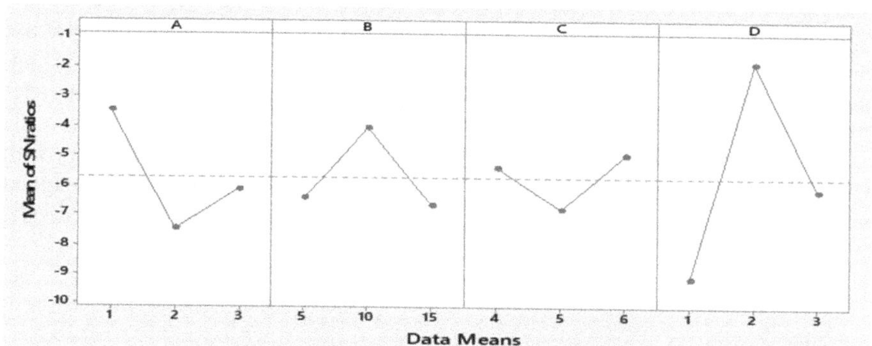

Fig. 21.6 Main effects plot for SN ratios

ratios are read opposite the mean graph which means the most significant values are optimum.

For calculating the cutting gap, the objective function "smaller the better" type was used, as shown in Table 21.8. When an ideal value is as minimum defined, then the difference between measured data and ideal value is expected to be as minimal as possible to achieve the goal of the experiment.

Table 21.8 Response for the signal to noise ratios

Level	A	B	C	D
1	−3.481	−6.415	−5.364	−9.131
2	−7.492	−4.028	−6.796	−1.847
3	−6.123	−6.653	−4.937	−6.119
Delta	4.011	2.625	1.860	7.284
Rank	2	3	4	1

Table 21.9 Confirmation test comparison between prediction and actual

	Prediction	Actual
Level	A1B2C3D2	A1B2C3D2
Mean	2.80386	2.24763
S/N Ratio	0.159	0.131

21.3.1 Confirmation Test

A confirmation test is required in Taguchi's parameter design methods to remove concerns about selecting control parameters, experimental design, or response assumptions. This test is to validate the set of control parameters identified as meaningful by Taguchi methods and to verify that the recommended control parameter settings are either optimal or near-optimal.

The confirmation test was done by the specific combination of parameters with their levels, which were predicted as A1B2C3D2. The new experiment is designed for conducting new experiments to get the best cutting gap performance. The results are tabulated as shown in Table 21.9.

21.4 Conclusion and Recommendation

From the experiment and data analysis that has been performed, the objective is achieved. The Taguchi approach to quality engineering, places a great deal of intensity on minimizing variation as the primary means of improving quality. The idea is to design products and processes whose performance is not affected by external conditions and to build this during the development and design stage through the use of experimental design.

An experimental study has been done to find the significant factor for the cutting gap. research, the influence of the nozzle gap, feed rate, oxygen pressure, and layer on the oxy-fuel cutting process of mild steel was studied. By using the Taguchi method, the optimum setting for obtaining the minimum cutting gap is A1 (nozzle gap, 1 mm), B2 (feed rate, 10 mm/min), C3 (oxygen pressure, 6 bar), and D2 (layer, 2). By experimenting, it can be observed that the layer is the more affecting parameter compared to the nozzle gap, feed rate, and oxygen pressure. Also, it shows that using the Taguchi method in the Minitab18 software is the best method to carry out the best optimization in this experiment.

References

Bologa, O., Breaz, R.-E., Racz, S.-G., Crenganiş, M.: Decision-making Tool for Moving from 3-axes to 5-axes CNC Machine-tool. Procedia Comput. Sci. **91**, 184–192 (2016)

Deng, C., Guo, R., Zheng, P., Liu, C., Xu, X., Zhong, R.Y.: From open CNC systems to cyber-physical machine tools: a case study. Procedia CIRP **72**, 1270–1276 (2018)

Minquiz, G.M., Borja, V., López-Parra, M., et al.: A comparative study of CNC part programming addressing energy consumption and productivity. Procedia CIRP **14**, 581–586 (2014)

Mustafa, N., Ibrahim, M.H.I., Asmawi, R., Amin, A.M., Masrol, S.R.: Green strength optimization in metal injection molding applicable with a Taguchi method L9 (3)4. In: International Integrated Engineering Summit. vol 773. Applied Mechanics and Materials. Trans Tech Publications Ltd, pp. 115–117 (2014)

Pang, J.S., Ansari, M.N.M., Zaroog, O.S., Ali, M.H., Sapuan, S.M.: Taguchi design optimization of machining parameters on the CNC end milling process of halloysite nanotube with aluminium reinforced epoxy matrix (HNT/Al/Ep) hybrid composite. HBRC J. **10**(2), 138–144 (2014)

Ramakrishna, C.S., Raghuram, K.S., Ben, B.A.: Process modelling and simulation analysis of CNC oxy-fuel cutting process on SA516 grade 70 carbon steel. Mater. Today Proc. **5**(2, Part 2):7818–7827 (2018). https://doi.org/10.1016/j.matpr.2017.11.461

Reddy, N.R., Satyanarayana, V.V., Kumar, P.S., Rao, V.R.: An investigation of oxy-fuel cutting of mild steel using principle component analysis and taguchi technique. Int. J. Mech. Prod. Eng. Res. Dev. **7**, 457–464 (2017)

Wakjira, M.W., Altenbach, H., Perumalla, J.R.: Analysis of CSN 12050 carbon steel in dry turning process for product sustainability optimization using taguchi technique. Balani K, ed. J Engn:7150157. https://doi.org/10.1155/2019/7150157

Chapter 22
Welding Parameter Response to Depth Penetration—An Experimental Study

Muhammad Hadzimi Md Ludin, Mohd Riduan Ibrahim, Tajul Adli Abdul Razak, and Pranesh Krishnan

Abstract To join ferrous and non-ferrous metals, metal inert gas welding (MIG) is an essential welding operation. The welding quality is affected by the MIG input welding parameters, and the weld bend geometry affects the weld quality. This paper describes the factors that affect the welding parameters (voltage, current, speed, and AISI 1020 steel penetration depth) during welding. The methodology involves the Taguchi technique for data acquisition and optimization of the welding parameters. Lastly, the conformations test results show the effectiveness of the analysis of penetration through the variation among the predicted and experimental values.

Keywords GMAW · MIG · Angular distortion · Taguchi method · L9

22.1 Introduction

Metal gas inert welding (MIG) is a metallurgical fusion process, used to join structural components together. MIG is also known as gas metal arch welding GMAW, as it uses oxygen to stabilize the welding arc (Prakash et al. 2016). The welding makes two structural components combine together such that heating and solidification result in the permanent joint. It is achieved by forming an arc between a metal electrode and a workpiece. Welding technology is commonly used in every type of manufacturing

M. H. M. Ludin · T. A. A. Razak (✉)
Mechanical Section, Malaysian Spanish Institute, Universiti Kuala Lumpur, 09000 Kulim, Kedah, Malaysia
e-mail: tajuladli@unikl.edu.my

M. R. Ibrahim
Manufacturing Section, Malaysian Spanish Institute, Universiti Kuala Lumpur, 09000 Kulim, Kedah, Malaysia
e-mail: mohdriduan@unikl.edu.my

P. Krishnan
Intelligent Automotive Systems Research Cluster, Electrical Electronic and Automation Section, Malaysian Spanish Institute Universiti Kuala Lumpur, Kulim Hi-Tech Park, 09000 Kulim, Kedah, Malaysia
e-mail: pranesh@unikl.edu.my

M. H. Abu Bakar et al. (eds.), *Progress in Engineering Technology III*,
Advanced Structured Materials 148, https://doi.org/10.1007/978-3-030-67750-3_22

Table 22.1 The parameter used in the experiment by Kumar et al. (Das et al. 2013)

Variables	Low level (1)	Medium level (2)	High level (3)
Current (A)	190	195	200
Voltage (V)	20	23	26
Welding speed (mm/min)	0.037	0.042	0.048

sector: shipbuilding, equipment in railroad, construction, pipelines, boilers, launch vehicles, nuclear reactors, aircraft, and automobiles (Kumar and Goyal 2018). It can be used in the fabricating industry or for repairing metal products.

The process finds its application in air, underwater, and space. Usually is a high current for the MIG welding starting at 80 A and may require more than 12,000 A in spot welding (Das et al. 2013). It is different for welding two razor blades, at which is used a gas tungsten arc only at five amps to obtain a good result. According to Zaidi and Madavi (2018), the electric current is used for variable welding parameter. The recommended current used for this experiment is 150 A (results at 100, 125, 150 A) whereas the recommended arc voltage is 25 V (results at 20, 25, 30 V). Besides, the rate of gas flow is set at 14 L/min (results at 12, 14, 16 L/min).

Based on the journal article by Tewari et al. (2010), the welding speed is defined as the rate at which the electrode travels along the seam or the rate at which it travels under the electrode along the seam. It is possible to make some general statements about travel speed. Increasing the travel speed and keeping constant arc voltage and current will reduce the bead width and also increase the penetration until the optimum speed is reached at which maximum penetration is achieved (Sheikh and Kamble 2018). It will result in decreasing penetration by increasing the speed beyond this optimum.

Based on the journal by Kumar et al. (2017), a total of 9 experiments are conducted, and each experiment is based on a combination of levels as shown in the Table 22.1. The third experiment, for example, is performed by keeping the independent design variable one at level 1, variable two at level 2, variable three at level 3, and variable four at level 4. The orthogonal arrays have the following selected properties, which reduce the number of experiments to be performed.

According to Chavda et al. (2014) in their finding, the Taguchi method is used to identify the process parameters with the maximum distribution of stress in the MIG for medium carbon steel. The levels of parameters by Zaidi and Madavi (2018) are listed in Table 22.2.

22.2 Methodology and Experimental Setup

See Figs. 22.1 and 22.2.

Table 22.2 Level of MIG welding process variable

Variables	Low level (1)	Medium level (2)	High level (3)
Current (A)	180	190	200
Voltage (V)	21.5	22.5	23.5
Gas flow rate (lit/min)	15	17	19
Welding speed (mm/min)	200	225	250

Fig. 22.1 Machine setup and nozzle gap to the workpiece

Fig. 22.2 Welding process and welding specimen

22.2.1 *Experimental Work*

In this experiment, there were three variable parameters selected. The variable setting is the feed rate, current, and nozzle gap. The three levels of the variable parameters are shown invariable parameters three levels are shown in the Table 22.3.

The design matrix is produced by a Taguchi orthogonal array; level L9 was selected. Table 22.4 displays the L9 Taguchi design (orthogonal array). L9 means

Table 22.3 Process parameters and level

Parameter	Low level (1)	Medium level (2)	High level (3)
Feed rate (mm/min) (A)	12	16	20
Current (Amp) (B)	120	150	180
Nozzle gap (mm) (C)	1	3	5

Table 22.4 L9 orthogonal array

Trial no.	Feed rate (mm/min)	Current (A)	Nozzle gap (mm)
1	12	120	1
2	12	150	3
3	12	180	5
4	16	120	5
5	16	150	1
6	16	180	3
7	20	120	3
8	20	150	5
9	20	180	1

nine runs. 3^3 involves three factors with three levels each. The table columns represent the control factors, the table rows represent the runs (a combination of factor levels), and each table cell represents the factor level for the experiment.

Table 22.5 Taguchi design of experiment

Experiment run	Process parameter			Designation	Result
	A	B	C		
1	12	120	1	A1B1C1	1.362
2	12	150	3	A1B2C2	2.760
3	12	180	5	A1B3C3	2.880
4	16	120	5	A2B1C3	2.292
5	16	150	1	A2B2C1	2.074
6	16	180	3	A2B2C2	2.726
7	20	120	3	A3B1C2	1.655
8	20	150	5	A3B2C3	1.7383
9	20	180	1	A3B3C1	0.901

22.3 Results and Discussion

According to the Taguchi method, the mean is the desired value for the output characteristic. There are 3 signal-to-noise ratios of common interest for optimization of static problems, smaller-the-better, larger-the-better, and nominal-the-best. Better quality is obtained by choosing the penetration for optimum welding performance.

In this experiment, the work is effect by the readings of the obtained penetration through the measuring instrument after cutting all the welded specimens perpendicular to the direction of welding, see Table 22.6 which includes also the variations. Results indicate that the welding speed has a significant effect. Increased welding speed and decreased current raise the ultimate tensile strength of the welded joint. The voltage does not add to the weld strength. Irrespective of the quality attributes, the higher signal-to-noise ratio correlates to a better quality. Hence, the optimal level of the process variables is the level with the highest signal-to-noise ratio.

Based on Fig. 22.3, the experiment shows that the higher rank that effect the depth penetration of MIG welding comes from the current. The value of factor B is 2.776 A. The second rank is the mean value for the feed rate is 2.364 mm/min. The mean value for the nozzle gap is 2.380 mm. The significant factor in this experiment comes from factor B, which is the current. From the graph in Fig. 22.3, the optimum parameter is A2B3C2. Besides, the significant factor is the biggest factor that affects the response in this experiment. It will change a large number of results compared with other elements (Fig. 22.4; Table 22.7).

22.3.1 Confirmation Test

The last step after the optimal combination of process parameters and their levels was obtained in the confirmation test. This confirmation test is required to run confirmation experiments to verify if the welding parameter setting produces the optimum performance. Thus, it evaluates the predictive ability of the Taguchi method for MIG welding performance. The optimum value of the experiment is A2B3C2 as shown in Table 22.8.

Table 22.6 Response table for mean

Level	A	B	C
1	2.319	1.754	2.037
2	2.364	2.191	2.380
3	2.039	2.776	2.304
Delta	0.325	1.022	0.343
Rank	3	1	2

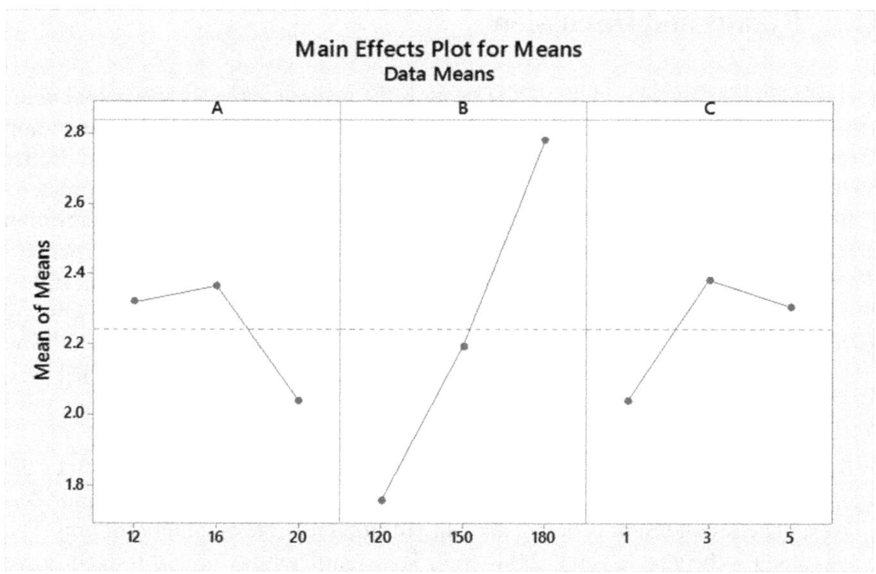

Fig. 22.3 Main effects plot for means

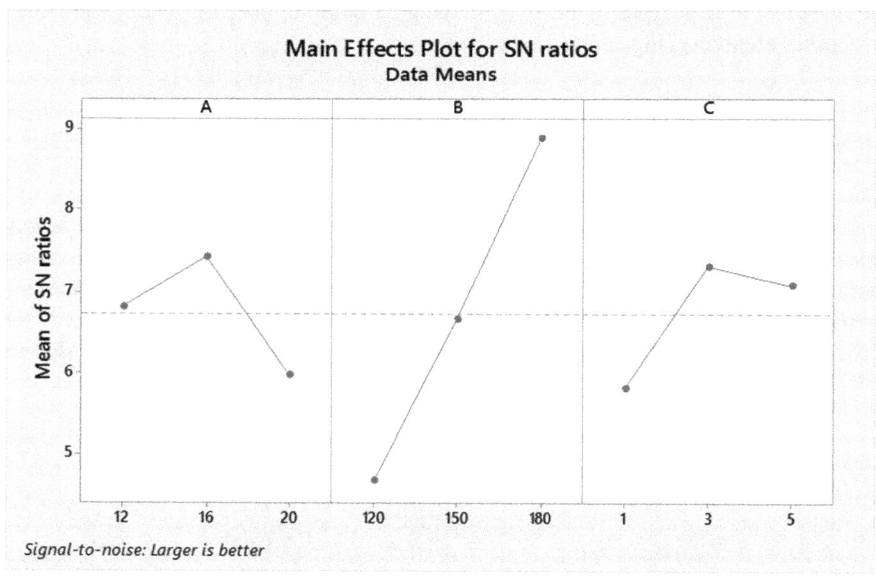

Fig. 22.4 Graph of main effects plot for S/N ratio

Table 22.7 Response for means

Level	A	B	C
1	6.797	4.655	5.806
2	7.417	6.655	7.302
3	5.962	8.865	7.068
Delta	1.455	4.210	1.495
Rank	3	1	2

Table 22.8 Optimum values from the experiment

	Prediction	Actual
Level	A2B3C2	A2B3C2
Mean	3.040	3.349
S/N ratio	10.134	11.654

22.4 Conclusion and Recommendation

Based on the experiment, all the data collected and analyses confirmed that the objective of this experiment was achieved. This experiment was conducted to find the significant factor for the depth penetration of MIG welding. Taguchi was applied as a method to find the optimal process parameter for penetration. The data collected in this experiment was analyzed with the Taguchi method and predicted by using the Minitab 18 Software. L9 orthogonal array, signal-to-noise (S/N) ratio, and analysis of variance were used to optimize of welding parameter. This experiment was conducted nine times with different setting parameter levels on the different workpieces. The three factors contribute to the response, and all have been considered for experimentation of MIG welding. During the study, it is found that the current is the significant factor that affects the depth penetration of welding. The other parameters, which are the feed rate and nozzle gap, is do not have a significant impact compared with the current setting.

Besides, before collecting the data, the fabrication process is one of the critical aspects that was considered in this experiment. The material used to make a frame machine is affecting the machine performance. High-quality content reduces the vibration of the device. The MIG welding machine process is suitable to join the mild steel plate while the parameter is considered. The data analyses using the Taguchi method and Minitab 18 software is essential instrumental in making this experiment successful.

References

Chavda, S.P., Desai, J.V., Patel, T.M.: A review on optimization of MIG welding parameters using Taguchi's DOE Method. Int. J. Eng. Manag. Res. **4**(1), 16–21 (2014)

Das, B., Debbarma, B., Rai, R.N., Saha, S.C.: Influence of process parameters on depth of penetration of welded joint in MIG welding process. Int. J. Res. Eng. Technol. **2**, 220–224 (2013)

Kumar, V., Goyal, N.: Parametric optimization of metal inert gas welding for hot die steel by using Taguchi approach. Mat. Sci. Res. India **15**(1) (2018). doi:http://www.materialsciencejournal.org/?p=7065

Kumar, C.L., Vanaja, T., Murti, K.G.K., Prasad, V.V.S.H.: Optimization of MIG welding process parameters for improving welding strength of steel. Int. J. Eng. Trends Technol. **50**(1), 26–33 (2017)

Prakash, A., Bag, R.K., Ohdar, P., Raju, S.S.: Parametric optimization of metal inert gas welding by using Taguchi approach. Int. J. Res. Eng. Technol. **5**(2), 176–182 (2016)

Sheikh, A.A., Kamble, P.D.: Optimization of welding process parameter to minimize defect in welding of sheet. Int. Res. J. Eng. Technol. **5**(05), 4310–4314 (2018)

Tewari, S.P., Gupta, A., Prakash, J.: Effect of welding parameters on the weldability of material. Int. J. Eng. Sci. Technol. **2**(4), 512–516 (2010)

Zaidi, A., Madavi, K.: Improvement of welding penetration in MIG welding. Int. J. Sci. Res. Sci. Technol. **4**(2), 1198–1203 (2018)

Chapter 23
Voice Activated Command for Automotive Applications Using a Raspberry PI

Muhammad Shafiq Rosli, Sazali Yaacob, and Pranesh Krishnan

Abstract Implementation of automation controls using voice commands is intended to increase the level of technology comparable to developing countries. Voice activation systems using user voice recognition are electronic devices installed to control the direction of a moving car without the use of a control device or switch. It only uses voice recognition to turn this system on. With this system, some applications can be used or activated by voice commands only. Therefore, users can focus on driving without interruption. The main objective of the project was to develop a system that would help users with limited mobility to control applications. However, the project can be divided into two parts which are the hardware and software development. Among the hardware used include the Raspberry Pi, motor driver, direct current motor, microphone and liquid crystal display. This project was developed using the Raspberry Pi as the primary control over the application. The driver motor will distribute the same voltage power to the direct current motor to function while receiving the signal from the Raspberry Pi. Besides, the microphone device is used to record the user's voice while giving instructions, and the liquid crystal display will display the direction of rotation of the motor when the motor is moving like moving forward, backward, left and right sides.

Keywords Voice command · Automotive applications · Voice recognition · Raspberry Pi

---

M. S. Rosli
Electrical Electronic and Automation Section, Malaysian Spanish Institute Universiti Kuala Lumpur, Kulim Hi-Tech Park, 09000 Kulim, Kedah, Malaysia
e-mail: shafiq.rosli04@s.unikl.edu.my

S. Yaacob (✉) · P. Krishnan
Intelligent Automotive Systems Research Cluster, Electrical Electronic and Automation Section, Malaysian Spanish Institute Universiti Kuala Lumpur, Kulim Hi-Tech Park, 09000 Kulim, Kedah, Malaysia
e-mail: sazali.yaacob@unikl.edu.my

P. Krishnan
e-mail: pranesh@unikl.edu.my

M. H. Abu Bakar et al. (eds.), *Progress in Engineering Technology III*,
Advanced Structured Materials 148, https://doi.org/10.1007/978-3-030-67750-3_23

23.1 Introduction

Nowadays, vehicles are the main means of transportation which are widely used to make our daily life more comfortable. Driving a car can make it easy for someone to move from one place to another quickly (Ni 2012). In sparsely populated areas, owning a car is even more critical since it provides the only opportunity for travelling long distances due to a lack of public transport. For older people, having more difficulties walking to the bus stop and cycling, driving is often the only option for independent mobility (Ellaway et al. 2003). However, some drivers find it difficult to control some applications while driving, especially the right and left signals. For example, many vehicles on the road do not give a signal when changing lanes or taking right or left directions. The drivers tend to forget to do so either by distraction in driving or by forgetfulness. When driving, the driver should focus on the driving to avoid any problems. For those who control the application in the vehicle by pressing the switch on and off may cause distraction to the drivers and lose focus on the driving. However, disabled drivers are no exception to this problem. Therefore, a vehicle needs a system that enables drivers to stay focussed and easy to control applications. This system also can help the people to overcome some limitations and can control few applications effortlessly using voice commands.

By developing voice activated commands, some applications that allow drivers to stay focussed on driving can be controlled automatically. By using voice recognition, they can control the applications smoothly and can concentrate on driving (van der Velde 2019). It also provides the benefit for disabled users who have trouble controlling the applications in the vehicle while driving. From this recognition, they are able to drive safely and focus on driving.

Several projects concerning the use of voice in the method of operation are present in the literature. Hidayat and Firmanda (2015), developed a home automation control system to provide the facility to control the home appliances residents to do homework. It proposes the idea of a home-based voice automation system. The system comprises over a smartphone, electrical applications, voice module and website. It uses the voice command function to control any electrical device while keeping the web-based application to control detailed instruction or configuration. The system has been determined by voice commands related to electrical applications, and the specifications are different from other electrical applications in each house. The cloud-based API, Wolfram Alpha, was used to record voice commands that turn them into text for processing by the Raspberry Pi.

Sharifuddin et al. (2019) developed an intelligent wheelchair movement control using convolutional neural networks (CNNs). The supporting software comprises of the convolution neural networks, the voice recognition module and voice data processing module. The system completes the functions of collecting the voice data, distilling character, voice recognition and voice data processing in terms of mel-frequency cepstral coefficients features extraction. The central processor of this system is the Raspberry pi 3B. The system will identify the word and send a signal to the motor driver. Pang and Chen (2018) established a voice recognition using

Google Home and Raspberry Pi for smart socket control. It works on a set of voice commands that are set to control electrical applications in a home via a Bluetooth socket. The system connects the API.ai functionality of Google Home to Google accounts. Google Home receives instructions to be written by API.ai assessment and response instructions. The proposed system consists of household appliance control, entertainment system, hall facilities monitoring and home security. This system also uses a service provided by Amazon, which is capable of capturing voice commands and processing them. The command will be sent to the local web server created by the Raspberry Pi microprocessor. This extension service is provided by Ngrok by creating a secure public uniform resource locator (URL) (Yue 2017).

The remainder of this paper deals with the development of the voice activated command to control automotive applications using the Raspberry Pi to identify the voice command of a specific user this is explained in the methodology section. The signal processing and classification model development is explained in the feature extraction section. The final section details the implementation methods. The data will be processed and classified for a specific action. If matching is established, the corresponding motor driver will turn on or off the desired electrical appliance following the voice command.

23.2 Methods

The system is to control the application automatically. Application automation systems typically work using wireless technology, but very few are controlled by voice. The proposed system is an innovation to this technology where it is introduced for convenience by simply giving voice commands to the microphone. It uses a Raspberry Pi, DC motors, motor driver, microphone, Google speech recognition for recognition and switch. Additional equipment like LCD is often added to expand the system's ability to automate applications and make them more user-friendly. A USB microphone is required to record sound. The Raspberry pi does not need modules to install microphones; instead, we need changes to the terminal. The voice command system is based on the use of voice and processing, which is then compiled as commands by this system. Google speech recognition is used to recognise the received voice to move motors. The speech recognition module (Mahesh 2016) is available on the Raspberry Pi, which connects the Raspberry Pi with Google for recognition via the internet. The microphone can also be connected directly to the USB port to convey verbal commands to the network.

23.2.1 Experimental Setup and Data Acquisition

During the data collection, several trials of speech signals were captured using a microphone device connected to a laptop. The words spelt during the trials were left, right, forward, backward and stop. The experiment was conducted in a quiet environment with negligible background noise, and 100 trials were given for each word. The speech signals were recorded at a sampling frequency of 16 kHz using a USB microphone. A Python script running the sound device recording function was used to record the voice. It recorded signal based on 8 bits per sample and mono channel so that the signal can be reproduced through several speakers, producing the same copy of the signal. The data is validated using data validation techniques such as ANOVA and t-test.

23.2.2 *Signal Preprocessing*

The signals received by the input device are processed in the Python software that allows the voice system to recognise the instructions given in controlling the application. In this system, the signal are processed to remove some unwanted components or features from a signal. The defining feature of the filter is the complete or partial suppression of the signal. This system uses butter low pass filters, which are filters that only pass low-frequency signals (Jirafe 2019). A low pass filter (LPF) attenuates a cut-off frequency, allowing lower frequencies to pass through the filter. It also complements the signal that is often higher than the cutting frequency. In other words, low pass filters help in eliminating short-term fluctuations and provide a smoother signal shape. These filters capture the signal while eliminating the noise. Figure 23.1 shows the filtered noise where the blue signal is the recorded voice and green for the filtered signal. With the filtering process, the system can recognise the voice more precisely. By using the python software, voice signals can be filtered more efficiently and produce smoother signals.

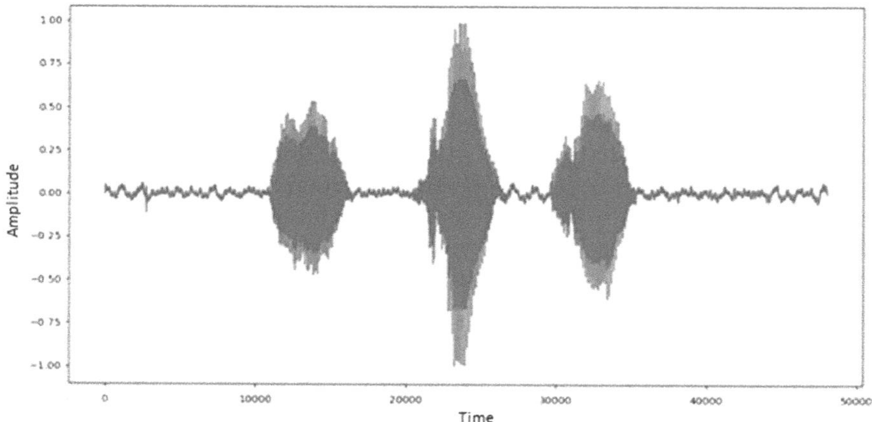

Fig. 23.1 System overview for the speech recognition system

23.2.3 *Signal Framing and Trimming*

The recorded signal produces many data depending on the signal length and is recorded at a sampling frequency of 16 kHz for 5 s. The data values generated by the signal will show negative, positive and zero values. The problem is because the signal is recorded in continuous form. The data in the signal is generated by the way the user speaks and the sound of the microphone. During the recording process, the microphone will not only record the voice of the user, but it will also record all the sounds produced in the environment, including the noise generated by the user or electronic device. However, the system will filter out the noise and leave only the user's voice. During filtering, there is also a data value generated, and this value is smaller than the data value generated by the user's voice.

Data generated from sounds other than commands will cause system clutter and systems will be challenged to identify. Therefore, the voice signal will be framed from a long signal to the correct signal for the voice command. With this process, the actual signal can be generated, and it will also generate real data for each command in the signal. So, Fig. 23.2 shows the raw signal and Fig. 23.3 shows the trimmed signal.

23.3 Feature Extraction

Feature extraction is an essential step in the construction of any pattern classification and aims at the extraction of the relevant information that characterises each class.

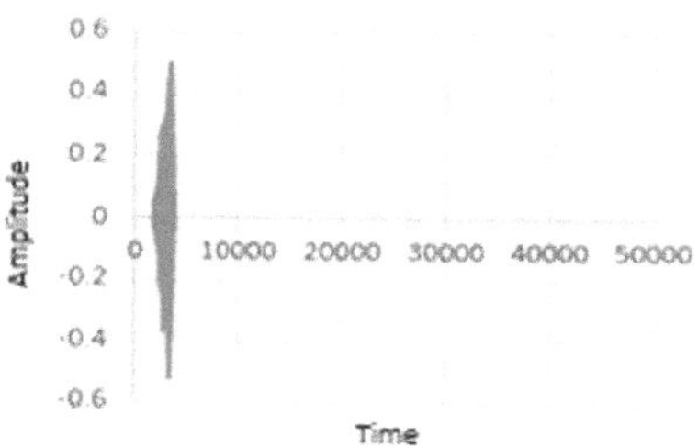

Fig. 23.2 A typical raw speech signal

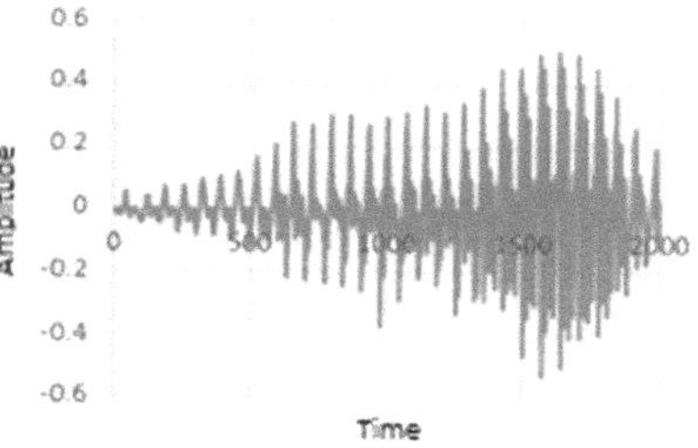

Fig. 23.3 A typical trimmed signal

Feature extraction is done after the pre-processing phase in the voice recognition system. Voice command signals that have been analysed and filtered will be extracted using features that have been programmed into the system. The primary task of pattern recognition is to take an input pattern and correctly assign it as one of the possible output classes.

23.3.1 Power Spectral Analysis

The time-domain voice signal recorded by the microphone was converted into the frequency domain using spectral analysis. Spectral analysis was used to measure the magnitude and frequency of signals as shown in Fig. 23.4. The measurements in the frequency domain are showing the value of energy available at each given frequency. Some measurements need to maintain the information about signal frequency, phase and amplitude.

The power spectral (PS) is defined as the square of the linear spectral magnitude:

$$\mathrm{PS} = (\mathrm{Mag})^2 \tag{1}$$

To relate the power in the time domain to the power in the frequency domain, the number for frequency window measurements need to be calculated according to the following integral formula:

$$\int |f(t)|^2 \mathrm{d}t = \left[N_T \cdot \frac{NF}{(\Delta F)^2} \right] \cdot \int |F(f)|^2 \mathrm{d}f \tag{2}$$

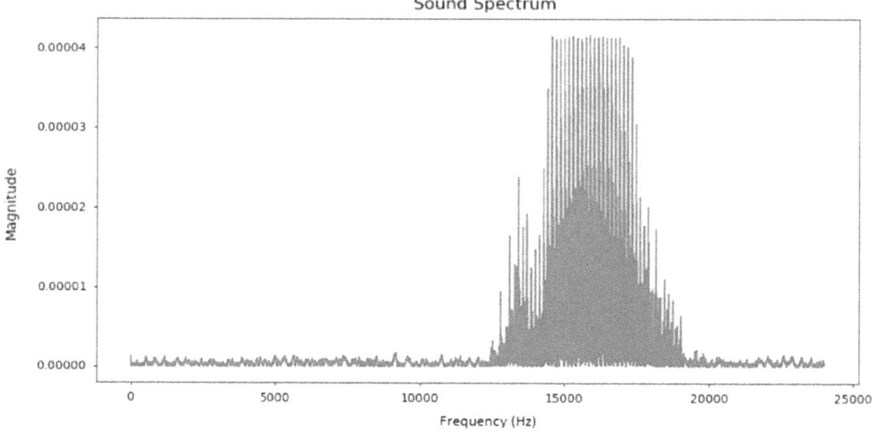

Fig. 23.4 Power spectral analysis

where N_T is number of samples selected in the time domain, N_F is total number of samples in the frequency domain and ΔF is total frequency range in the frequency domain.

23.3.2 *Fast Fourier Transform*

By using the fast Fourier transform (FFT), the frequency components of a signal buried in a noisy time-domain signal can be found easily. It is an efficient algorithm for computing discrete Fourier transforms, and it was useful in solving problems in various applications. Fourier transform inverse fast is to convert signals from the frequency domain to the time domain. Non-periodic FFT signals will cause the resulting frequency spectrum to leak. Originally provided some windows to implement FFT to prevent leaks. The frequency-domain signal will be converted into the time domain after analysis. The Fourier transform will break the time signal and will return information about the frequency of all sine waves needed to simulate the time signal (Vink 2017). For the same sequence of distance values, the discrete Fourier transform (DFT) is defined as:

$$X_k = \sum_{n=0}^{N-1} x_n \cdot e^{-\frac{i2\pi kn}{N}} \tag{3}$$

where N is number of samples, n is current sample, X_n is value of the signal at time n, k is current frequency (0 Hz to N − 1 Hz) and X_k is result of DFT.

23.3.3 *Classification*

In the classification process, the system uses the energy and average values contained in the recorded voice signal data. The process also classifies each voice signal received into five directions such as front, right, left, back and stop. When a voice command is delivered via a microphone, the system displays the programmed directions in the Python software. The system will also display different directions or words if the system cannot recognise the given voice. To determine the value of energy contained in a sound signal, a sum formula is used.

$$E = \text{np.sum}(x * *2) \tag{4}$$

where E is energy and x is voice signal data.

Voice signals are computed using different average values for each signal. Hence, each programmed command can be classified using only the average value. The formula for determining the average value used in the python software is as follows.

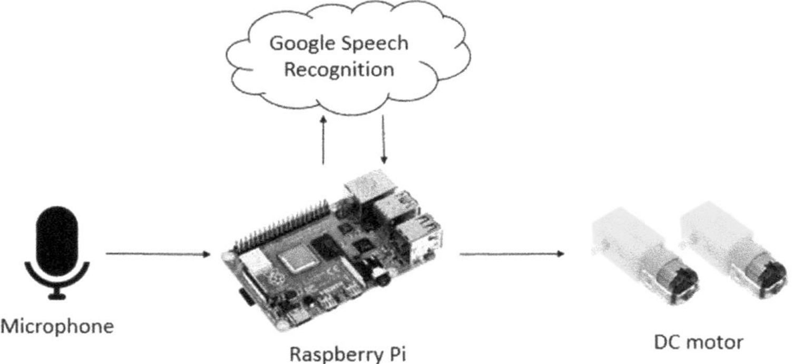

Fig. 23.5 Block diagram of the Google speech recognition

23.3.4 Recognition

The last process in this voice command system is the recognition process as shown in Fig. 23.5. With this process, the system will recognise the recorded sound to control DC motors. With this process, the system can determine the instructions received to control the application properly. This recognition process requires an internet connection to work. If the system is turned on without an internet connection, the recognition will not be possible, and the system will display an error. This is because every voice signal data will be sent to Google Speech Recognition to convert voice signals to text (Jordan 2020). However, the system will display the wrong word if it does not recognise the word provided by the user. If the text that Google recognises matches the programmed instructions, then the DC motor will move following the instructions.

23.3.5 Hardware Interface

As an implementation, an android application had been developed by using MIT App Inventor. The primary function of this application is to monitor the user. In this application, the average of drowsiness and the duration of eye closured are displayed. On behalf of the user, they can log into their account to supervise their driving behaviour.

Figure 23.6 shows the overall interface of the components used in this system. The hardware implementation used in this system has been successfully installed and works perfectly. Each interface for hardware used in systems has its role. The Raspberry Pi acts as a data processing centre and sends signals to the motor driver. Therefore, the Raspberry Pi processes the data received from the microphone and sends it to Google speech recognition to recognise instructions for moving the application. The motor driver works to distribute power and instructions to the DC motor

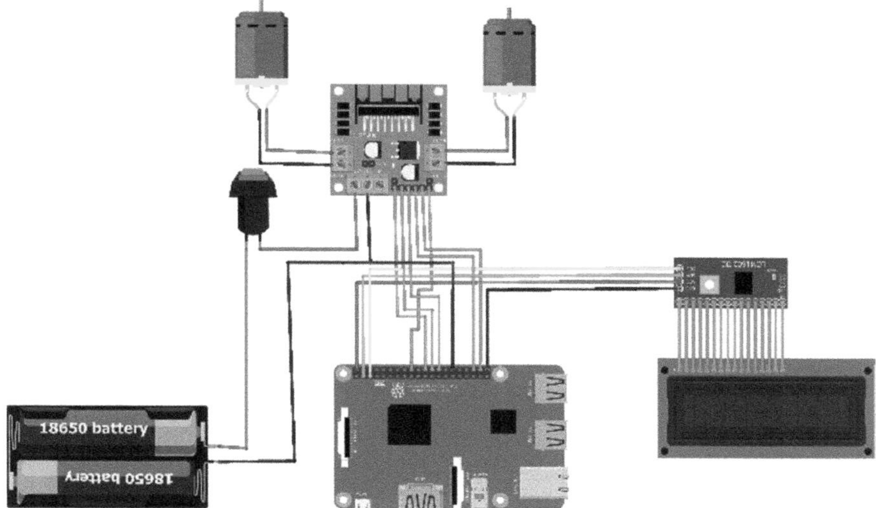

Fig. 23.6 Hardware connection diagram

to move. When the motor driver receives a signal from the raspberry, it will control the direction of movement for the motor. DC motor, as an output application moves in a specific direction and stops when receiving a signal. 16 × 2 LCDs serve as informants to users about the processes that are being performed in this system. The LCD will tell the user to start voice recording, recognition time and text that can be converted from the voice signal data. DC motors and motor driver are supplied with 12 V by 18,650 Li-Ion batteries. Besides, the switch is used in this system to cut and connect power from the battery to the components used.

23.4 Results and Discussion

23.4.1 Analysis of Speech Recognition

Speech recognition was analysed using different voices where the first voice is the owner's voice and the second voice is another individual voice. Feedback of motor movements after instruction was also analysed in this topic. With two different voices when directing through a microphone can be compared to assess the level of recognition of this system. It uses speech recognition to convert speech to text when using a microphone. It will provide easy packing for various popular public speech recognition APIs. The Raspberry Pi also needs to be connected to the Wi-Fi and use Google Speech Recognition to process the recognising commands from the user. In this process, the duration parameter is set in the recording function to stop reading after

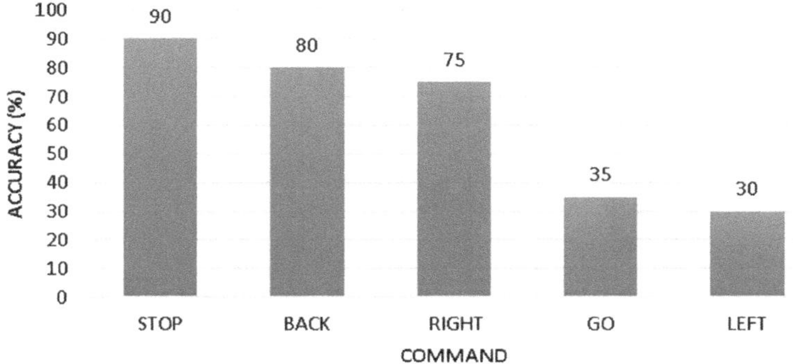

Fig. 23.7 Speech accuracy analysis

5 s. Then, the audio data will be uploaded to Google for output text. The responding times for recognising the command have been recorded by twenty times. The average response time recorded was 6.475 s, and most often recorded time in these trials was 6.6 s. It has a slight delay in recognising voice commands from users.

The bar graph shown in Fig. 23.7 shows the difference in accuracy by the recognition system to recognise different voice commands. It seems that when a voice command is given through a microphone, the accuracy of each type of command also varies depending on the way the user is speaking. This makes sense since each of these different expressions will produce different values of accuracy. Besides, the tone coming out of the user's mouth also plays a role in the accuracy of the recognition. Another thing to keep in mind is the grouping of graphs. The categories for the percentage in determining the direction of motor movement through speech recognition are not equal. Each of the first three bars represents the voice commands of stop, back and right to achieve more than 50% accuracy, but the next two groups drop below 50% making their values 35% and 30%, respectively.

The bar graph in Fig. 23.8 shows the percentage value in system accuracy to recognise what others are saying. An analysis is done using the voice of another person in giving voice commands to the system using a microphone. Voice commands recognised by the system were recorded for accuracy values in twenty trials. The result of this analysis is that the 'stop' voice commands provided by the user are of the highest value. Also, the instructions given by the user are of the lowest value in accuracy compared to other instructions. The accuracy value in the percentage of 'stop' command provided by the user is 75%. The lowest accuracy value for this voice is 25%. The second most common command detected by the system is the following command, where it achieves 65% accuracy. The third voice command most detected by the system is the right command. The accuracy value for these instructions is 45%. Voice commands detected by less accurate systems are left commands. The accuracy value for this command is 35%.

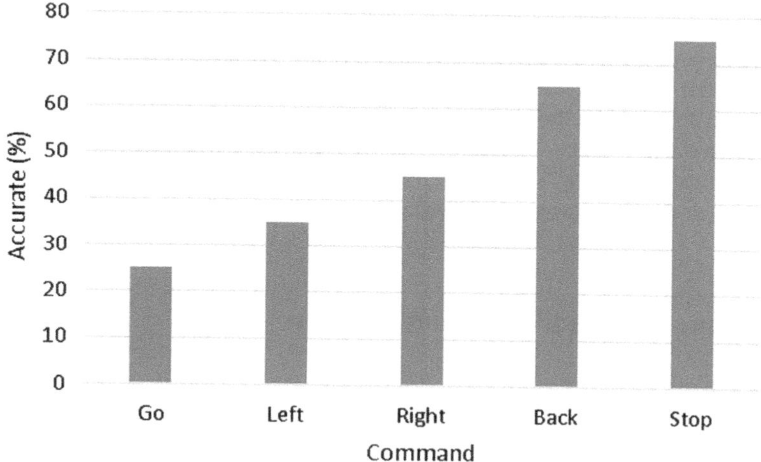

Fig. 23.8 Speech accuracy percentage analysis

23.4.2 Comparison of Speech Accuracy Analysis

According to the bar chart in Fig. 23.9, the voice recognition system can detect the first voice more accurately than any other voice. This is because the language and voice styles used are clear and easy to recognise by the system. The system can more accurately recognise other voices in the left direction than the first voice. For the left word, the first voice is less clear than the other voice. Therefore, the system cannot recognise the first voice more accurately than the other voices. Only three commands are most accurate and detected by this system, namely stop, back and

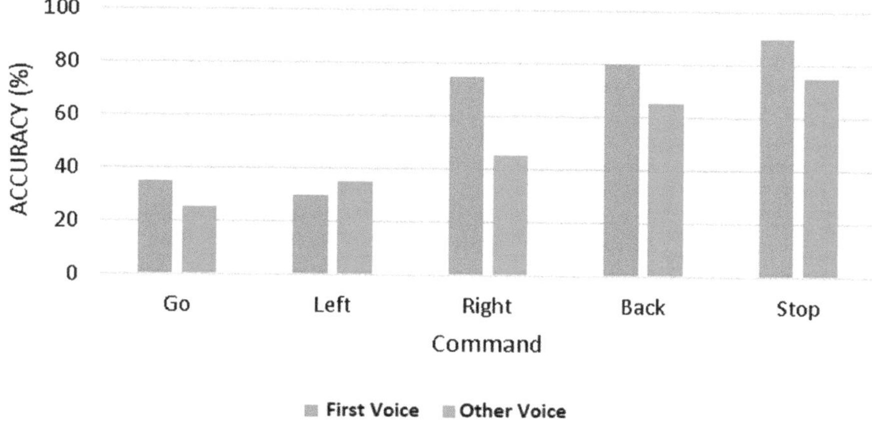

Fig. 23.9 Comparison of speech accuracy analysis

right. The other two directions of go and left are very inaccurate and less detectable when voice commands are given via a microphone. The graph bar indicates that the system cannot accurately identify both words because the pronunciation and word styles are different from those already programmed in Google Speech Recognition. However, these five voice commands can be traced and translated from speech to text. DC motor also works well when the text recognised by the recognition system is the same as that programmed in Python.

23.5 Conclusion and Recommendation

This work shows that an automated voice control system can be built using Raspberry Pi which is relatively cheap and a widely available single board computer. The main contribution of this research provides an automatic control system and voice command operation. The system offers a whole new experience in people's lives. The user will need to use a microphone connected to the Raspberry Pi to give instructions. This system is widely tested with different sounds to determine the accuracy of the system in the process of recognising voice signals. The prototype works well and runs smoothly with the programmed correctly. The project can be improved with external devices like android to give voice commands to Raspberry via Bluetooth or Wi-Fi. However, for a superior quality, a robust microphone with noise cancellation can be used.

References

Ellaway, A., Macintyre, S., Hiscock, R., Kearns, A.: In the driving seat: psychosocial benefits from private motor vehicle transport compared to public transport. Transp. Res. Part F Traffic Psychol. Behav. **6**(3), 217–231 (2003)

Hidayat, S., Firmanda, S.F.: Scheduler and voice recognition on home automation control system. In: 2015 3rd International Conference on Information and Communication Technology (ICoICT), pp. 150–155 (2015)

Jirafe, N.: How to filter noise with a low pass filter - Python. Analytics Vidhya (2019). https://medium.com/analytics-vidhya/how-to-filter-noise-with-a-low-pass-filter-python-885223e5e9b7. Accessed 4 April 2020

Jordan, K.: (2020) How to convert speech to text in Python. About Hacker Noon. https://www.thepythoncode.com/article/using-speech-recognition-to-convert-speech-to-text-python. Accessed 10 Oct 2020

Mahesh, S.: Voice recognition using Raspberry Pi 3. Rhydolabz-wiki (2016). https://www.rhydolabz.com/wiki/?p=16234. Accessed 4 April 2020

Ni, R.: Driving under reduced visibility conditions for older adults. In: Meyers, R.A. (ed.) Encyclopedia of Sustainability Science and Technology, pp. 3135–3150. Springer, New York (2012). https://doi.org/10.1007/978-1-4419-0851-3_476

Peng, C., Chen, R.: Voice recognition by Google Home and Raspberry Pi for smart socket control. In: 2018 Tenth International Conference on Advanced Computational Intelligence (ICACI), pp. 324–329 (2018)

Sharifuddin, M.S.I., Nordin, S., Ali, A.M.: Voice control intelligent wheelchair movement using CNNs. In: 2019 1st International Conference on Artificial Intelligence and Data Sciences (AiDAS), pp. 40–43 (2019)

van der Velde, N.: How Does Speech Recognition Technology Work? Summa Linguae Technologies. (2019). https://www.globalme.net/blog/how-does-speech-recognition-technology-work. Accessed 4 April 2020

Vink, R.: Understanding the fourier transform by example. Ritchie Vink Blog (2017). https://www.ritchievink.com/blog/2017/04/23/understanding-the-fourier-transform-by-example/. Accessed 21 May 2020

Yue, C.Z., Ping, S.: Voice activated smart home design and implementation. In: 2017 2nd International Conference on Frontiers of Sensors Technologies (ICFST), pp. 489–492 (2017)

Chapter 24
Out-of-Plane Surface Deformation Measurement of Advanced High Strength Steel Using 3D DIC

Nurrasyidah Izzati Rohaizat, Christophe Pinna, Hassan Ghadbeigi, Dave N. Hanlon, Ishak A. Azid, and Sharmiwati M. Sharif

Abstract This paper presents the out-of-plane deformation analysis on a dual-phase steel with a UTS of 1000 MPa (DP1000) by using three-dimensional (3D) optical digital image correlation (DIC) technique on a laboratory scale punch test. The laboratory scale punch test being used in this study has been built to deform sheet steel specimens in a manner that resembles the forming process in automotive industry. The results obtained from conducting the punch test using three-dimensional (3D) optical digital image correlation on the investigated DP1000 materials showed that a uniform deformation takes place until a strain values of 10%. The deformation begins to localise around the sides of the punch as the maximum strain value at the area of necking reaches 19.7% just before fracture. DIC results also showed that the laboratory scale punch tests are equi-biaxial, where the strain distribution throughout the test are uniform. The mechanical response from the punch test shows that the punch test has good repeatability.

N. I. Rohaizat (✉) · I. A. Azid · S. M. Sharif
Universiti Kuala Lumpur, Malaysian Spanish Institute Kulim Hi-Tech Park, 09000 Kulim, Kedah, Malaysia
e-mail: nurrasyidah.izzati@unikl.edu.my

I. A. Azid
e-mail: ishak.abdulazid@unikl.edu.my

S. M. Sharif
e-mail: sharmiwati@unikl.edu.my

N. I. Rohaizat · C. Pinna · H. Ghadbeigi
Department of Mechanical Engineering, The University of Sheffield, Mappin Street, Sheffield S1 3JD, UK
e-mail: c.pinna@sheffield.ac.uk

H. Ghadbeigi
e-mail: h.ghadbeigi@sheffield.ac.uk

D. N. Hanlon
Tata Steel Research, Development and Technology, PO Box 10.000, 1970 CA Ijmuiden, The Netherlands
e-mail: dave.hanlon@tatasteel.com

M. H. Abu Bakar et al. (eds.), *Progress in Engineering Technology III*, Advanced Structured Materials 148, https://doi.org/10.1007/978-3-030-67750-3_24

Keywords Advanced high strength steels · Digital image correlation · Dual phase steels · Failure behaviour · Strain analysis

24.1 Introduction

The global demands for lighter and safer cars have led to great advancement in the automotive steel industries, especially in the development of advanced high strength steels (AHSS) (Balliger and Gladman 1981). A common type of AHSS is usually found in the automotive industry is dual-phase steel, which is known as or DP steel. DP steel exhibits high tensile strength and good ductility, which allow weight reduction achievable in automobiles components, thus becoming popular in the automotive industry (Sarwar et al. 1996). The excellent mechanical properties of DP steel can be attributed to the unique microstructure which consists of a hard martensite phase embedded in a soft ferrite phase. However, due to the presence of both ferrite and martensite phases, the damage initiation and propagation happening in DP steels is not well understood and needs to be investigated further.

Most procedures used to study the damage mechanisms in dual-phase steels have been done either through micro-tensile testing inside a SEM or through post-mortem investigation on broken specimens. Although tensile testing inside a SEM provides useful insight into the deformation of phases leading to damage initiation, the extent of damage at the surface of the specimen is very limited as most damage development takes place at the specimen sub-surface.

There is a need to develop an improved and controlled test procedure to allow crack propagation to be investigated. Moreover, the conditions for damage development should be more representative of industrial forming operations, i.e. involving biaxial or out-of-plane deformation of the specimen. In addition, the forming test allows for a higher degree of deformation of the specimen, thus promoting of damage nucleation.

The aim of this study is to develop a controlled experimental procedure using a laboratory scale punch test which resembles the forming process in automotive industry and to be combined with a 3D digital image correlation technique to allow deformation analysis on the DP steel studied.

24.2 Literature

In the attempt to understand the damage mechanisms and deformation of the dual-phase steels, the digital image correlation technique or also known as the DIC technique has gained a lot of attention from researchers especially to quantify displacements and deformations on deforming surfaces, hence allowing researchers to further implementing this technique from large scale to micro-scale and nano-scale mechanical testing applications (Kang et al. 2006; Ghadbeigi et al. 2010; Kapp et al. 2011).

In many studies which have conducted a micro-tensile test to study damage initiation in DP steel at the scale of microstructure have reported that damage begins in the mid-thickness (sub-surface) of a tensile specimens due to high stress triaxiality. An investigation using X-ray tomography during in situ (micro-scale) tensile test found that voids formation is heterogenous and the highest void fraction is found to be at the centre of the specimen. It is explained that the high density of void fraction at the location because of the outer shape of the specimen after necking, which induces a high stress triaxiality (Maire et al. 2008). A study involving the application of in situ tensile testing in a SEM on DP600 steel has successfully captured crack propagation during the onset of fracture by interrupting the micro-tensile test during necking. Crack development is found to initiate at the centre where the void volume fraction and stress triaxiality are the highest. In terms of crack propagation path, microstructural observations showed that the crack propagated mainly occurs in the ferrite phase with occasional deflection when martensite was in the way (Ghadbeigi et al. 2010).

Tensile testing inside a SEM provides useful insight into the deformation of phases leading to damage initiation, however the extent of damage at the surface of the specimen is very limited as most damage development takes place below the surface. Thus, in this paper this developed procedure will also ensure the enough deformation on the studied DP steel generated crack at the surface, hence allowing crack propagation to be studied at microstructural scale.

24.3 Methodology

24.3.1 Material

For this research, the material used for investigation is a dual-phase steel with a UTS of 1000 MPa. The DP1000 steel contain a uniform distribution of the ferrite and martensite phase. The investigated DP1000 steel shows a 50–50% composition of ferrite and martensite phase. The received materials are cold rolled DP1000 sheet steels with a production thickness of 1.6 mm. The mechanical properties and the chemical composition of the investigated DP1000 steels are shown in Tables 24.1 and 24.2, respectively.

The DP1000 steel sheets are then cut using an electrical discharged machine (EDM) into 90 mm diameter circular blanks specimens. The specimen diameter has been chosen so that the centre of the specimen, which will be in contact with punch, is not affected from the heat of the EDM cutting process.

Table 24.1 Mechanical properties of the investigated material

Material	Galvanization	$YS_{0.2\%}$ (MPa)	TS (MPa)	Gauge (mm)
DP1000	No	729	1051	1.6

Table 24.2 Chemical composition

Material	Galvanization	$YS_{0.2\%}$ (MPa)	TS (MPa)	Gauge (mm)
DP1000	No	729	1051	1.6

In order for the image correlation of the optical 3D DIC system to work, specimens are required to have a unique speckle pattern on the surface. Random speckling on specimen surface are done using matte black and white spray paints. A white matte layer serves as a contrasting background and random tiny black speckles pattern created using black spray paint is uniformly distributed on the entire specimen.

24.3.2 Punch Test and DIC Set up

The punch test rig consists of die sets to secure the 90 mm circular specimen blanks. The die sets have clearance in the centre to allow specimens to deform when loaded with a 20 mm diameter conical punch.

The punch test rig is set up together with DIC cameras on a 100 kN Electric Mayes machine as shown in Fig. 24.1. The general method of assembling the whole rig starts with the clamping of the blanks. The blank is secured in between the dies and tightened before being put on the punch. The die set is supported with a pair of stands from the bottom for a little clearance between the punch tip and the specimen to avoid the punch from damaging the specimen before the start of the experiment. Supporting bars are then placed on top of the die set to provide a hold down force

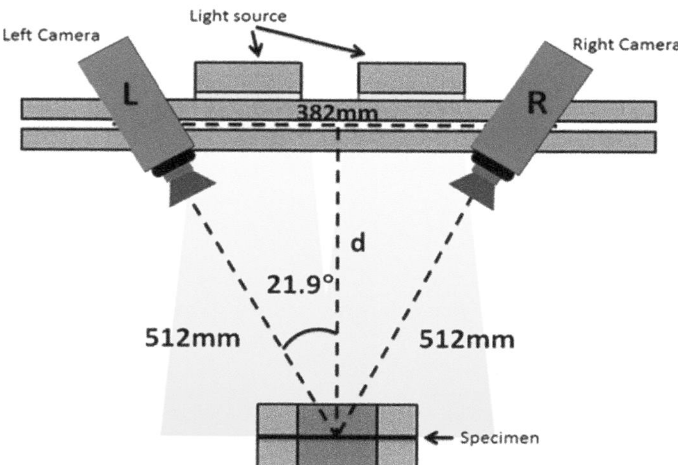

Fig. 24.1 3D DIC (stereo vision) cameras set up at a measured distance relative to the position of specimen

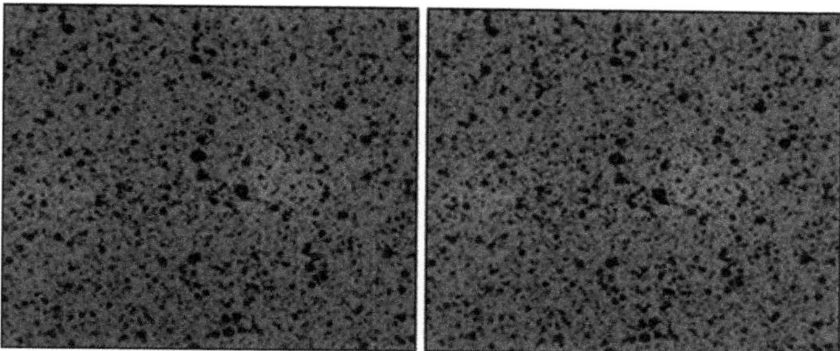

Fig. 24.2 Views from each cameras which should be focussed on the same location on the sample, preferably at the centre. (Left) view from camera 1 and (Right) view from camera 2. Together, the two images will be processed to enable stereo vision

during the test and is tightly bolted at the connection which secures the position of both top supporting bar and bottom supporting bar of the Mayes machine to secure the die set.

The 3D DIC set up is done by mounting cameras on the boom at the top of the test rig and connecting both cameras to the computer to view the area of interest on the specimen. On the computer, the software being used for the pattern recognition is based on the principles of digital image correlation is the VIC-3D by Correlated Solution. Every pair of images continuously captured by both cameras is processed into stereo images. Cameras orientation is adjusted until the area of interest is at the centre of camera views, then the best focal length is found by adjusting the lenses. Figure 24.1 shows the measurements of cameras set up relative to the specimen. Cameras are adjusted until they are in focus at the same location on the specimen preferably at the centre region. Examples of speckle images with a good focus are shown in Fig. 24.2.

Once both cameras are in focus, the calibration process is carried out by using a pre-determined dot-pattern calibration plate. At least 25 images are captured for the calibration process at which later only 15 images with good calibration score are chosen to make sure the calibration error between the two cameras is below 0.03. During the capturing of calibration images, the calibration plate is moved under the cameras view to produce images involving translation, rotation and in- and out-of-plane movements. Punch test can be started as soon as 3D DIC set up is ready.

24.3.3 Source of Errors in DIC Technique

It is important to ensure the reliability of the measured data when using the DIC system. Identifying the source of errors before running the DIC system can be useful to ensure the errors throughout using the DIC technique can be controlled from the beginning as setting up the system to the end when analysing DIC results.

In DIC, one of the error sources is classified as correlation. The correlation errors can be further divided into statistical and systematical errors. There are several influences from the hardware system and environment conditions such as noise from cameras, different illumination conditions for the cameras, photon shot noise, contrast in image intensity and stochastic pattern of the specimen surface. The main statistical errors are usually due to the noise in camera image. By using different types of cameras, the reduction of camera noise can greatly reduce the resulting correlation errors (Neumann and Krupka 2007).

Therefore, the setting up and calibration procedure should be carried out carefully to ensure the measurements obtained from DIC analysis are correct.

24.4 Result and Discussion

In this investigation, the received 1.5 mm thickness DP1000 steel sheets of different tempering conditions are machined into 90 mm diameter blanks and tested until the specimen fails. From running this punch test, it is expected to observe the crack formation on the surface of the specimen which later will be inspected in the mid-thickness to provide information on the damage mechanisms occurring in a condition similar to the forming process in industry. In conjunction with that, the effect of different heat treatments applied on the formability of the studied DP1000 steels can also be compared.

Three sets of 90 mm diameter specimens are produced to allow punch tests to be repeated. The first batch of DP steel samples is tested using the non-contacting optical method, 3D DIC coupled with the punch test to allow deformation and strains to be measured. The second and third batch are intended to validate the obtained results which will prove if the developed punch test has good repeatability and the collected data from the tests are statistically reliable.

24.4.1 Force-Displacement Response

In this experiment, the punch test rig along with two stereo vision DIC cameras are securely fixed on a MAYES 100 kN electric machine. For each test, the machine is set to load the specimens at constant displacement rate of 2.5 mm per minute in compression mode. Before testing begins, it is ensured that there is no contact between the specimen and the tip of the punch. The DIC system begins the stereo image capturing as soon as the Mayes machine is set to run. Data acquisition device of the machine will display the real-time force-displacement response of the loading state during the test. The test is stopped once the force response drops abruptly which can also be heard from the loud sound produced by the broken sample. Figure 24.3 shows the crack appearing on the through the specimen thickness after loading the specimen until fracture.

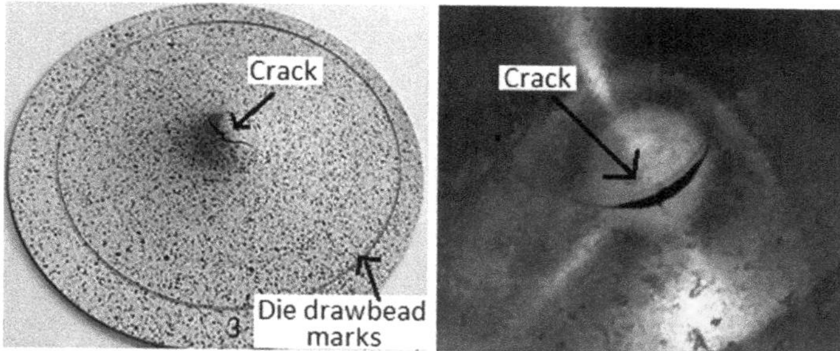

Fig. 24.3 Examples of broken punch test specimens. The broken specimen is coated in paint for DIC (left) and the revealed crack after the speckled coating is removed from the specimen surface (right)

Force-displacement data from all three specimens of the punch test have been successfully collected during the test and are shown in Fig. 24.4. Data from each tests are labelled as aq-1, aq-2 and aq-3 as shown in the force-displacement curves. In these plots, the displacement is re-aligned to 0, when the load is at 1 kN. This is because it is difficult to identify the start point of contact between punch and specimen due to low noise in machine reading below 1 kN. The increment in load reading becomes stable as it gets close to 1 kN for all tests.

Figure 24.4 shows that the result for every tested DP1000 specimen steel shows very close similarity which suggest that the designed punch test has good repeatability from the produced results. The maximum values for both force and displacement from plots in Fig. 24.4 for all three specimens are presented in Table 24.3. Mean values are calculated based on the data obtained from the three batches.

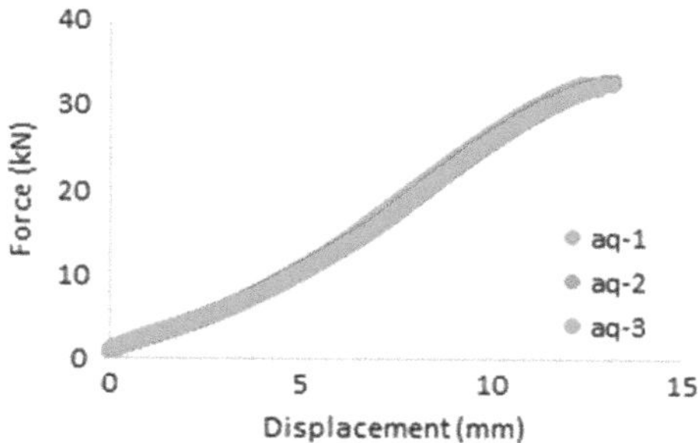

Fig. 24.4 Force-displacement response obtained from conducting punch test plotted according the tempering conditions

Table 24.3 Force (F) and out-of-plane displacement values before fracture of specimens

aq-1		aq-2		aq-3		Mean, μ	
F_{max} (kN)	W_{max} (mm)	F_{max} (kN)	W_{max} (mm)	F_{max} (kN)	W_{max} (mm)	F_{max} (kN)	W_{max} (mm)
32.8	13.1	32.8	12.8	33.1	13.2	32.91	13.03

F_{max}: Maximum force, W_{max}: Maximum out-of-plane displacement

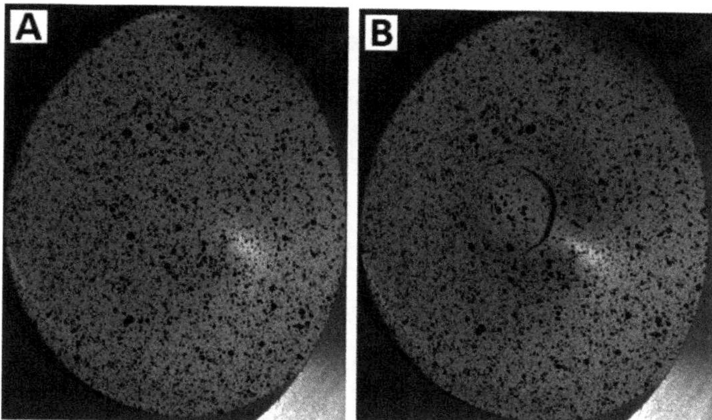

Fig. 24.5 Example of punch test specimen before (left) and after (right) testing

24.4.2 3D Strain Distribution in Punch Test

The optical 3D DIC method being used alongside in this test provides further information on the out of plane deformation happening on the surface of the specimen. After the test, the captured images from the DIC device are processed in order to provide information on the strain localisation and deformation of the specimen.

For the optical 3D DIC method to work, two cameras are attached to the punch test rig and are focussed on the centre of a well-lit specimen. Hence, allowing the 3D DIC system to analyse out-of-plane deformation of the specimen. The image captured by the DIC system is shown in Fig. 24.5, where image (A) is the undeformed specimen taken before test begin and image (B) is the failed specimen taken after the test is stopped.

Once experiments are completed, the captured images are processed using a DIC commercial software, Correlated Solutions VIC-3D to extract deformation and strain results. The full deformation development happening when the punch test took place can be observed in the series of images in Fig. 24.6.

From the DIC, deformation of tested specimens can be analysed in terms of their strain distributions. Referring to Fig. 24.6, the series of four images (1–4) shows the strain development in the as-quenched specimen from the beginning of the test to the maximum loading state before the specimen failed. In image (1), the specimen is still intact and there is no contact yet between the punch and specimen. Hence, result from the strain plot before punch contact is the noise from the correlation with a maximum value of 0.03%. From image (2) to (3), the strain begins to grow most noticeably at the centre region from 1.72% to 10.80%. Approaching the final stage of test before the specimen failed as shown in image (4), deformation in the sample seemed to localise around the outside of the centre region as indicated by the arrow with a high strain concentration with an average value of 19.7%. A similar pattern of result is observed in all other specimens where the strain first accumulates at the

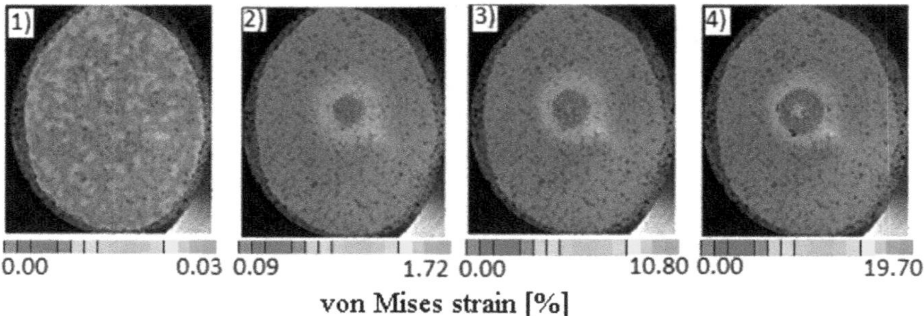

Fig. 24.6 Plots (1)–(4) showing the increment in von Mises strain distribution of as-quenched specimen being loaded during the punch test. High strain of localised necking is as indicated in (4)

centre and later with increasing loading, strain increment begins to localise around the outside of centre region (resulting from the large surface contact between punch and specimen).

In results shown in Fig. 24.6 (4), it is clearly seen that specimens are close to the equi-biaxial stretch state. This is shown by the red contour around the necking area which shows uniform straining at around 19.7%. Findings are also in agreement with results reported by Chen et al. (2012).

24.5 Conclusions

The punch test using 3D DIC was capable to exert out-of-plane deformation with strain values above 50% on the studied DP1000 steels. The deformation achieved using the punch test is higher compared to the maximum deformation achieved from other methods such as a standard tensile test. The 3D DIC technique proved to be very useful in providing information surface strain measurement. The use of 3D DIC technique when running the punch test allows the strain to be measured at the instance of crack propagating in the punch test.

Acknowledgements Author would like to thank Tata Steel, Ijmuiden for providing the research materials. Special thanks to The University of Sheffield for allowing this project to happen. This project will not be possible without the sponsorship from Majlis Amanah Rakyat (MARA) and Universiti Kuala Lumpur, Malaysia.

References

Balliger, N.K., Gladman, T.: Work hardening of dual-phase steels. Met Sci. **15**(3), 95–108 (1981)

Chen, X., Xie, X., Sun, J., Yang, L.: Full field strain measurement of punch-stretch tests using digital image correlation. SAE Int. J. Mater. Manuf. **5**(2), 345–351 (2012)

Ghadbeigi, H., Pinna, C., Celotto, S., Yates, J.R.: Strain evolution measurement at the microscale of a dual phase steel using digital image correlation. Appl. Mech. Mater. **24–25**, 201–206 (2010a). https://doi.org/10.4028/www.scientific.net/AMM.24-25.201

Ghadbeigi, H., Pinna, C., Celotto, S., Yates, J.R.: Local plastic strain evolution in a high strength dual-phase steel. Mater. Sci. Eng., A **527**(18–19), 5026–5032 (2010b). https://doi.org/10.1016/j.msea.2010.04.052

Kang, J., Jain, M., Wilkinson, D.S., Embury, J.D.: Microscopic strain mapping using scanning electron microscopy topography image correlation at large strain. J. Strain Anal. Eng. Des. **40**(6), 559–570 (2006). https://doi.org/10.1243/030932405X16151

Kapp, M., Hebesberger, T., Kolednik, O.: A micro-level strain analysis of a high-strength dual-phase steel _Kapp(2011).pdf. Int. J. Mat. Res. **102**(06), 687–691 (2011)

Maire, E., Bouaziz, O., Di Michiel, M., Verdu, C.: Initiation and growth of damage in a dual-phase steel observed by X-ray microtomography. Acta Mater. **56**, 4954–4964 (2008). https://doi.org/10.1016/j.actamat.2008.06.015

Neumann, I., Krupka, R.: Error estimations in digital image correlation technique Thorsten Siebert 1a, Thomas Becker, Karsten Spiltthof. Adv. Exp. Mech. V. 265 (Published online 2007)

Sarwar, M., Priestner, R., Materials, M., Centre, S.: Influence of ferrite-martensite microstructural morphology on tensile properties of dual-phase steel. J. Mater. Sci. **31**, 2091–2095 (1996)

Chapter 25
Small Scale Prototype Development of Vehicle to Vehicle (V2V) Communication Using the Light Fidelity (LiFi) Technology

Norzalina Othman, Lutfil Hadi Ideris, and Zainal Abidin Che Hassan

Abstract Vehicle to vehicle, also known as a V2V communication using the light fidelity (LiFi) technology is about transmitting and receiving data wirelessly. The data is transmitted by modulating the light intensity given by a light source of the front vehicle, for instance from the vehicle's tail lamp. The development of the V2V communication prototype is a proposed method on transmitting the data via optical communication in order to prevent the collision between two vehicles: front and rear vehicle. In this project, the LiFi transmitter has been developed which consist of the light emitting diode (LED) that is used as a source for data transmission and the LiFi receiver that was designed with the photodiode to receive the data. For this time, the experiments have been done in order to study the transmission rate of single character with multiple characters. The colors of the transmitter also effects the obtained result. The proposed communication system based on LiFi for vehicle to vehicle (V2V) is a cost-effective solution with high data rate capabilities.

Keywords Vehicle to vehicle (V2V) · Light fidelity (LiFi) · Vehicular communication

25.1 Introduction

The field of science and technology is moving swiftly toward its progress. The human being uses this technological change for the saving of time and comfort. Development ranges from wired to wireless communication; and human beings have taken one step upstairs in wireless communication and developed the light fidelity technology, LiFi.

N. Othman (✉) · L. H. Ideris · Z. A. C. Hassan
Universiti Kuala Lumpur, Malaysian Spanish Institute Kulim Hi-Tech Park, 09000 Kulim, Kedah, Malaysia
e-mail: norzalina@unikl.edu.my

L. H. Ideris
e-mail: lutfil.ideris17@s.unikl.edu.my

Z. A. C. Hassan
e-mail: zainalabidin14@s.unikl.edu.my

M. H. Abu Bakar et al. (eds.), *Progress in Engineering Technology III*,
Advanced Structured Materials 148, https://doi.org/10.1007/978-3-030-67750-3_25

Harald Hass is known as LiFi's father from Edinburgh University who in his TED talks (Sarkar et al. 2015) told of the existence of this technology. The essence of this technology, according to Hass, is the strength and ability of diodes emitting light. In the near future, LiFi is an emerging technology that uses visible light bandwidth for data transmission and is 10,000 times greater than the band used in Wi-Fi technology (Sarkar et al. 2015). Therefore, when vast numbers of users demand Wi-Fi, RF bandwidth is increasingly being used, resulting in a clogged signal. Usage of the LiFi system is the answer. The idea is to use light bulbs as a source for data transmission at our homes as shown in Fig. 25.1.

Road collisions lead to loss of human lives. Such deaths are due to car accidents. Studies show that most collisions are due to the vehicle following being unaware of the vehicle's actions ahead. Based on the Malaysian Institute of Road Safety Research (MIROS), almost 81% the road accidents happen are caused by the human behavior, 13% is because of road condition and only 6% is caused of vehicle problem. The most common accident that happens in Malaysia is the rear end accident.

If the vehicle ahead may interact with the rear vehicle, as seen in Fig. 25.2, the collision can be avoided. There are several techniques to incorporate such connectivity technology, i.e., 5.9 GHz dedicated short range connectivity (DSRC) wireless in which two vehicles can communicate at the 5.9 GHz frequency and the vehicular

Fig. 25.1 LiFi technology

Fig. 25.2 V2V communication

ad hoc network which is the implementation of MANETs in which two vehicles can communicate by wireless fidelity (Yang et al. 2004). Using LiFi is intended to introduce a network which is cost-effective and has a high data rate. Because high-intensity LED lights are already present in automobiles these lights can be used as LiFi transmitters. In vehicles using LiFi technology, the collision can be avoided by adding only cheap circuitry.

25.2 Literature

The visible light communication protocol applied to the V2V (vehicle to vehicle) network was developed by Ferraz and Santos (2015). The VLC makes low cost use of the white LED. LED has less energy required so replaced with tail and headlights and reduces the thermal damage, LED is preferred as high light due to its high bandwidth and immunity to electromagnetic source interference. The technique of Manchester Coding is used for transmission, because it is safe and reliable. The distance from the vehicles was 20 m.

Ergul et al. (2015) presented the VLC and its challenges. Therefore, VLC has a large bandwidth and a high data carrying capacity thousand times greater than the radio frequency spectrum. The paper stated that the device is effective in carrying up to 300 Mbps of data rate within 25 ft range.

Using visible light communication, Kim et al. (2012) examined the outdoor environmental situation faced by vehicle to vehicle contact. Head light and rear light used for saturation of the transmitter and a picture diode was used for receiver of the light signal. In daytime outdoor environmental conditions the total distance is 20 m range. Outdoor communication problems faced during observation, such as sunlight glare, distortion of the photo diode; increase contact range during the day. In this way, transmitter and receiver are implemented with filter design, error correcting and improving signal power.

Nachimuthu et al. introduced the design methodology of vehicle to vehicle contact using VLC (Abdulsalam et al. 2015). The transmitter is mounted on the front vehicle and a receiver on the rear vehicle is attached. The data received may be helpful in taking further action such as tracking the vehicle speed or preventing collision. The system consists of two main transmitter sections, and the receiver. Modulation is used to modulate the input signal and data is transmitted in 0's and 1's type (bulb flashes, i.e., on and off).

25.3 Methodology

The system comprises of a transmitter section and a receiver section as shown in Fig. 25.3.

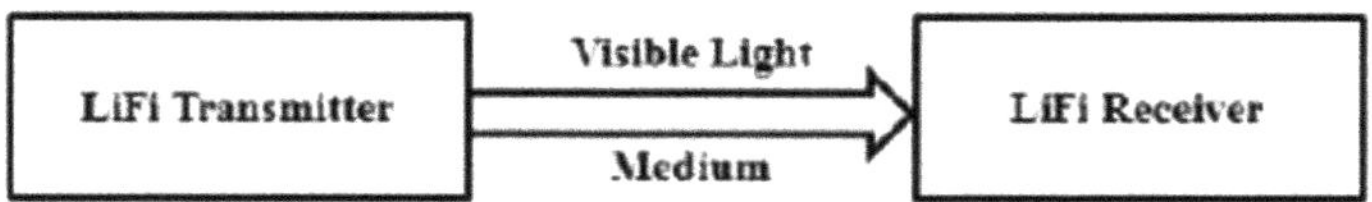

Fig. 25.3 Block diagram of LiFi communication system

Vehicle A's front lights act as transmitter at behind spot and send pulses of 0's and 1's. The flickering of LEDs should be done very quickly, so that the human eye cannot visualize it. The photodiode in the rear of vehicle B receives the data transmitted in current form at the front position. The system applies to scenario when vehicle A is within 5 m, front light transmits speed data to vehicle B to be processed and estimate the distance. The transmitter and receiver block are diagram shown in Figs. 25.4 and 25.5.

For experiment sake, this project is conducted to find the time elapsed between the data during transmission. Serial print is used to define what kind of data will be sent. The data is processed by Arduino ATMEGA and is sent to the LED driver which provides a constant current to the LED. Serial print is a part of Arduino Window which can display when the data is transferred. The transmitted data from the LED is received by the photodiode at receiver section in the form of a current pulse. The received pulse is undetectable and very small. On detection, the received data is displayed on the Arduino window.

The transmitter design in the Fritzing software is shown in Fig. 25.6. The cable/USB is connected to the Arduino when the data is confirmed and be entered then the LED will be on and transmits the signal of single-character data. The purpose of the final transmitter circuit is to test the circuit design output before implementing in a real design.

The final receiver design in the Fritzing software is shown in Fig. 25.7. The character is shown by the Arduino serial monitor with time impulse. When the photodiode detects the photon, it produces the pulses of the current. It is tested using an input square wave supposedly.

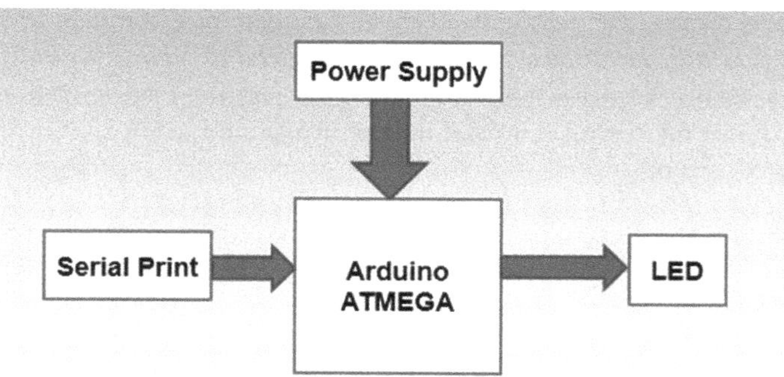

Fig. 25.4 Block diagram of transmitter section

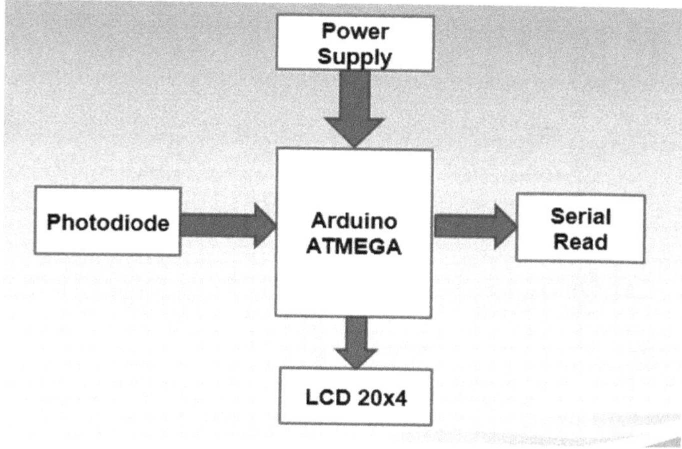

Fig. 25.5 Block diagram of receiver section

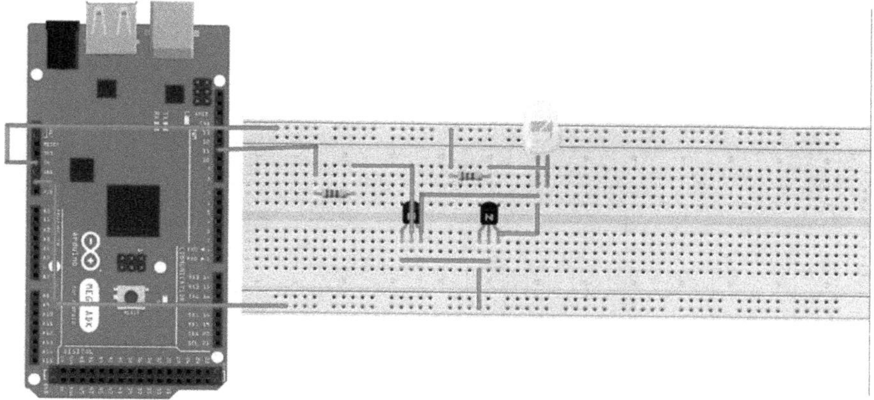

Fig. 25.6 Transmitter circuit design in Fritzing

25.4 Result and Discussion

This section is an elaboration on the result and the discussion of the analysis. In each experimental analysis, all the analysis transmission is being run with a fast flick of the transmitter with a certain data (character), and the data is given into 2 kind of methods which are 2 data at the same time and 3 data at the same time. Then by using 2 type of transmitter sources, 2 of them are assigned with a parameter which is the distance from 30, 25, 20, 15, 10 and 5 cm for comparison. The data have been obtained from the Arduino serial monitor as shown in Fig. 25.8. The data at Arduino serial monitor were exported into a Microsoft excel data sheet to do a further analysis.

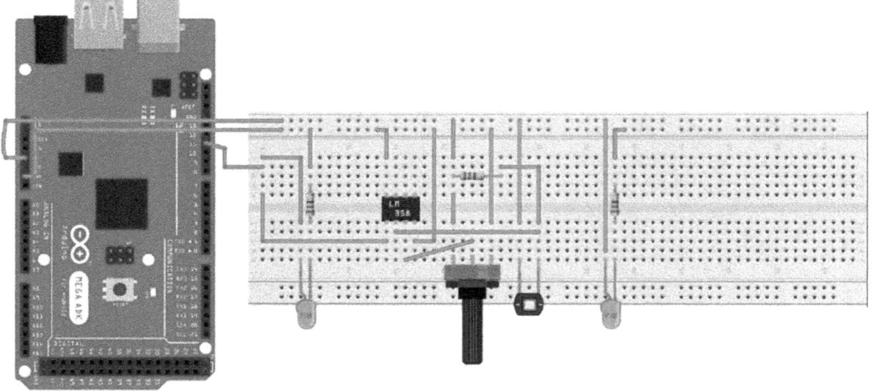

Fig. 25.7 Receiver circuit design in Fritzing

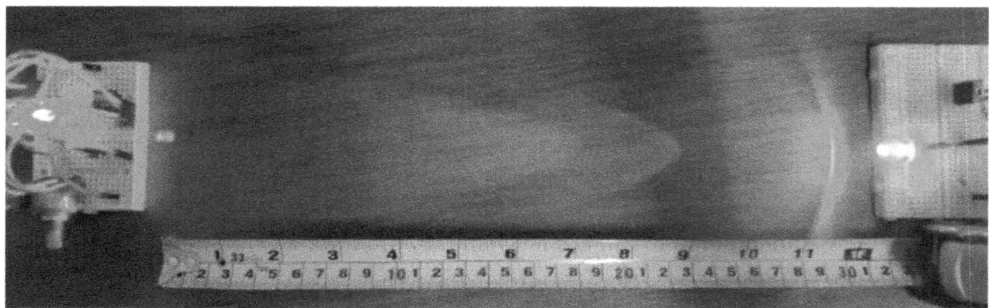

Fig. 25.8 Data transmission of LiFi system in certain distance

Figure 25.9 is the result of the data that have been extracted from the spreadsheet and plotted on the graph for data analysis.

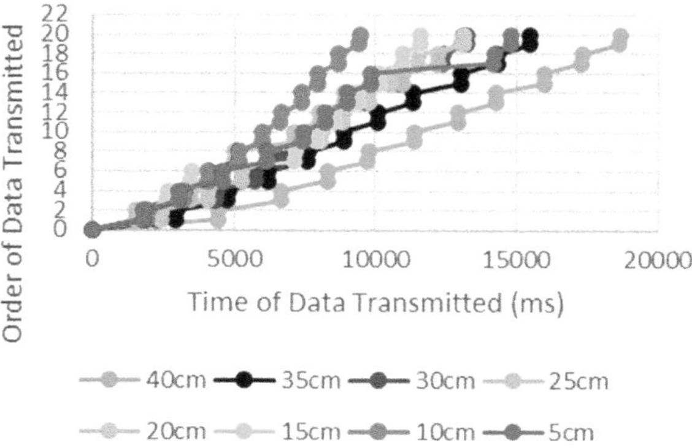

Fig. 25.9 White receiver of 2 input at same time

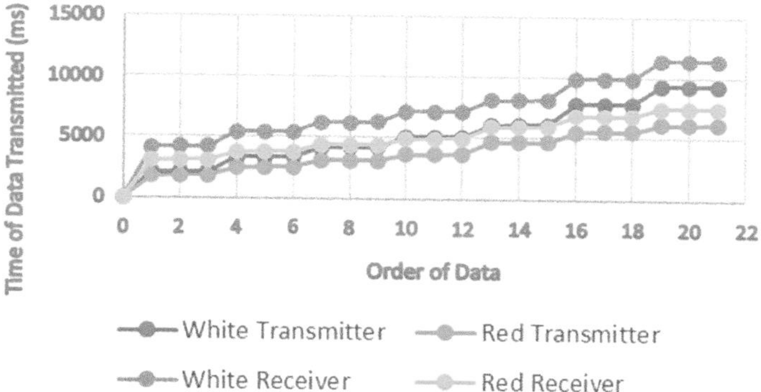

Fig. 25.10 Distance comparison of 3 data

Figure 25.10 displays the result of the extracted data from the table data sheet. The aim of this constructed graph is to study the gap of the time when the info or data that will be transmit to another party. Besides, to investigate the different color of the transmitter if there will be any clear difference. From that both graphs, the pattern between transmitter and receiver are very smooth and almost accurate based on the ordered data given. For example, the gradient between data 1 and 2 are very small, while the gap between data 2 and 3 are ramping in a short time because the data that being transmitted is not given continuously due to voltage fluctuating at LED and cause of the very fast flick in the LED itself. The pattern of the graph is quite similar even the source of the transmitter is of different color.

25.5 Conclusions

As a conclusion, the result obtained is achieving the necessity that should have in the transmission communication and the data that were sent by the transmitter to receiver had no data loss. Based on the result, the delay that happened during the data are being transmitted is in range of 0.011–0.014 s. As from result and discussion, the most suitable and recommended type is the white LED. The data shows that it has slightly higher accuracy than the red and the data will be more accurate when the distance is getting shorter. As for the recommendation, it is recommended for a further analysis to increase the value of data that can be given in the same time by do a study about LED driver to prevent voltage fluctuation. In addition, by increasing the voltage of the transmitter (LED) for getting a better and accurate data in a short time delay. An advanced array coding must be applied in order to prevent the data losses while processing the transmission.

References

Abdulsalam, N.A., Hajri, R.A., Abri, Z.A., Lawati, Z.A., Bait-Suwailam, M.M.: Design and implementation of a vehicle to vehicle communication system using Li-Fi technology. In: 2015 International Conference on Information and Communication Technology Research (ICTRC), pp. 136–139 (2015)

Ergul, O., Dinc, E., Akan, O.B.: Communicate to illuminate: State-of-the-art and research challenges for visible light communications. Phys. Commun. **17**, 72–85 (2015)

Ferraz, P.A.P., Santos, I.S.: Visible light communication applied on vehicle-to-vehicle networks. In: 2015 International Conference on Mechatronics, Electronics and Automotive Engineering (ICMEAE), pp. 231–235

Jamali, A.A., Rathi, M.K., Memon, A.H., Das, B., Ghanshamdas, Shabeena, : Collision avoidance between vehicles through LiFi based communication system. IJCSNS **18**(12), 81–87 (2018)

Kim, D., Yang, S., Kim, H., Son, Y., Han, S.: Outdoor visible light communication for inter-vehicle communication using controller area network. In: 2012 Fourth International Conference on Communications and Electronics (ICCE), pp. 31–34 (2012)

Sarkar, A., Agarwal, S., Nath, A.: Li-Fi technology: Data transmission through visible light. IJARCSMS **3**(6), 1–10 (2015)

Yang, X., Liu, L., Vaidya, N.H., Zhao, F.: A vehicle-to-vehicle communication protocol for cooperative collision warning. In: The First Annual International Conference on Mobile and Ubiquitous Systems: Networking and Services. MOBIQUITOUS, pp. 114–123 (2004)

Chapter 26
Design of an Electromagnetic Vibration Energy Harvester Using the Vehicle Suspension

Fazidah Saad, Muhammad Najib Abdul Hamid, and Ahmad Nabil Khan Ahmad Zairini

Abstract The vibration energy dissipated from the oscillation and damping of the vehicle suspension system is usually wasted. This vibration energy can be converted into voltage output to power up other suitable electrical components in a vehicle. The wasted vibration energy can be harvest by using many types of energy harvester, and the electromagnetic energy harvester is one of the interesting harvesters to explore. The electromagnetic energy harvester is a system that captures the wasted energy from the ambient source and converts them into other usable and useful energy. The main objective of this paper is to design an study of electromagnetic vibration energy harvester using the vehicle suspension. The main idea is to design an electromagnetic energy harvester that can fit the vehicle suspension and harvest the maximum power output through the vibration of the suspension. The components of a basic electromagnetic energy harvester include the coil, magnets, wires, and the mounting case to keep the components together that can be attached to the vehicle suspension. The design is to be studied based on two proposed designs of the electromagnetic vibration energy harvester. Each design has a different mechanism and parameters set up that resulted in varying voltage output values. The electromagnetic energy harvester design was simulated using the input frequency of 50 Hz with magnet permeability, and the voltage output value discussed. The design has the highest collected simulation data of voltage output that can be used as a guideline for future research.

Keywords Electromagnetic energy harvester · Vibration energy · Vehicle suspension

F. Saad (✉) · M. N. Abdul Hamid · A. N. K. Ahmad Zairini
Universiti Kuala Lumpur, Malaysian Spanish Institute Kulim Hi-Tech Park, 09000 Kulim, Kedah, Malaysia
e-mail: fazidah@unikl.edu.my

M. N. Abdul Hamid
e-mail: mnajib@unikl.edu.my

A. N. K. Ahmad Zairini
e-mail: nabil.zairini@s.unikl.edu.my

M. H. Abu Bakar et al. (eds.), *Progress in Engineering Technology III*,
Advanced Structured Materials 148, https://doi.org/10.1007/978-3-030-67750-3_26

26.1 Introduction

The automotive industry and technology have evolved since the beginning of the automobile invention over the past centuries. Over time, many technological inventions and advancements were developed to ensure vehicles achieved a progressive efficiency in terms of energy usage and power output and fuel consumption (Li et al. 2013). The automobile consists of many complex components that make up together to become a system. In today's scenario, most manufacturers are pursuing the maximum efficiency and simultaneously reduce the wasted energy formed by the vehicle systems such as engines, transmission systems, and suspensions (Li et al. 2013).

Energy harvesting is a process of converting the output of the waste energy from the ambient source into other useful energy (Kumar et al. 2016). The energy is captured and stored in the harvester and converted into other forms of energy, depending on the implemented output type. An energy harvester offers a small amount of power for a low energy system, and the stored electricity can be used as the power for the vehicle electronics (Patil and Gawade 2012). Types of energy harvester include piezoelectric, electromagnetic, solar, etc. (Wei and Jing 2017). For the electromagnetic energy harvester type, the basic principle on which almost all electromagnetic generators are based is the Faraday's law of electromagnetic induction. It states that the voltage, or electromotive force (EMF), induced in a circuit is proportional to the time rate of change of the magnetic flux linkage of the circuit (Bakhtiar and Khan 2019).

The vibration of the suspension system, together with its components, produces kinetic and potential energy. This energy is dissipated and wasted, which can be benefited by reusing them to supply the energy toward other systems. The energy gained from the vibration of the suspension system can be harvest into other usable energy in terms of voltage output (Wei and Jing 2017).

The common design and structure of the electromagnetic energy harvester typically produced on a large scale and are not practical to be fitted in the automotive components, such as the suspension. The scope of this study is to design an electromagnetic energy harvester that can be adapted to the suspension, which is based on previous studies from other researchers. The design and simulation of the electromagnetic energy harvester for the vehicle suspension, are being done using the SOLIDWORKS and ABAQUS software.

26.2 Literature

In the kinetic energy recovery process, the motion or oscillations of a system, mainly vibrations are converted into electrical energy. At present, various techniques for converting vibration energy of the suspension system to electrical energy were proposed. The most used transfer mechanisms are by using piezoelectric, hydraulic, and electromagnetic approaches (Georgiev and Kunchev 2018). In the electromagnetic system, the electromagnetic induction law of Faraday states that, when a conductor put in a changing magnetic field, EMF is induced, which is called the induced EMF. If the conductor circuit also induced, the current induced is called the induced current (Bakhtiar and Khan 2019). Flux associated with the coils varies by wheel motion, cutting magnetic flux lines and producing a voltage. The benefit of this method is that due to friction, there is no heat generation.

The dissipation of vibration energy in the vehicle suspension system is one type of energy losses in vehicles. The potential of harvestable power from the suspension system has been studied for decades (Zuo and Zhang 2013). The vibrations at the suspension are produced when a vehicle moves on a road, and it increases during moving on a rough road. Essentially, the shock absorbers act as an energy dissipating device during the vehicle movement, and the vibration energy is dissipated out of the vehicle in the form of heat. Several scholars had extensively explored the energy harvesting potential based on vehicle suspensions and its effect on fuel-saving (Abdelkareem et al. 2018). Zuo and Zhang (Zuo and Zhang 2013) demonstrated that the potential energy of a typical passenger car is between 100 and 400 W, considering that the car is traveling on good and average roads with a 97 km/h approximately.

Currently, many kinds of regenerative suspension systems with different strategies have been proposed and studied to recover the dissipated vibration energy in vehicle suspension. The energy harvesting mechanisms for suspension systems have also been investigated to improve fuel efficiency and thus reduce energy consumption (Li et al. 2013). Substantially, the harvested energy from suspension vibrations could be used for another electrical usage in the vehicle, such as charging batteries and alternators. Besides that, it also might be able to provide energy for semi-active or active suspensions used in the vehicle to achieve better ride quality and road handling (Wang et al. 2018).

Although there are already many electromagnetic energy harvesters that have been designing and proposed in regenerative vehicle suspension due to the power potential, there is still room for new ideas to foster. Many parameters can be manipulated to achieve huge power energy conversion.

26.3 Design and Analysis

There are many new designs and development of electromagnetic regenerative suspension systems being designed and studied (Jamil et al. 2015), but each design has its pros and cons. There are two conceptual designs of vibration electromagnetic energy harvester designed in this paper based on the availability of space in the vehicle suspension system. It is very crucial that the main idea of the designs is to attach perfectly to the vehicle suspensions and can harvest the wasted vibration that occurs during the vehicle movement.

26.3.1 Designs of Vibration Energy Electromagnetic Energy Harvester and Working Principle

These two designs are using the systems that use a barrel-like principle (Abdullah and Jamil 2015), as in Fig. 26.1a. Basic components in the designs can be termed as in Fig. 26.1b (Abdullah and Jamil 2015). The main advantage of this design is the possibility to separate the housing from the top and bottom cover that makes this concept easy to set the parameter of the device. By disassembling the part of the system, the coil and magnet can be modified, which made the parameter setting simpler.

Both conceptual designs can refer to as in Fig. 26.2a,b, where Fig. 26.2a shows Design 1, while Fig. 26.2b shows Design 2.

For design 1, a set of two coils is placed outside the magnet. All the linkages are connected to each top and bottom mount. The main shaft is from the bottom linkage connected to the magnet, and the magnet's motion is controlled by the spring, which is

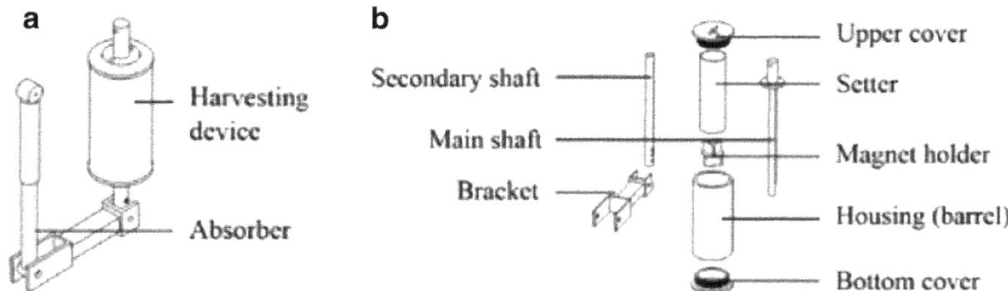

Fig. 26.1 **a** Idea of attaching the energy harvester with vehicle shock absorber. **b** Basic components in the designs

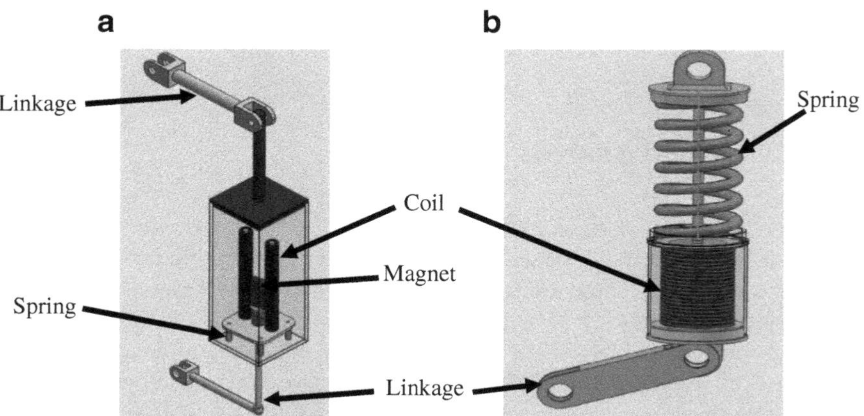

Fig. 26.2 **a** Design 1. **b** Design 2

Table 26.1 Summary of design parameters for design 1 and design 2

Parameters	Design 1	Design 2
Magnet height, mm	200	200
Magnet radius, mm	100	100
Wound coil height, mm	220	220
Number of coil turn	20	10
Coil internal diameter, mm	140	140
Distance between magnet and coil, mm	10	10
Wire diameter, mm	2	2

attached below the magnet. As the magnet moves up and down because of the absorber oscillation, it will create a magnetic field. The speed of suspension oscillation affects the movement of the magnet in the coil. The dimension and measurement for magnet and also coils are based on the overall dimension of the vehicle's absorber length and spring length. The design's consideration is also based on the confined space available in the vehicle frame near the suspension system of the car. The design intended to attach with fitment that will ensure that the vibration can be captured efficiently to be transferred to the energy harvester.

The parts in design 2 are almost similar in the arrangement of magnet and coils as in design 1, but the design is much simpler with less moving parts and components. There is one coil with a magnet placed inside it, and it is mounted inside the spring. The magnet motion is controlled by the spring movement that is mounted directly on the suspension linkages. Here the linkage is made shorter compared to design 1. A summary of the design parameters for each design is shown in Table 26.1.

26.3.2 Design Analysis

The Abaqus software is used to simulate the magnetic field and magnetic flux change inside of the electromagnetic energy harvester and both electromagnetic energy harvester design focused on the direction of the magnet motion, attachment to the suspension absorber, and the linkage of the device. Each design has a different magnet motion between the coils as the attachment is different.

The Abaqus Electromagnetic Model software has the simplicity of material assignation and direct alternation between property selection, such as electromagnetic flux density (EMB) and electric field vector (EME). The material assigned as electrical behavior in this section is Nickel, with a 596X105 conductivity (Raju and Venkatachalam 2013). The magnet permeability is set up to 1000, respectively, with isotropic type (Raju and Venkatachalam 2013)—the section assigned as electromagnetic, homogeneous solid. For the simulation, only the inside component that makes up the electromagnetic energy harvester is simulated, as shown in Fig. 26.3. This

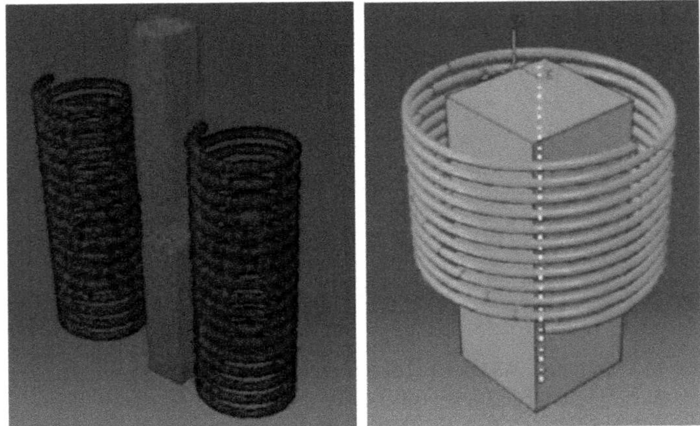

Fig. 26.3 Assembly of coil and magnet in simulation for analysis simplicity in design 1 and design 2

excludes the housing and linkage of the final product design to reduce the complexity during the analysis.

The output variable is selected to simulate the electrical and magnetic properties of the model, and they are magnetic flux density vector (EMB), magnetic field vector (EMH), and electric field vector (EME). The vibration frequency assigned is 50 Hz (Georgiev and Kunchev 2018) in all the simulation and the simulation has been done at four surface distances between the magnet and the coil, which is at 5 mm, 10 mm, 15 mm, and 20 mm. An example of load setup and meshing of design 2 in the software shown in Fig. 26.4a, b.

26.4 Result and Discussion

The results gained from the simulation in Abaqus is as in Table 26.2.

The changes in the magnetic flux density versus the distance of the surface magnet and the coil for each design can be seen in the graphs in Fig. 26.5a, b.

Both designs are having decrement of the magnetic flux density value due to the increasing distance of the magnetic surface and coil during the magnet oscillation inside the coil. This has proven that any further distance of the magnet and coil will give a low value of EMB. Nevertheless, design 1 is having a lower value of EMB if compared to design 2. This is due to the fact that design 2 is having a bigger area of magnetic field line arise as the magnet located inside the coil itself, which makes the magnet fully surrounded by the coil.

The output voltage of the electromagnetic energy harvester is calculated as (Bakhtiar and Khan 2019)

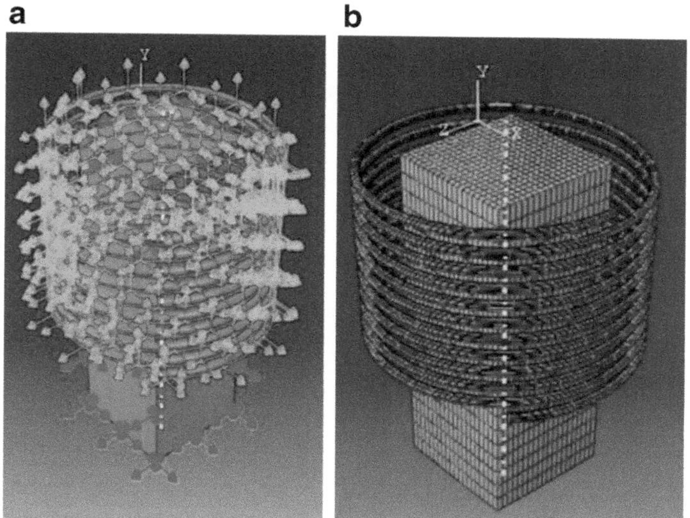

Fig. 26.4 **a** Load setup. **b** Meshing

Table 26.2 Simulation result of magnetic flux density for both design

	Magnetic Flux Density (EMB), T			
Surface distance of magnet and coil	5 mm	10 mm	15 mm	20 mm
Design 1	1.804E−11	1.35E−11	1.20E−11	1.15E−11
Design 2	5.846E−10	5.52E−10	5.43E−10	5.17E−10

$$V_l = -U\left(\frac{dB_z}{dz}\right)_l \sum_{i=1}^{n} S_i \tag{26.1}$$

where l is the voltage induced in a single layer of a coil, and it depends on the relative velocity U of the membrane (coil) with respect to the magnet, the magnetic flux density gradient dB_z/d_z over the layer, and the area sum S_i of turns n of the layer in the wound coil. For a single layer, a wound coil having n number of turns and internal diameter d_p, the area sum is estimated as (Bakhtiar and Khan 2019)

$$\sum_{i=1}^{n} S_i = \sum_{i=1}^{n} \frac{\pi}{4}(d_i)^2 = \frac{\pi}{4}\sum_{i=1}^{n}\left(\frac{d_p}{2} + (i-1)d_w\right)^2 \tag{26.2}$$

where $d_i = \frac{d_p}{2} + (i-1)d_w$ and d_w is the coil wire diameter. Relative velocity of the membrane (coil) with respect to the magnet can be calculated as being equal to the frequency (F) multiplied by the magnet length (L), $U = FL$.

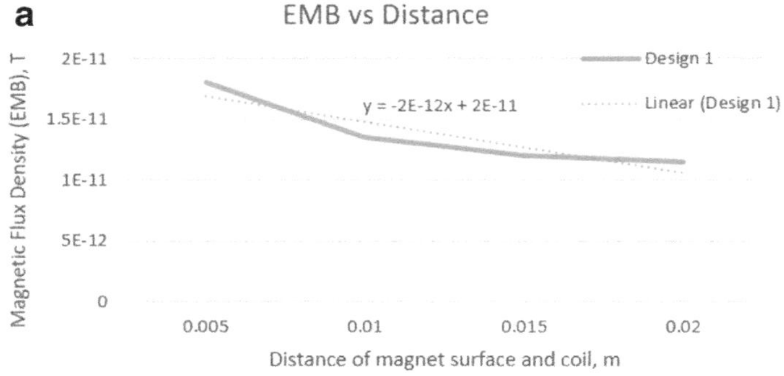

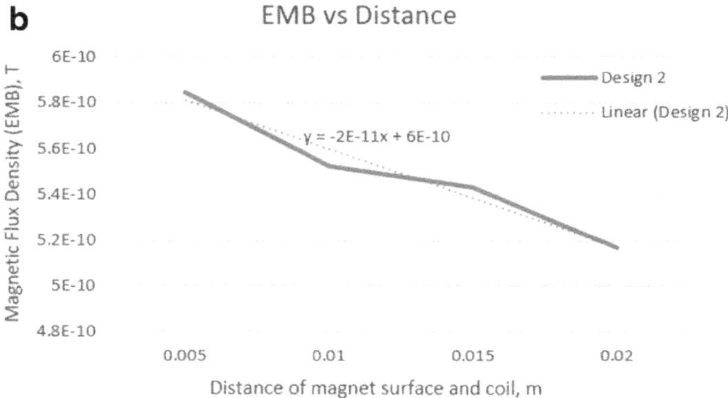

Fig. 26.5 **a** Magnetic flux density versus the distance of surface magnet and the coil for design 1. **b** Magnetic flux density versus the distance of surface magnet and the coil for design 2

For multiple layers, the total voltage, V_T can be calculated by substituting (26.1) in (26.2) (Bakhtiar and Khan 2019).

$$V_T = -U\left(\frac{dB_z}{dz}\right)_l \frac{\pi}{4} \sum_{i=1}^{n} \left(\frac{d_p}{2} + (i-1)d_w\right)^2 \tag{26.3}$$

From the graphs in Fig. 26.5a, b, the magnetic flux density gradient *dBz*/*dz* can be calculated as the graph gradient. All parameters used to calculate the total voltage are shown in Table 26.3.

Results show the voltage output gain from each design of the electromagnetic energy harvester. Although design 1 used a double coil, the internal diameter of the coil designed to be the same as design 2, and it shows that the number of coils used does not affect much the total output voltage. But the design of the coil and magnet's internal diameter will be affected because design 1 does not have the magnet fully surrounded by the coil. In contrast, the magnet in design 2 is entirely surrounded by the coil.

Table 26.3 Parameters used in total voltage calculation

	Design 1	Design 2
Frequency, F	50 Hz	50 Hz
Magnet Length, L	0.2 m	0.2 m
U	10 m/s	10 m/s
$\frac{dB_z}{dz}$	−2E−12 T	−2E−11 T
Number of coil turn, n	20	10
Coil internal diameter, d_p	0.14 m	0.14 m
Wire diameter, d_w	0.002 m	0.002 m
Total Voltage, V_T	1.832E−13 V	1.217E−12 V

26.5 Conclusions

Based on the theory of electromagnetic induction and reference of previous electromagnetic energy harvester design, two design outputs were generated. This study focused on the magnet motion in different parameters and direction, in terms of current-induced and magnetic field strength sourced from the vehicle suspension. Design 1 uses the upward and downward movement of the magnet that is surrounded by two coils. Design 2 uses a single coil with a magnet placed inside the coil diameter. Simulation of movement and electromagnetic behavior of the energy harvester has been carried out for the design, and each design has different magnetic flux outputs with different parameters.

In this paper, the proposed design of vibration electromagnetic energy harvester successfully harvests the oscillation of the shock absorber to electrical energy although the obtained output voltage is rather a small value with the maximum value of 1.217E−12 V and design 2 that has a simple design but has the largest contact surface contact of the magnet and coil surface where the coils entirely surrounded the magnet had proved to get the higher voltage output than design 1 that has more coils but less contact surface between the magnet and coils.

References

Abdelkareem, M.A.A., Xu, L., Ali, M.K.A., et al.: Vibration energy harvesting in automotive suspension system: a detailed review. Appl. Energy. **229**, 672–699 (2018)

Abdullah, M.A., Jamil, J.F.: Harvesting energy from the vibration of suspension of a passenger vehicle. Recent Adv. Mech. Eng. 128–133 (2015)

Bakhtiar, S., Khan, F.U.: Analytical modeling and simulation of an electromagnetic energy harvester for pulsating fluid flow in pipeline. Sci. World J. 5682517 (2019). [Laude V (ed)]. https://doi.org/10.1155/2019/5682517

Georgiev, Z., Kunchev, L.: Study of the vibrational behaviour of the components of a car suspension. In: MATEC Web Conference, vol. 234 (2018). https://doi.org/10.1051/matecconf/201823402005

Jamil, J.F., Abdullah, M.A., Tamaldin, N., Mohan, A.E.: Fabrication and testing of electromagnetic energy regenerative suspension system. J. Teknol. **77**(21), 97–102 (2015)

Kumar, A., Balpande, S.S., Anjankar, S.C.: electromagnetic energy harvester for low frequency vibrations using MEMS. Procedia Comput. Sci. **79**, 785–792 (2016)

Li, Z., Zuo, L., Luhrs, G., Lin, L., Qin, Y.: Electromagnetic energy-harvesting shock absorbers: design, modeling, and road tests. IEEE Trans. Veh. Technol. **62**(3), 1065–1074 (2013)

Patil, R.H., Gawade, S.S.: Design and static magnetic analysis of electromagnetic regenerative shock absorber. IJAET **3**(2), 54–59 (2012)

Raju, A.B., Venkatachalam, R.: Analysis of vibrations of automobile suspension system using full-car model. Int J Sci Eng Res. **4**(9), 2105–2111 (2013)

Wang, R., Ding, R., Chen, L.: Application of hybrid electromagnetic suspension in vibration energy regeneration and active control. J. Vib. Control. **24**(1), 223–233 (2018)

Wei, C., Jing, X.: A comprehensive review on vibration energy harvesting: Modelling and realization. Renew. Sustain. Energy Rev. **74**, 1–18 (2017)

Zuo, L., Zhang, P.S.: Energy harvesting, ride comfort, and road handling of regenerative vehicle suspensions. J. Vib. Acoust. **135**(1), 011002 (2013). https://doi.org/10.1115/1.4007562

Chapter 27
Application of Clean Agents in Fire Suppression Systems: An Overview

Abdul Shukor Jum'azulhisham, Abdul Razak Muhammad Al-Hapis, Hassan Azmi, and Jamian Rahim

Abstract The application of clean agent for fire extinguishing evolves from time to time. Lately, there is an increasing number of applications of F-gaseous in refrigeration, air-conditioning, and blowing agent as well. This includes the use of fluorine gas as its extinguishing agent. The application of clean agents in the fire suppression systems is better than before. These agents are reviewed in this paper together with other gaseous. The awareness of greenhouse effects, climate change, ozone depletion and global warming enables innovative green technologies to grow more rapidly for the sustainability of the future generation. This study aims to choose the needs of the new fire suppression clean agent to comply with the standards and regulations. The study of fluorinated gaseous (F-gaseous) in this paper brings to the needs of a new study of new gaseous focused to counter global warming. The study will also include related computational models and experimental investigations conducted for the research related in the gas type clean agent fire systems to suit to the effectiveness of its application. This article presents a review of these studies, the trend of studies and suggests a direction for future research developments.

Keywords Kigali amendment · Fire suppression systems · Halon · Clean agent · Hydrofluoroolefin

A. S. Jum'azulhisham (✉) · A. R. Muhammad Al-Hapis · H. Azmi
Universiti Kuala Lumpur Malaysian Spanish Institute, Kulim Hi-Tech Park, 09000 Kulim, Kedah, Malaysia
e-mail: jumazulhisham@s.unikl.edu.my

A. R. Muhammad Al-Hapis
e-mail: alhapis@unikl.edu.my

H. Azmi
e-mail: azmi_hassan@unikl.edu.my

J. Rahim
Faculty of Engineering Technology, Universiti Tun Hussain Onn Malaysia, Pagoh Higher Education Hub, 84600 Muar, Johor, Malaysia
e-mail: rahimj@uthm.edu.my

M. H. Abu Bakar et al. (eds.), *Progress in Engineering Technology III*,
Advanced Structured Materials 148, https://doi.org/10.1007/978-3-030-67750-3_27

27.1 Introduction

Fire is a process in which substances combine chemically with three elements in a form of a chain reaction between oxygen, heat, and fuel. Each key element must be present simultaneously to create fire. The fire can be only extinguished when either one of these elements is sufficiently removed. Choy and Fong described that fires can be instantly extinguished with attribution of three possible actions, by breaking off or removal of heat, separation of the fuel and the oxygen, or the removal of the oxidizer (Suryoputro et al. 2018; Yeturu et al. 2016) as described in Fig. 27.1.

Most of the gaseous type of fire suppression systems extinguish the fire by lowering the oxygen level below 18.5% or by reducing the temperature. This paper will generally discuss the literature of fire suppression systems and finally, focus on the current trend of the application of clean agents.

The modern fire extinguishing is typically separated into four main methods as illustrated in Fig. 27.2. It can be divided based on the material used in extinguishing fire including dry powder, water-based, foam, and gaseous including the clean agent.

Gaseous fire suppression agents begin with the use of halons in the 1930s. halon 1301 (CF_3Br) and halon 1201 were found to be very effective in suppressing fire in those days (Pagliaro and Linteris 2017). It was claimed that the halon has less toxic, non-corrosive, nonflammable properties and it is compatible with other materials (Choy and Fong 2003). However, due to the negative environmental impact of chlorofluorocarbons (CFCs) in halons, the Montreal Protocol had decided to ban its usage in the 1980s (Godin-Beekmann et al. 2018; Newman 2018). Since then, few agents were developed to replace them including hydrocluorofluorocarbons (HCFC) (Kumar and Rajagopal 2007; Daviran et al. 2017; Yang and Wu 2015) hydrochlorofluoroolefin (HCFO), fluorinated ketone (FK) (Linteris et al. 2018), perfluorocarbons (PFCs), hydrofluorocarbons (HFCs) (Wang et al. 2014), and hydrofluoroolefin (HFOs) (Daviran et al. 2017; Wang et al. 2019; Nilsson et al. 2017; Arora et al. 2018)

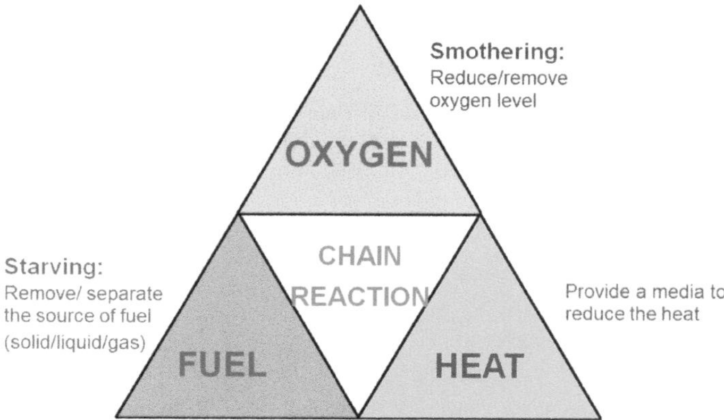

Fig. 27.1 Fire triangle

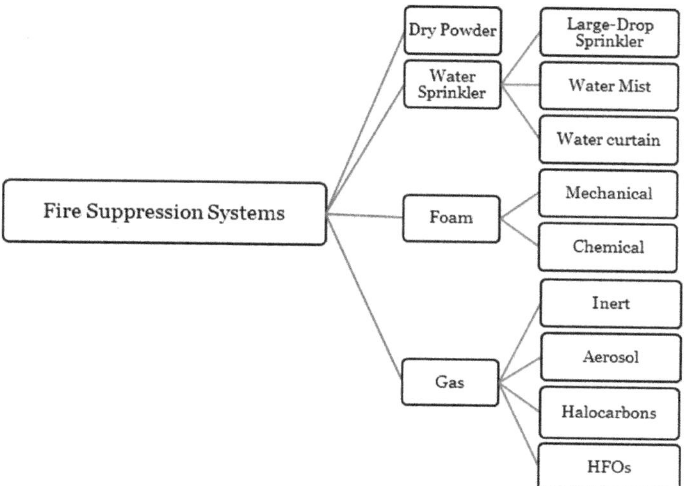

Fig. 27.2 Methods of fire extinguishing and its substance

as well. The generation of the gas fire suppression systems as shown in Fig. 27.3 expands with the use of halons.

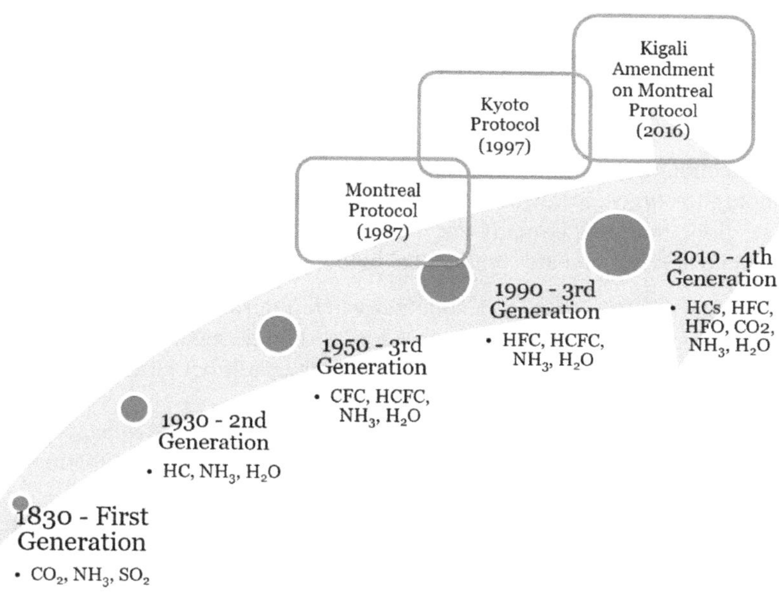

Fig. 27.3 The evolution of gaseous fire suppression systems

Generally, there are two methods of applying an extinguishing clean agent:

1. Total flooding system.
2. Local application.

The UL 2166 - Standards for Safety for Halocarbon Clean Agent Extinguishing System Unit defines a total flooding system as the system to supply the clean agent arranged to discharge the extinguishing agent into the intended protected volume and to fill the volume to an effective extinguishing agent consideration.

The total flooding applications are not only applied in gas type fire suppression systems (Li 2008) but also applicable for water mist and foam fire extinguishing system. The total flooding gas protection system applications was established since the era of halons (Choy and Fong 2003) until now.

The gas type of fire suppression systems is currently claimed to be the most effective fire extinguishing systems compared to water sprinkler, water mist, and foam. It leaves no residue to the affected area.

The study primarily aims to highlight the clean agents to extinguish flames and the new alternative toward reducing environmental negative impacts. The following sections will include the issue of greenhouse effects, the overview of the range of gaseous, studies on the material used in fire suppression including fluorine. The studies conducted by researchers and possible studies to be conducted. Ultimately, this paper concludes with a summary of findings and future recommendations.

27.2 Issue of Greenhouse Effects Based on the Production and Use of These Gaseous

The selection of future gaseous for the production processes, keep and safe use including for fire suppression must take into consideration the effect of the greenhouse, global warming issues (Gencer and Oru 2015), and ozone depletion for future generation survival and are summarized below:

1. The greenhouse effect is a situation where the surface of the earth and lower atmosphere become warmer caused by carbon dioxide and water vapour. This permits the sun's energy flow through to the earth but blocks energy from the earth to return back into space.
2. Ozone depletion potential (ODP) is a measure of potential substances to reduce the amount of ozone in the atmosphere which blocks harmful radiation including ultraviolet (UV) from the sun penetrating the earth.
3. Global Warming Potential Index has been established to put the contribution of all the non-carbon dioxide (CO_2) greenhouse gaseous (GHG) to the climate change over the normalized CO_2 scale. Table 27.1 shows the classification of GWP.

Table 27.1 GWP classification

GWP	Max
Ultra-low	< 30
very low	31–100
Low	101–300
Moderate	301–1000
High	1001–3000
Very high	3001–10,000
Ultra-high	10,001–30,000

27.3 Overview of Gaseous Fire Suppression Systems

Within a century, a wide range of fire extinguishing technologies was developed, including powdered based extinguishers, water-based, water mist, water curtain technology, high expansion foam. and gaseous. Currently, gaseous type of fire suppression systems including clean agents is actively being studied.

The National Fire Protection Association defined a "Clean Agent" as an electrically non-conducting, volatile, or gaseous fire extinguishing that does not leave a residue on surfaces upon evaporation (Linteris 2011).

The demand for efficiency, effectiveness during and once it has been applied in extinguishing fire and productivity makes gaseous fire suppression systems highly demanded. There is a broad range of gaseous used in fire suppression systems. The function of gas in suppressing fire is to reduce the oxygen level and eliminate fire from heating up and spread up in the work system.

Halons was phase-off in the 1990s. It was then changed to the aerosol type of gas which is much more reliable (Zhang et al. 2015a). The evolution of fluorinated gaseous produces better results then. The fluorinated gas is also known as "F-gas" involved in refrigeration, air-conditioning, blowing agent, fire suppression systems, etc. (Abas et al. 2018). These man-made gaseous float in the atmosphere for centuries which contributes to global warming.

Below, the evolution of gaseous specifically for suppressing fires that been reviewed.

27.3.1 Aerosols

An aerosol consists of fine solid particles used in a fire suppression system. Aerosols have been used in the fire suppression systems for the application in the forms of dust, fume, smoke, mist, and fog. The hot aerosol fire extinguishing technology was developed based on pyrotechnics. Hot aerosol does not need to be driven out by pressurized gaseous. The hot aerosol is capable to extinguish class A, B, C, D, and K fires (Zhang et al. 2015a). Cold aerosol extinguishing technology based on dry

Table 27.2 Composition of inert gaseous to extinguish the fire

Commercial name		Chemical formula	Substance content
Inergen	IG-451	N_2ARCO_2	Nitrogen (52%), Argon (40%), CO_2 (8%)
Nitrogen	IG-100	N_2	Nitrogen (100%)
Argonite	IG-55	N_2AR	Argon (50%), Nitrogen (50%)
Argon	IG-01	Ar	Argon (100%)

powders or particles dispersed in a solution is called cold aerosol technology. These aerosol leave residue after the fire is extinguished.

27.3.2 Inert Gas

An inert gas is generally non-reactive with other substances. In other words, their atoms do not combine with other atoms in chemical reactions. Some molecular gaseous in inert gaseous which are suitable for extinguishing fire are nitrogen and carbon dioxide. The inert gaseous include inergen (IG-541), arganite (IG-55) argotech (IG-01), and N100 (IG100) were identified as shown in Table 27.2 and had been commonly used in fire extinguishing systems. These gaseous respond against fire by the suffocation method. It has been claimed as non-toxic, stable at high temperature, zero ODP, and GWP.

27.3.3 Halocarbons

Halocarbons consist of a combination of carbon and one or more halogens such as bromine, chlorine, iodine, and fluorine. The combination of these chemical compounds creates halons, hydrochlorofluorocarbons (HCFCs) and, hydrofluorocarbons (HFCs) which have been applied in extinguishing the fire. However, the application of halocarbons is not limited for fire suppression system but they also have been used for air conditioning, refrigeration, blowing agent, solvent, and others. Some of these agents have been identified as contributors to the environmental issue.

27.3.3.1 Halons

There were various types of halons produced, including halon 1301 (CF_2Br-bromotrifluoromethane) and halon 1202 ($CBrCIF_2$—dibromodifluoromethane). Halon was made of a combination of hydrogen, carbon, chlorine, and bromine atoms (Kim 2001). Initially, it was introduced to protect vehicles, structures, and equipment from fire. The application of halon is not limited in fire suppression systems, but also applied in air conditioning and refrigeration. Due to global ozone-depleting awareness, the production and application of these gaseous were ceased phase-by-phase in 1996 in developed countries and developing countries in 2010 as stated in article 5 (Godin-Beekmann et al. 2018).

27.3.3.2 Hydrocluorofluorocarbons

Hydrocluorofluorocarbons (HCFCs) compound consist of hydrogen, fluorine, chlorine, and carbon atoms. Although HCFCs ODP is zero, they are less potential at destroying the stratospheric ozone than CFCs (Godin-Beekmann et al. 2018). Previous studies have reported on HCFCs in the application in refrigeration (Kumar and Rajagopal 2007; Yang and Wu 2015), feedstock, heating elements (Yang and Wu 2015), solvent, and fire suppression systems (Huo et al. 2017). Among all 38 HCFCs listed under the United Nation Environmental Programme (Ozone Secretariat 2006), HCFC-123 (Kumar and Rajagopal 2007) and HCFC-124 were suitable for fire suppression systems replacing halon 1301. Comparisons of HCFCs on ODP, GWP, and atmospheric lifetime are presented in Table 27.3.

The GWP of HCFC is classified high, ranging from 3500 to 11,700. A strong decision was made by the Montreal Protocol in the process of phasing out the ozone-depleting substances (ODS) (Kumar and Rajagopal 2007), resulting in a group of CFCs and HCFC refrigerants have been banned. A decision was made and action of replacement of HCFC to heptafluoropropane took place.

27.3.3.3 Hydrofluorocarbons

There is a wide range of hydrofluorocarbons with some specific applications including refrigeration, air-conditioner, blowing agent, solvent, fire suppression agent (Wang et al. 2014) etc. Based on experimental studies, heptafluoropropane (HFC-227ea) was found as the best in extinguishing the fire (Choy and Fong 2003) compared to its family, because of less toxic, shorter lifetime, and better electrically insulating proposition (Wang and Duan 2016). The use of pentafluoroethane, known as HFC-125, was a fire extinguishing agent (Ting et al. 2016). Some studies state that the GWP is slightly higher than HFC-227ea and slightly lower in the atmospheric lifetime. HFC-236fa was used in extinguishing fires. It was found that the HFC-236 isomers have low acute and sub-chronic toxicity potential, although the HFC-236ea isomer tends to be somewhat more toxic on a sub-chronic basis (Brock et al. 2000).

Based on F-gas regulations, subject to the provisions of the United Nations Framework Convention on Climate Change (UNFCCC), HFCs were characterized by zero ODP but high in GWP as illustrated in Fig. 27.4. Even though HFC-227ea is safe in occupied area, because of Kigali amendment on the Montreal Protocol, it was stated that these gaseous have to be phased-down (Godin-Beekmann et al. 2018; UNEP 2016).

Table 27.3 Some of the HCFCs controlled under the Kyoto Protocol

Name	Refrigerant No	Chemical name	Chemical formula	ODP	GWP	Atm lifetime (years)
HCFC-21	R-21	Dichlorofluoromethane	$CHFCl_2$	0.04		1.7
HCFC-22	R-22	Monochlorodifluoromethane	CHF_2Cl	0.055	1810	12.0
HCFC-123	R-123	Dichlorotrifluoroethane	$C_2HF_3Cl_2$	0.02–0.06	77	1.3
HCFC-124	R-124	Monochlorotetrafluoroethane	C_2HF_4Cl	0.02–0.04	609	5.8
HCFC-141b	R-141b	Dichlorofluoroethane	CH_3CFCl_2	0.11	752	9.3
HCFC-142b	R-142b	Chlorodifluoroethane	CH_3CF_2Cl	0.065	2130	17.9
HCFC-225ca	R-225ca	Dichloropentafluoropropane	$CF_3CF_2CHCl_2$	0.025	122	1.9
HCFC-225cb	R-225cb	Dichloropentafluoropropane	CF_2ClCF_2CHClF	0.033	595	5.8

Source United Nation Environment Programme

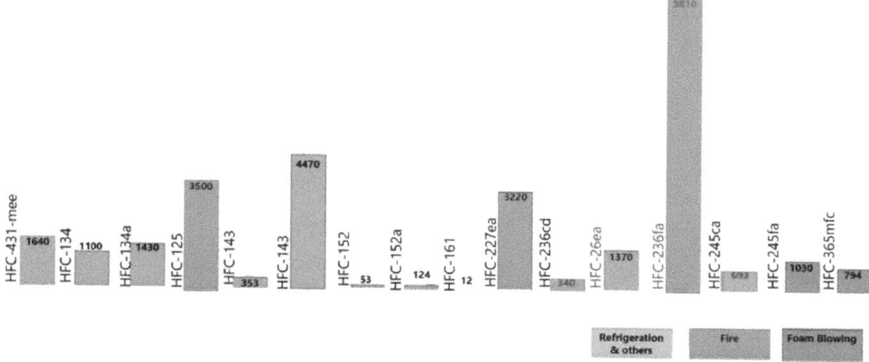

Fig. 27.4 Global warming potential of HFCs

27.3.4 Fluorinated Ketones

FK-5–1-12 known as $C_6F_{12}O$ is a fluorinated ketone with chemical formula of $CF_3CF_2C(O)CF(CF_3)_2$ as illustrated in Fig. 27.5. A colorless liquid, electrically non-conductive vapor and considered as a clean agent. The unsaturated organic compound, FK-5–1-12, does not remove oxygen and is safe for humans trapped in the area. It is not identified as contributing to the ozone layer depletion. FK-5–1-12 has been seen in experimental and numerical studies including in testing multiple fire performance including the cup burner test, insulation strength and decomposition characteristics of gas mixture (Zhang et al. 2017). The numerical simulations were performed using a detailed kinetic mechanism (Linteris et al. 2018; Xu et al. 2016). The experimental and numerical studies have been conducted to determine the unstretched laminar burning velocity of premixed CH4-air and C3H8-air flames with this gas (Pagliaro and Linteris 2017).

Fig. 27.5 The molecular structure of $CF_3CF_2C(O)CF(CF_3)_2$ $C_6F_{12}O$

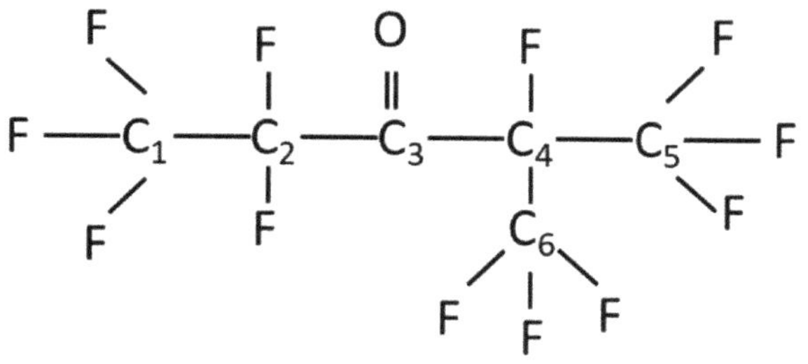

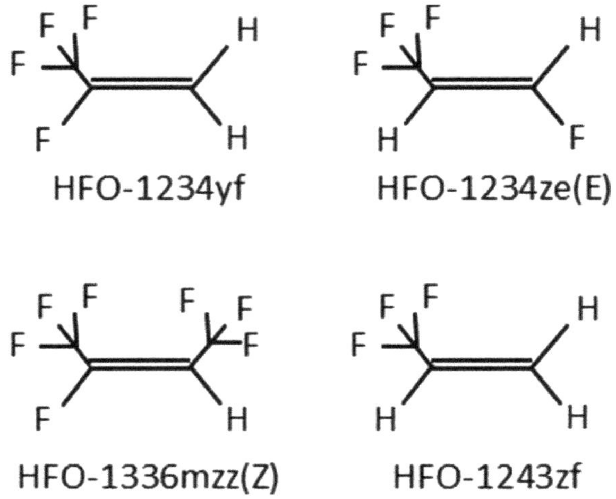

Fig. 27.6 Structural formula of some HFOs

27.3.5 *Hydrofluoroolefin*

Hydrofluoroolefin (HFOs) contains hydrogen, fluorine, and carbon just like HFCs, but they are specifically different. Most of HFOs are built in a double covalent bond as shown in Fig. 27.6. HFOs have a low toxicity level, do not contain chlorine and pose no threat to the ozone layer (Zeng et al. 2020). The application of HFOs is similar to other gaseous discussed. Until today, forty-eight HFOs (Minor and Spatz 2015) and blends were systematically analysed including HFO-1224yd, HFO-1224yf, HFO-1234yf (Zeng et al. 2020; Wu et al. August 2019; Leck 2009), and HFO-1336mzz(Z) (Nilsson et al. 2017; Minor et al. 2010). Some are applied purposely for refrigeration (Arora et al. 2018; Leck 2009; Fluorochemicals 2014), air conditioning (Daviran et al. 2017; Gencer and Oru 2015), solvent, foam, heat pump (Nilsson et al. 2017), blowing agents (Wu et al. 2019), and fire suppressant but some can be used for more than one applications through some tests. In terms of the possibility for the flammable situation, the likelihood of ignition is low (Leck 2009). The range of ODP is ultra-low, supporting with the GWP range below very low and shorter atmospheric lifetimes (Gil and Kasperski 2018) makes them the most effective gaseous suitable to replace halons, HCFCs, and HFCs.

Not many studies found HFOs suitable to extinguish the fire. The most recent finding shows that HFO-1336mzz(Z) has been tested using the cup burner test. As a result, HFO-1336mzz(Z) shows better suppression performance with a lower concentration than HFCs extinguishing agents (Nielsen et al. 2007a; Wang et al. 2020a).

Focused on the type of clean agent fire suppression systems and its impact on climate change, a few researches were conducted and there were not as many as other applications in air-conditioning and refrigeration.

Some of the identified extinguishing agents highlighted reduce the level of oxygen which may contribute to bad effects on living things trapped in the fire including humans. The proposal of new suppression agent which reduces the temperature of the fire is highly recommended.

As mentioned earlier supported with data obtained from United Nation Environmental Plan (UNEP), it is summarized that the best way in selecting the type of gas for further study is based on the environmental effects including the lowest issues on global warming criteria including ODP, GWP, and atmospheric lifetime. It is clear and the recommendation to be made that more studies on HFOs as an alternative of HFC-227ea to be conducted.

27.4 Studies on Clean Agent Gaseous Used as Fire Suppressors

There are numerous techniques used to assess the effectiveness of the fire suppressant gas. The use of computational fire modelling (Wang et al. 2020b) with assistance by the software simulation approach has eased the analysis (Cournoyer 2014). The verification method through experimental tests confirmed the success of the study. There are investigations with a combination of computation and experimental studies on these gaseous fire suppressors (Kumar and Rajagopal 2007). Previous studies on computation and the experimental investigations highlighted below:

- Modelling of state and thermodynamic cycle properties of HFO using a cubic equation of state. In this study, Peng and Robinson cubic equation of state was applied. As a result, specific enthalpy, specific entropy, and specific volume were generated (Yang et al. 2018).
- Modelling of fire suppression by fuel cooling with water spray was conducted using fire dynamic simulator (FDS). FDS simulation has also been applied in fireball modelling to predict thermal effects in the gas industry (Neto 2010; Jenft et al. 2017).
- Fire modelling was used to simulate the active fire and gas temperatures, heat release rate, finite element analysis (FEA) Wang et al. 2020b), at which the heat of reaction and energy was constant (Sellami et al. 2017). The purpose of this modelling was to predict fire characteristic of various materials including the size of the flame, ignition time, flashover, and fire growth. The experiments and computations were conducted accordingly. The data from the computational modelling of fire phenomena help to identify the limitations of self-contained fire extinguishers.
- The model-computed atmospheric lifetime, radiative forcing, and GWP for HFCs were implemented. This model is being considered as a non-ozone depletion replacement. This model was implemented in the study of hydrofluorocarbons (HCFC) (Biteau and Steinhaus 2016). It is recommended to apply this study for HFOs.

- A numerical modelling of fire extinguishing gas retention was identified. The possible use of CFD assists the researchers in analysing the omission of the discharge gas systems (Magid and Bray 1999) and the flow of liquid and gaseous through a pipe and T junctions (Kubica and Czarnecki 2019). This study can be conducted on another corresponding type of fire suppression novel agents.
- The Martin-Hou equation of state (MH EoS) was implemented in the analysis of HFO-1234yf and comparison with R-134a gas was done (Wu et al. 2019). This tool generates tables of thermodynamic properties and generating accurate pressure-enthalpy (PH) diagrams. Further, it can be refined as more accurate measurement data for the HFO-1234yf becomes available.
- The computational method using a time-dependent, axisymmetric numerical code (UNICORN) is applicable on the cup burner flame to show the combustion inhibition and improvement by halon 1301 and selected substitute fire extinguishing agents. This particle method may also be employed to test more potential agents related to fire suppression systems and better comparison data can be instantly publicized (Athulya and Miji Cherian 2016).
- A cup burner test was conducted through the experimental method to identify and examine the extinguishing concentration of Class B fire. The principle of the test is that the diffusion flames of fuels burning in a round reservoir (burner as the cup), centrally located in a coaxially flowing air stream and it is extinguished by the presence of HFC-227ea to the air (Choy and Fong 2003). The same test was also conducted for HFO-1336mzz(Z) (Wang et al. 2020a). It was proven that HFO-1336mzz(Z) has a very good potential to replace HFCs but still not enough convincing without supporting more research.
- Computational fluid dynamics (CFD) is one of the engineering tools (Wang and Duan 2016), i.e., the most widely used modelling method in solving most mathematical problematic issues in fluid mechanics based on the eulerian-eulerian formulations (Takahashi et al. 2017). A model for predicting suppression within CFD codes using the vulcan fire-field model (Swain and Mohanty 2013) was also identified. CFD is applicable for:

  (a) Turbulence models (Takahashi et al. 2017) (k-epsilon, k-omega, reynolds stress model Hewson et al. 2003), large eddy simulation (Basavarajappa and Miskovic 2019)
  (b) Combustion models (mixture fraction, eddy dissipation, flamelet)
  (c) Thermal radiation models (discrete transfer, monte carlo).

- The CFD method is typically a well-established tool that helps in design, prototyping, testing, and analysis. The use of CFD simulations was also used to simulate leakage of the refrigerant HFO-1234yf through the air-condition ventilation of a vehicle cabin. The simulation was important to understand whether the lower flammability limit could be exceeded during inadvertent leakage of the refrigerant into the passenger compartment of a vehicle (Leck 2009).
- Some comparative study based on CFD was performed to analyse and obtain the most effective and reliable results. FDS, Ansys Fluent, and Ansys CFX are the most common computational packages (Zhiyin 2015). The use of abaqus, ansys

fluent as part of CDF software eases the user to design, process, simulate, and obtain required data in efficient time.

- It is important to understand the environmental matters before selecting the best gas for study. Experimental studies were conducted to determine the global warming potential and atmospheric lifetime of one of the related gaseous (Yang et al. 2017).
- ReaxFF simulations developed by van Duin were implemented to study the pyrolysis mechanisms of HFO-1336mzz(Z) Huo et al. 2017).

Based on the findings, many researchers focused on the modelling on fluids, study of suppressant gaseous thermodynamics, mechanical properties. However, the study of alternatives of HFCs including HFC-227ea has huge potential as discussed for the applications in refrigeration (Arora et al. 2018; Wu et al. August 2019), application in heat pump (Nilsson et al. 2017), as well as based on a comparative study with another type of gas in automotive air conditioning systems (Nielsen et al. 2007b). The latest study of the new suppressant has not been identified until now.

The use of CFD in analysing and detection and suppression gaseous, as well as fires in power plant (Siu et al. 2017) were identified. So far, there has been little discussion about the specific type of clean agent applied in the CFD simulations. More studies are required on specific clean agents and simulations using CFD.

27.5 Future Recommendations

Based on GWP, ODP, life cycle climate performance, toxicity, and physical properties of the reviewed fire suppression gaseous, it can be concluded that the potential research must be continued. The recent work of literature focused on the reduction of global warming and atmospheric lifetime. The ultimate GWP_{100} target of production of new gases must be below zero. Considering the aim of the atmospheric lifetime, the production of new gaseous will improve from a few years to a few days. Today, we can see the increasing numbers of works that can be found compared to the past.

There are several combinations of modelling and experimental analyses were identified and can be done including the cup burner test, simulations using CFD, CFX, experimental studies on the effectiveness of flame response and comparative studies with studied gaseous.

However, it is difficult and challenging to find any research material on the spreading of this gas in a specific space, the effectiveness of the distribution nozzle with the existing clean agent distribution nozzle, the quantity of the gas, size, pressure, flow rate of the piping system toward the outlet, the capability of the system to extinguish different sizes of flames or type of combustion materials, comparison of its effectiveness with other HFOs and many more.

Based on the findings, the authors are motivated to conduct further investigation, especially on HFOs.

The results presented here may facilitate further studies on the application of the clean agent fire suppression system for the sustainability of the future generation. The implications of the results of this research will be recommended for future works.

References

Abas, N., Kalair, A.R., Khan, N., Haider, A., Saleem, Z., Shoaib, M.: Natural and synthetic refrigerants, global warming: a review. Renew. Sustain. Energy Rev. **90**, 557–569 (2018). https://doi.org/10.1016/j.rser.2018.03.099

Arora, P., Seshadri, G., Tyagi, A.K.: Fourth-generation refrigerant : HFO 1234yf. Curr. Sci. **115**(October), 1497–1503 (2018). https://doi.org/10.18520/cs/v115/i8/1497-1503

Athulya, A.S., Miji Cherian, R.: CFD modelling of multiphase flow through T junction. Procedia Technol. **24**:325–331 (2016). https://doi.org/10.1016/j.protcy.2016.05.043

Basavarajappa, M., Miskovic, S.: CFD simulation of single-phase flow in flotation cells: effect of impeller blade shape, clearance, and Reynolds number. Int. J. Min. Sci. Technol. 1–13 (2019). https://doi.org/10.1016/j.ijmst.2019.05.001

Biteau, H., Steinhaus, T., Christopher Schemel, A.S., Guy Marlair, N.B.A.J.L.T.: Calculation Methods for the heat release rate of materials of unknown composition. Fire Saf. Sci. 1165–1176 (2016) . https://doi.org/10.3801/IAFSS.FSS.9-1165. [January 2009]

Brock, W.J., Kelly, D.P., Munley, S.M., Bentley, K.S., McGown, K.M., Valentine, R.: Inhalation toxicity and genotoxicity of hydrofluorocarbon (HFC)-236fa and HFC-236ea. Int. J. Toxicol. **19**(2), 69–83 (2000). https://doi.org/10.1080/109158100224881

Choy, W.M., Fong, N.K.: an introduction to clean agents heptafluoropropane. Int. J. Eng. Perform. Based Fire Codes **5**(4), 181–184 (2003)

Cournoyer, M.E.: Fire modeling of an emerging fire suppression system. J. Chem. Health Saf. **21**(5), 8–13 (2014). https://doi.org/10.1016/j.jchas.2014.05.011

Daviran, S., Kasaeian, A., Golzari, S., Mahian, O., Nasirivatan, S., Wongwises, S.: A comparative study on the performance of HFO-1234yf and HFC-134a as an alternative in automotive air conditioning systems. Appl. Therm. Eng. **110**, 1091–1100 (2017). https://doi.org/10.1016/j.applthermaleng.2016.09.034

Fluorochemicals, D.: HFO-1336mzz-Z as a low GWP working fluid for transcritical rankine power cycles (2014)

Gencer, A., Oru, V.: Characteristics of some new generation refrigerants with low GWP. Energy Procedia **75**, 1452–1457 (2015). https://doi.org/10.1016/j.egypro.2015.07.258

Gil, B., Kasperski, J.: Efficiency evaluation of the ejector cooling cycle using a new generation of HFO/HCFO refrigerant as a R134a replacement. Energies **11**(8), 1–18 (2018). https://doi.org/10.3390/en11082136

Godin-Beekmann, S., Newman, P.A.., Petropavlovskikh, I.: 30th anniversary of the montreal protocol : from the safeguard of the ozone layer to the protection of the Earth's climate. Comptes Rendus Geosci. **350**, 331–333 (2018). https://doi.org/10.1016/j.crte.2018.11.001

Hewson, J.C., Tieszen, S.R., Sundberg, W.D., Desjardin, P.E.: CFD modeling of fire suppression and its role in optimizing suppressant distribution. NIST Special Publication (2003)

Huo, E., Liu, C., Xu, X., Dang, C.: A ReaxFF-based molecular dynamics study of the pyrolysis mechanism of HFO-1336mzz(Z). Int. J. Refrig. (2017). https://doi.org/10.1016/j.ijrefrig.2017.07.009

Jenft, A., Boulet, P., Collin, A., Trevisan, N., Mauger, P., Pianet, G.: Modeling of fire suppression by fuel cooling. Fire Saf. J. (2017). https://doi.org/10.1016/j.firesaf.2017.03.067

Kim, A.: Recent development in fire suppression systems. In: 5th AOSFST, Newcastle, Aust 12–27 (2001)

Kubica, P., Czarnecki, L.: Numerical modelling of the fire extinguishing gas retention in small compartments. Appl. Sci. **9**(663), 1–22 (2019). https://doi.org/10.3390/app9040663

Kumar, K.S., Rajagopal, K.: Computational and experimental investigation of low ODP and low GWP HCFC-123 and HC-290 refrigerant mixture alternate to CFC-12 Bureau of Indian Standards. Energy Convers. Manag. **48**, 3053–3062 (2007). https://doi.org/10.1016/j.enconman.2007.05.021

Leck, T.J.: Evaluation of HFO-1234yf as a potential replacement for R-134a in refrigeration applications. In: 3rd IIR Conference on Thermophysical Properties and Transfer Processes of Refrigerants, vol 155, pp 1–9 (2009)

Li, P.Y.: Total flooding gas protection system with FM-200. Int. J. Eng. Performance-Based Fire Codes (1):14–20 (2008)

Linteris, G.T.: Clean agent suppression of energized electrical equipment fires. Fire Technol. **47**, 1–64 (2011)

Linteris, G.T., Babushok, V.I., Sunderland, P.B., Takahashi, F., Katta, V.R., Meier, O.: Unwanted combustion enhancement by C6F12O fire suppressant. Proc. Combust. Inst. **34**(2), 2683–2690 (2018). https://doi.org/10.1016/j.proci.2012.06.050

Magid, H., Bray, R.G.: Atmospheric lifetime and global warming potential of HFC-245fa. J. Geophys. Res. **104**, 8173–8181 (1999)

Minor, B., Spatz, M.: HFO-1234yf low GWP refrigerant update. Int. Refrig. Air Cond. Conf., 14–17 (2015)

Minor, B.H., Herrmann, D., Gravell, R.: Flammability characteristics of HFO1234yf. Am Inst Chem Eng **29**(2), 150–154 (2010). https://doi.org/10.1002/prs

Neto, M.A.M.: Modeling of state and thermodynamic cycle properties of HFO-1234yf using a cubic equation of state. J. Braz. Soc. Mech. Sci. Eng. (2010). https://doi.org/10.1590/S1678-58782010000500005

Newman, P.A.: The way forward for montreal protocol science. C. R. Geosci. 2–7 (2018). https://doi.org/10.1016/j.crte.2018.09.001

NFPA 2001 World Fire Safety Congress. Can. Consult. Eng. (2001)

Nielsen, O.J., Wallington, T.J., Singh, R.: Atmospheric chemistry of CF3 CF=CH2: kinetics and mechanisms of gas-phase reactions with Cl atoms, OH radicals, and O3. Chem. Phys. Lett. **439**, 18–22 (2007). https://doi.org/10.1016/j.cplett.2007.03.053

Nielsen, O.J., Wallington, T.J., Singh, R.: Atmospheric chemistry of CF3CF=CH2: kinetics and mechanisms of gas-phase reactions with Cl atoms, OH radicals, and O3. **439**:18–22 (2007). https://doi.org/10.1016/j.cplett.2007.03.053

Nilsson M, Nes H, Kontomaris K (2017) Measured performance of a novel high temperature heat pump with HFO-1336mzz (Z) as the working fluid. 12th IEA Heat Pump Conf.

Ozone Secretariat, U.: Handbook for the Montreal Protocol on Substances That Deplete the Ozone Layer, 7th edn. Nuited Nation Environment Programme, Nairobi (2006)

Pagliaro, J.L., Linteris, G.T.: Hydrocarbon flame inhibition by $C_6F_{12}O$ (Novec 1230): unstretched burning velocity measurements and predictions. Fire Saf. J. **87**, 10–17 (2017). https://doi.org/10.1016/j.firesaf.2016.11.002

Sellami, I., Manescau, B., Chetehouna, K., De, I.C., Nait-said, R.: BLEVE fireball modeling using Fire dynamics simulator (FDS) in an Algerian gas industry. J. Loss Prev. Process Ind. **54**, 69–84 (2018). [September 2017]. https://doi.org/10.1016/j.jlp.2018.02.010

Siu, N., Apostolakis, G., Siu, N., Apostolakis, G.: A methodology for analyzing the detection and suppression of fires in nuclear power plants. Nucl. Sci. Eng. **5639**, 213–226 (2017). https://doi.org/10.13182/NSE86-A17264

Suryoputro MR Buana FA Sari AD Rahmillah FI (2018) Active and passive fire protection system in academic building KH. Mas Mansur, Islamic University of Indonesia. In: MATEC Web of Conferences https://doi.org/10.1051/matecconf/201815401094

Swain, S., Mohanty, S.: A 3-dimensional Eulerian–Eulerian CFD simulation of a hydrocyclone. Appl. Math. Model. **37**, 2921–2932 (2013). https://doi.org/10.1016/j.apm.2012.06.007

Takahashi, F., Katta, V.R., Linteris, G.T., Babushok V.I.: A computational study of extinguishment and enhancement of propane cup- burner flames by halon and alternative agents. Fire Saf. J. 1–7 (2017, February). doi:https://doi.org/10.1016/j.firesaf.2017.04.010

Ting, W., Ying-jie, H., Pin, Z., Ren-ming, P.: Study on thermal decomposition properties and its decomposition mechanism of penta fl uoroethane (HFC-125) fire extinguishing agent. J. Fluor. Chem. **190**, 48-55 (2016). https://doi.org/10.1016/j.jfluchem.2016.08.006

UNEP (2016) The Kigali Amendment to the Montreal Protocol: HFC Phase-Down https://www.unep.fr/ozonaction/information/mmcfiles/7809-e-Factsheet_Kigali_Amendment_to_MP.pdf.

Wang, Q., Duan, Q., Sun, J.: The efficiency of heptafluoropropane fire extinguishing agent on suppressing the lithium titanate battery fire. Fire Technol. (2015) [April 2016]. https://doi.org/10.1007/s10694-015-0531-9

Wang, X., Wu, R., Cheng, L., Zhang, X., Zhou, X.: Suppression of propane cup-burner flame with HFO-1336mzz(Z) and its thermal stability study. Thermochim. Acta **683**, 178463 (2020) [September 2019]. https://doi.org/10.1016/j.tca.2019.178463

Wang, X., Wang, W., Song, B., Lv, S., Liu, Z.: Measurement and correlation of viscosity of HFC227ea and HFC236fa in the vapor phase. Int. J. Refrig. **46**, 152–157 (2014). https://doi.org/10.1016/j.ijrefrig.2014.05.006

Wang, S., Guo, Z., Han, X., et al.: Experimental evaluation on low global warming potential HFO-1336mzz-Z as an alternative to HCFC-123 and HFC-245fa. J. Therm. Sci. Eng. Appl. (2019). https://doi.org/10.1115/1.4041881

Wang, Y., Wang, X., Zhang, X., Fu, H., Tan, Z., Zhang, H.: Theoretical and experimental studies on the thermal decomposition and fire-extinguishing performance of cis-1,1,1,4,4,4-hexafluoro-2-butene. Int. J. Quant. Chem. (2020). https://doi.org/10.1002/qua.26160

Wu, X., Dang, C., Xu, S., Hihara, E.: State of the art on the flammability of hydrofluoroolefin (HFO) refrigerants. Int. J. Refrig. (August 2019). https://doi.org/10.1016/j.ijrefrig.2019.08.025

Xu, W., Jiang, Y., Ren, X.: Combustion promotion and extinction of premixed counterflow methane/air flames by $C_6F_{12}O$ fire suppressant (2016). https://doi.org/10.1177/0734904116645829

Yang, Z., Wu, X.: Retro fits and options for the alternatives to HCFC-22. Energy **59**(2013), 1–21 (2015). https://doi.org/10.1016/j.energy.2013.05.065

Yang, R., Khan, F., Yang, M., Kong, D., Xu, C.: A numerical fire simulation approach for effectiveness analysis of fire safety measures in floating liquefied natural gas facilities. Ocean Eng. **157**, 219–233 (2018). https://doi.org/10.1016/j.oceaneng.2018.03.052

Yang, R., Khan, F., Yang, M., Kong, D., Xu, C.: A numerical fire simulation approach for effectiveness analysis of fire safety measures in floating lique fied natural gas facilities. Ocean Eng. **157** 219–233 (2018). https://doi.org/10.1016/j.oceaneng.2018.03.052. [December, 2017]

Yeturu, S.K., et al.: Assessment of knowledge and attitudes of fire safety—an institution based study. J. Pharm. Sci. Res. **8**(11), 1281–1284 (2016)

Zeng, F.P., Lei, Z.C., Miao, Y.L., Yao, Q., Tang, J.: Reaction thermodynamics of overthermal decomposition of $C_6F_{12}O$. In: Lecture Notes in Electrical Engineering. (2020). https://doi.org/10.1007/978-3-030-31680-8_5

Zhang, X., Halim, M., Ismail, S., Hee, C.: Hot aerosol fire extinguishing agents and the associated technologies: a Review. Braz. J. Chem. Eng. **32**(03), 707–724 (2015)

Zhang, X., Ismail, M.H.S., Ahmadun, F.R.B., Abdullah, N.B.H., Hee, C.: Hot aerosol fire extinguishing agents and the associated technologies: a review. Braz. J. Chem. Eng. **32**(3), 707–724 (2015). https://doi.org/10.1590/0104-6632.20150323s00003510

Zhang, X., Tian, S., Xiao, S., Deng, Z., Li, Y.: Insulation strength and decomposition characteristics of a $C_6F_{12}O$ and N_2 gas mixture. (2017). https://doi.org/10.3390/en10081170

Zhiyin, Y.: Large-eddy simulation: past, present and the future. Chin. J. Aeronaut. **28**, 11–24 (2015)

Chapter 28
Design and Analysis of Fire Extinguisher Nozzle Spray Using Design of Experiment Method

Abdul Shukor bin Jum'azulhisham, Abdul Razak bin Muhammad Al-Hapis, A. Rhaffor Kauthar, Hassan Azmi, and Krishnan Pranesh

Abstract A fire extinguisher is the first-line defense system against fire eruption in the early stage. The conventional nozzle that has been used with the current fire extinguisher is not fully effective as the user needs to get closer to the fire sources to extinguish it. This will bring potential hazards to the user himself as the fire sources are closer. The objective of this paper, are to design, test a new fire extinguisher nozzle, and configure the diameter and length of the nozzle. Affected parameters such as velocity and pressure at the nozzle outlet also have been studied. The optimal parameter configuration of the nozzle is the final aim of this project. In this study, besides experimentation method, flow simulation and analysis in SolidWorks 2016 software were used alongside with the Taguchi method (L9) to find the optimum parameter for the nozzle design. Data shows for the Taguchi method study, in terms of pressure the best combination is level 1 with a factor of A (50 mm diameter) and level 3 factor of B (50 mm length) as it produces the lowest reading of 10.10 bar and 10.110 bar, respectively. Meanwhile, as for the velocity data, the best combination is level 1 with a factor of A (50 mm diameter) and level 1 with a factor of B (40 mm length) which obtained a reading of 54.34 m/s and 31.6 m/s, respectively, greater compared to others. Thus, this study concluded with the optimum configuration for

A. S. Jum'azulhisham (✉) · A. R. M. Al-Hapis · A. R. Kauthar · H. Azmi
Manufacturing Section, Universiti Kuala Lumpur Malaysian Spanish Institute, Kulim Hi-Tech Park, 09000 Kulim, Kedah, Malaysia
e-mail: jumazulhisham@s.unikl.edu.my

A. R. M. Al-Hapis
e-mail: alhapis@unikl.edu.my

A. R. Kauthar
e-mail: kauthar@unikl.edu.my

H. Azmi
e-mail: azmi.hassan@unikl.edu.my

K. Pranesh
Intelligent Automotive Systems Research Cluster, Electrical Electronic and Automation Section, Universiti Kuala Lumpur Malaysian Spanish Institute, Kulim Hi-Tech Park, 09000 Kulim, Kedah, Malaysia
e-mail: pranesh@unikl.edu.my

M. H. Abu Bakar et al. (eds.), *Progress in Engineering Technology III*,
Advanced Structured Materials 148, https://doi.org/10.1007/978-3-030-67750-3_28

the new nozzle outlet 2 diameter of 50 mm in diameter and a nozzle length of 50 mm in length as this configuration produces the best results.

Keywords Dry powder fire extinguisher · CAD · Angular distortion · Design of experiment

28.1 Introduction

A fire extinguisher or fire expressant is an active fire protection tool used to put out or control small fires, usually in emergency situations. Due to the size, it increases the portability as to be carried around and stored in many places where it does not consume a lot of space. The fire extinguisher has a wide range of application based on the type of fire that needs to be handled as it has a specific type for the different types of fire needed to be handled. Although, it is not intended for use on an out-of-control fire, which can bring harm and endangers the user, or otherwise requires the expertise of a fire brigade. The researcher found that the design of the nozzle of the fire extinguisher needs to be studied. Hence, it is believed that effectiveness and efficiency can be improved. This project focuses on the study of effective nozzle spray by simulating and experimenting using a robust statistical method using the design of experiment (DOE) with the nozzle design and using the dry powder agent type. The study will emphasize the development of conceptual design, simulation, and experimental study to suggest the best parameters or factors that can affect the outcome of the spray effectiveness. Factors affected in fire extinguisher nozzle design may include nozzle length, angle, and diameter. The prototype will then to be set for an experiment to observe its effectiveness.

Dry Powder: This is also known as ABC dry powder fire extinguisher. The main constituents of ABC powders are ammonium dihydrogen phosphate (Li et al. 2019) and ammonium sulphate.

The ammonium dihydrogen phosphate is a white crystal, and its molecular formula is $NH_4H_2PO_4$, soluble in water, and slightly soluble in ethanol. It is mainly used for producing fertilizer and extinguishers. The ABC powder is often referred to as general-purpose or multi-purpose extinguisher powder and is capable of fighting class A, B, C, D, and Electrical fires except Class F the cooking oil and fat.

Taguchi Method: The Taguchi technique was applied in this study, where an orthogonal experimental array reduces much more variance and minimizing the process parameters (Lodhi and Agarwal 2014; Athreya and Venkatesh 2012). This array was set up the way of conducting the minimal number of experiments by minimizing it, which then results in the full information of all the parameters. The parameters are plugged into the orthogonal arrays in sequence, and then experiments will be run step-by-step by covering all of the factors of parameters. There are so many types of arrays that can pick from all of the arrays collection. The orthogonal array was selected by determining the parameter table and its variable option is shown in Table 28.1.

Table 28.1 Standard L9 orthogonal array

Experiment no	Factor A	Factor B
1	1	1
2	1	2
3	1	3
4	2	1
5	2	2
6	2	3
7	3	1
8	3	2
9	3	3

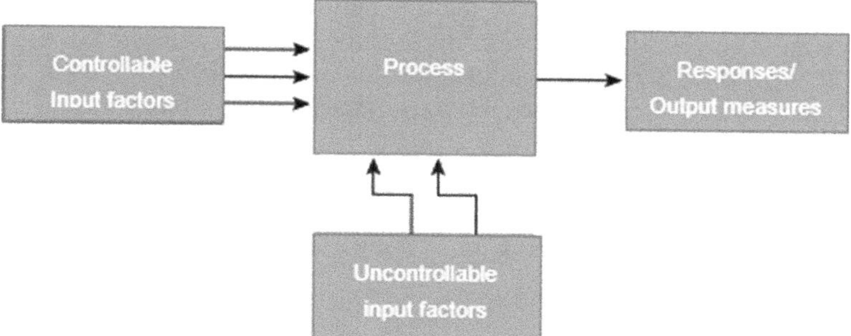

Fig. 28.1 Machine flow

Design of experiments: Design of experiments (DOE) is a systematic method to determine the relationship between factors affecting a process and the output of that process (Islam and Pramanik 2016). In other words, it is used to find cause-and-effect links. This information is needed to manage process inputs to optimize the output. It is used to determine the relationship between factors affecting the process and process responses. Figure 28.1 shows the flow of machine that is involved in DOE.

28.2 Methodology and Experimental Etup

28.2.1 Initial Experiment

The fire extinguisher nozzle experiment was conducted in a closed huge environment to ensure the data obtained covered the whole space and avoid wind, temperature and pressure disturbance. The large space allows the researchers to closely monitor the dust particles coverage. It also allows the researchers to measure the distance should be because the particles can be seen and covered the large area. The safety and health issues is also important because the fine dry powder particles can be hazardous to human respiratory system. These factors can give a significant impact on the experiment result.

28.2.2 Design Stage

The bi-nozzle system operates in dual mode. Mode one, as shown in Fig. 28.2(a), works with only one outlet, which is the center outlet with 12 mm of diameter. The working principle behind this mode is when the nozzle is fully closed, the pathway for the dry powder to travel through is only limited to one. The aim is to create a high pressure and velocity stream of the fine particles flow out through the outlet, thus creating a narrow spray distance. As for mode two, as presented in Fig. 28.2. Figure 28.2(b), the dry powder particles for this mode flow two direction through outlet 1 and outlet 2. For the current nozzle, outlet 2 has a bigger diameter with 52 mm. The working principle of these outlets requires the nozzle to turn counter clockwise to allow the particles travel through two pathways. This is to allow the fluid to pass through a more significant diameter outlet (outlet 2) that create a wider spray angle with a slow velocity. Figure 28.3 illustrate the evolution of the bi-nozzle design to the existing one.

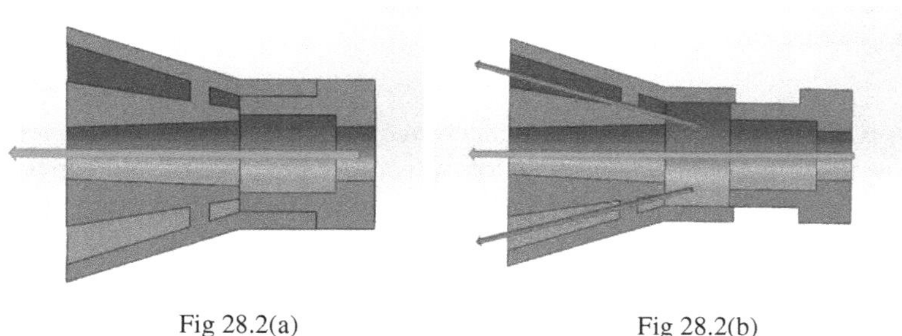

Fig 28.2(a) Fig 28.2(b)

Fig. 28.2 (a) and (b) Mode 1 Fig. 28.2(a) and Mode 2 Fig. 28.2(b) working principles

Fig. 28.3 Nozzle exterior design improvement

28.2.3 Simulation

The design begins with the setting up the project in SolidWorks. In the study on the nozzle, an internal analysis type was selected to simulate an internal flow. A list of pre-defined fluid already exists in SolidWorks but ammonium dihydrogen phosphate (Senthilkumaran et al. 2012) was not included in the library. This needs to be added manually. The wall was assumed to be adiabatic. The roughness of the internal nozzle is also assumed to be zero. The lid, which is an option that has on 'Flow Simulation' is added at first to the project inlet and outlet to allow SolidWorks to recognize the flow as internal. The boundary condition was set in the condition of fluid flowing out. The inner face of the lid was selected because the face needs to be laying on the boundary between the solid and fluid region where the mesh was also set. Selecting a higher mesh number will result better accuracy of the reading also increases as a higher number of computational points is used. On the contrary, a lower number of mesh will lower the computational time, but a lower accurate result will be obtained.

28.2.4 Design of Experiment Using the Taguchi Method

The Taguchi Design method was selected for the optimization (Athreya and Venkatesh 2012). For this study 3-Level design was selected. Two factors which are the outlet two diameters and length of the nozzle L9 orthogonal array was chosen and it produces 3^2 column. Numerical data is set on the control factor column (C1 and C2) and the response data column (C3). As shown in Fig. 28.4, the Taguchi design was selected and two control factor are chosen to be studied. The analyzed Taguchi design was selected and proceeded with result selection (C3). This generates a plot of main effects and interaction in the model for signal-to-noise ratios. For the purpose of this study, S/N Ratio for this function η, the Smaller-the-Better was selected as illustrated in Eq. 28.1.

$$\eta = -10\log_{10}\left(\frac{1}{n}\sum_{i-1}^{n} y^2 i\right) \tag{28.1}$$

It displays the main effects plot and interaction plots for means. Select the signal-to-noise goal for the response. In this study, the goal is to minimize the pressure (smaller is better) and to maximize the velocity (larger the better).

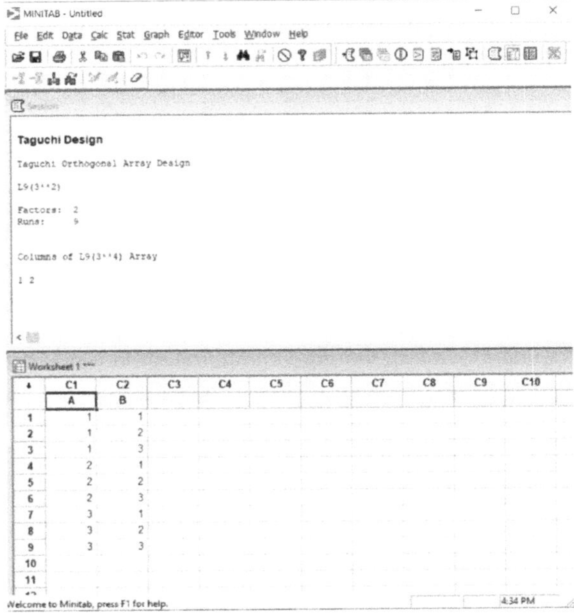

Fig. 28.4 Taguchi DOE setup with L9 orthogonal array using Minitab software

28.3 Results and Discussion

28.3.1 Initial Experiment Results

The experiment was run three times for each mode to find the average spray angle and spray distance. The total, 9 units of fire extinguisher are used loaded with 14 bar pressure as per standard requirement. This is to ensure that any pressure variance from the fire extinguisher tank can be eliminated. An image processing software called ImageJ was used to determine the spray angle of both modes while the distance was measured manually. For the spray angle, the software is able to detect the highest pixel point to the source on outlet then back to the lowest pixel point for the chosen spray as presented in Fig. 28.5 and Fig. 28.6.

In this experiment, there were three variable parameters selected. The variable setting are the distance, angles and nozzle modes. The three levels of variable parameters; low, medium and high) are shown in Table 28.2.

28.3.2 Nozzle Simulation Result

A total of 9 experiments consists of combinations of different outlet 2 diameter and nozzle length. Concept design 2 is chosen to run the simulation as it is the latest

Fig. 28.5 Experimenting nozzle design with mode 1

Fig. 28.6 Experimenting nozzle design with mode 2

improvement from the other two designs. The simulation took place under a velocity of 100 m/s discharge speed and 14 bar pressure at the inlet of the nozzle as fixed by the manufacturer. The simulation was implemented to all nine steps in order to find the impact toward the outlet 2 pressure and velocity that affected the spray angle and distance.

Figure 28.9 presents the pressure curve that is affected by the length of the nozzle. It clearly can be seen that the pressure at outlet 2 is inversely proportional to the

Table 28.2 Process parameters and level of two modes with 3 trials each

No. of trial	Mode 1 (at 14 bar)		Mode 2 (at 14 bar)	
	The distance of coverage (m)	Spray angle	The distance of coverage (m)	Spray angle
Trial 1	12	8.1°	9	14.2°
Trial 2	10	10.2°	8	13°
Trial 3	11	11.7°	10	15.8°
Average	11	10°	9	14.3°

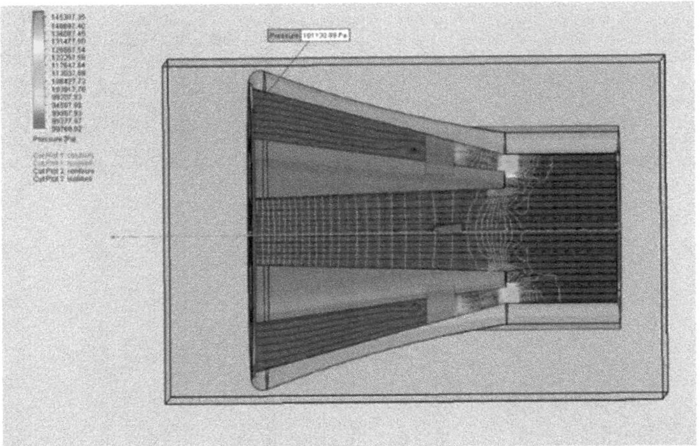

Fig. 28.7 Outlet 2—pressure reading of Mode 2 nozzle

nozzle length. As the nozzle length increases, the pressure will be decreased. The graph pattern is in a gradually decreasing manner. This is because, as the powder goes through along the nozzle, the pressure will continue to decrease due to the loss of potential energy. Figure 28.7 shows the pressure reading during Mode 2 nozzle was set. While, the velocity reading of Mode 2 nozzle set as illustrated in Fig. 28.8.

Presented in Fig. 28.10 is the velocity of the outlet 2 against the nozzle length. It shows that as the length of the nozzle increases the velocity happening on outlet two also increases. This is because the velocity is directly proportional to the length of the nozzle. Based on this graph, it is concluded that the higher the length and the diameter of the outlet, the faster the velocity.

28.3.3 Taguchi Method

The Taguchi method was conducted using the Minitab software (Khavekar et al. 2017; Arnold 2006; Zaman et al. 2019). The parameters and the results for these

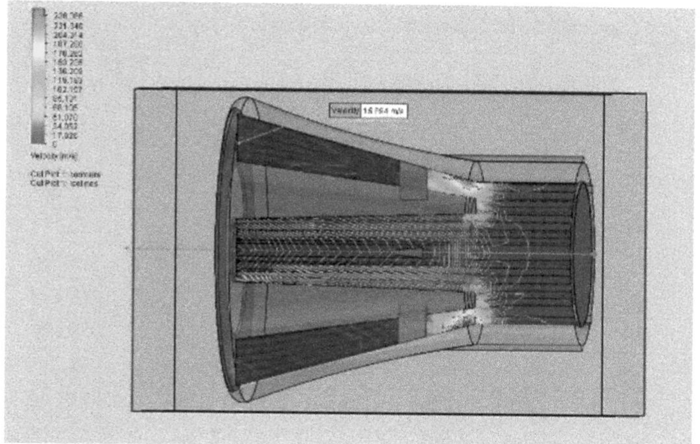

Fig. 28.8 Outlet 2—velocity reading of Mode 2 nozzle

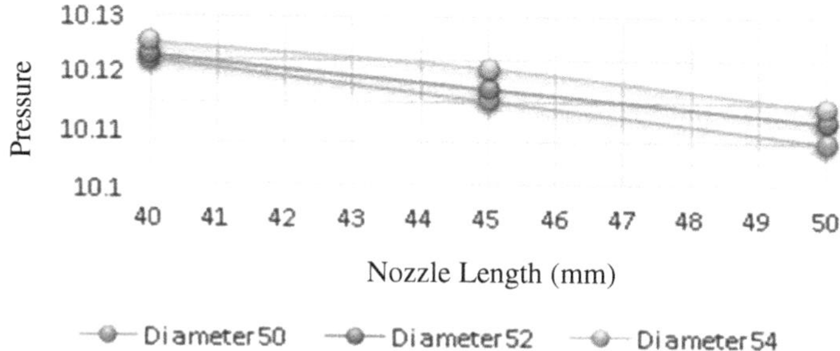

Fig. 28.9 Outlet 2 pressure curve

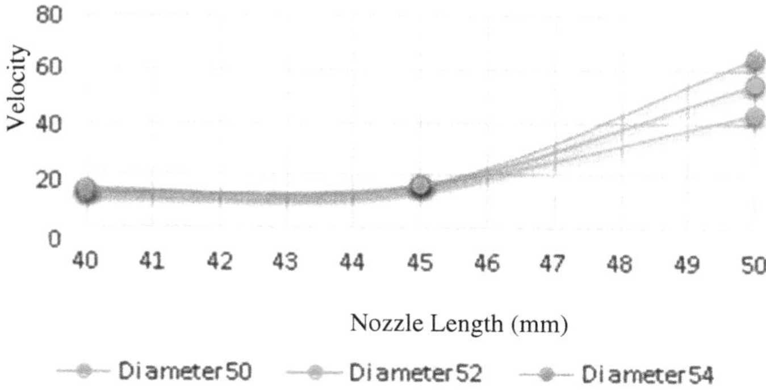

Fig. 28.10 Outlet 2 velocity curve

Table 28.3 Nine experiment with 2 parameters set and results

Experiment	Parameters		Outlet 2 pressure (bar)	Outlet 2 velocity (m/s)
	Outlet 2 diameter (mm)	Nozzle length (mm)		
1	50	40	10.12	12.58
2	50	45	10.11	15.954
3	50	50	10.10	43.481
4	52	40	10.121	13.386
5	52	45	10.113	15.894
6	52	50	10.105	54.860
7	54	40	10.124	14.233
8	54	45	10.118	15.899
9	54	50	10.1087	64.678

experiments were keyed in, a table of parameters setting and the result were produced. The parameters are the two nozzle outlet diameters and the nozzle length. These parameter are studied to find how it affects the results, which are the pressure and velocity at outlet 2. Table 28.3 shows all component required in the Taguchi method.

28.3.4 Taguchi Analysis: Result Pressure A, B

Based on the means graph in Fig. 28.11, it can be concluded that a low level of nozzle length would give a greater impact toward the result factors compare to low levels of nozzle diameter. In order to get the best effect based on the main effect data, the lower the nozzle length, the better. This is because the nozzle length factor produces the best effect toward the value of pressure on the outlet two and the best out of it was the level 3 (10.10 bar) compared to level 1 (10.12 bar) and level 2 (10.11 bar) referring to Table 28.5. For the diameter of outlet, two factors bring not so much impact to response: level 1 (10.110 bar) gives the lowest impact compared to level 2 (10.113 bar) and level 3 (10.117 bar). To summarize the optimum parameters to minimize the pressure is level 3 for the A factor and level 1 for the B factor. Referring S/N Ratios in Fig. 28.11, it shows that the B factor is the higher value of the signal-to-noise ratio (S/N) that has minimized the effect of the noise factors S/N highest value, which is a factor of B (means = −20.09) compared by a factor of A (mean = −20.095). Smaller-is-better was chosen for this study because it reduces the pressure on outlet 2. Equation 28.1 was applied in the signal-to-ratio. Smaller-is-better is implemented for outlet two pressure because a smaller value is needed. Thus the best combination is level 1 for the factor of A (50 mm) and level 3 for the factor of B (50 mm) as illustrated in Table 28.4.

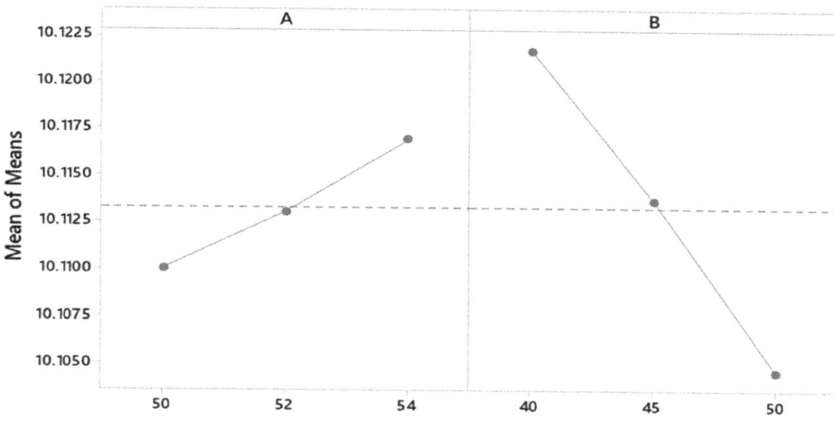

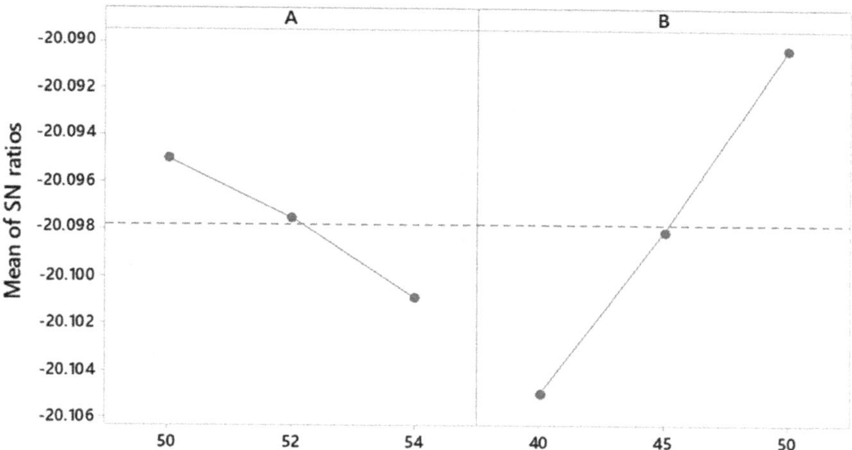

Fig. 28.11 Main effects plot for means and main effect plot S/N ratios

Table 28.4 Response table for the signal-to-noise ratio

Level	A	B
1	−20.10	−20.11
2	−20.10	−20.10
3	−20.10	−20.09
Delta	0.01	0.01
Rank	2	1

Table 28.5 Response table for mean

Level	A	B
1	10.11	10.12
2	10.11	10.11
3	10.12	10.10
Delta	0.01	0.02
Rank	2	1

28.3.5 Taguchi Analysis: Result Velocity A, B

Meanwhile based on Fig. 28.12 it has been concluded that the low level of nozzle length would give greater impact toward the result factors compare to low levels of nozzle diameter. To get the best effect based on the main effect data: the lower the nozzle length, the better. This is because the nozzle length factor produces the best effect toward the value of velocity on the outlet two and the best out of it was the level 3 (54.34 bar) compared to level 1 (13.40 bar) and level 2 (15.9 bar) referring to Table 28.7. For the diameter of outlet two, factors which bring not so much impact to response are: level 1 (24.00 bar) give the lowest impact compare to level 2 (28.05 bar) and level 3 (31.6 bar). To summarize the optimum parameter of the S/N ratio of Factor B from Table 28.6 and response for mean of factor B shown in Table 28.7 are the best.

Figure 28.12 shows the B factor gives the highest value of the signal-to-noise ratio (S/N) that has minimized the effect of the noise factors. S/N highest value is factor of B (means = 34.59) compared by factor of A (mean = 27.77). The smaller-the-better was chosen for this study because the goal os to reduce the pressure on outlet 2. The larger-is-better was implemented to the outlet two pressures because larger the value is needed for the velocity. Thus the best combination is level 1 factor of A (50 mm) and level 1 factor of B (40 mm). Although the observation from the velocity of outlet two may not be correct as it is to be opposite from the means graph.

28.4 Conclusion and Recommendation

In a nutshell, this study has concluded that in term of pressure at the outlet with two nozzles, the configuration of 50 mm diameter and 50 mm nozzle length is the best option. This is because the parameters set up, produce the lowest pressure read at outlet 2 in the simulation. The pressure recorded was 10.10 bar, slightly lower than the pressure of the current nozzle configuration, 10.113 bar. With 10.13 bar of pressure at the Outlet 2, average spray angle is 14.3°. The best configuration determined from using the Taguchi method were 50 mm diameter and 50 mm nozzle length. By using the Taguchi's method also in term of pressure, stated that the factor of nozzle length is the biggest distributor of pressure reduction compared to nozzle

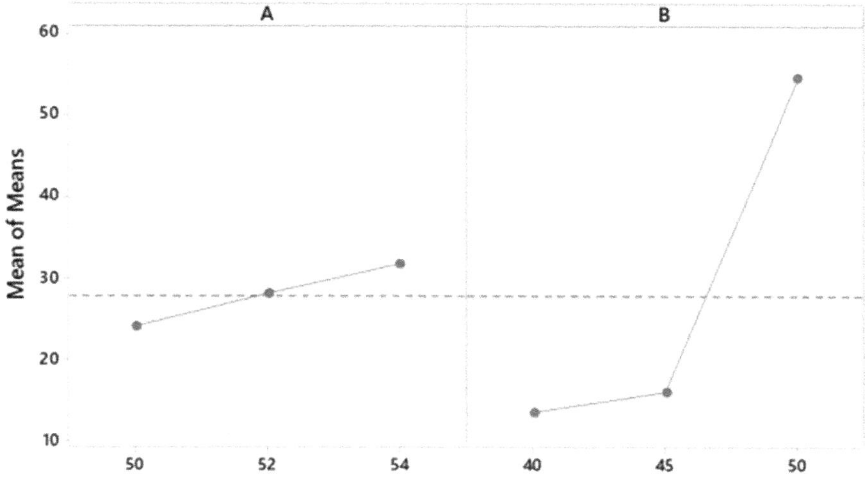

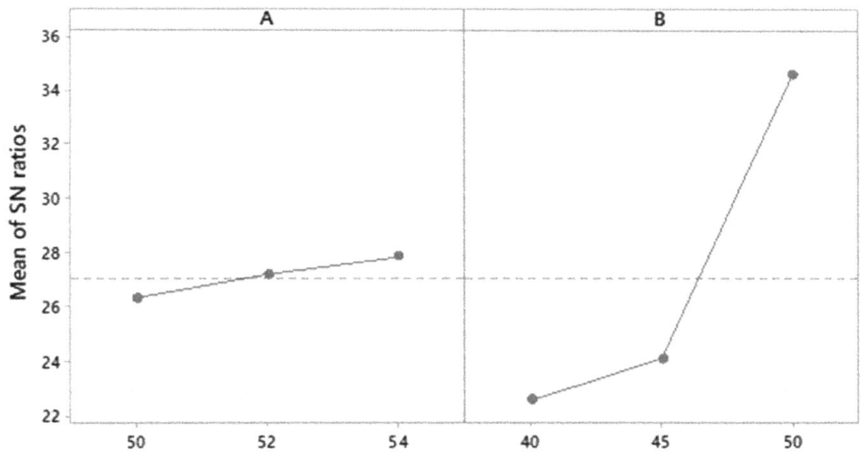

Fig. 28.12 Main effect plot for means and main effect plot SN ratios

Table 28.6 Response table for the signal-to-noise ratio

Level	A	B
1	26.27	22.53
2	27.11	24.04
3	27.77	34.59
Delta	1.50	12.06
Rank	2	1

Table 28.7 Response table for mean

Level	A	B
1	24.00	13.40
2	28.05	15.92
3	31.60	54.34
Delta	7.60	40.94
Rank	2	1

diameter. However, in terms of velocity at the outlet, it still needs to be studied as it defies one crucial aspect which was as the diameter increases, the surface area will also increase, thus it would make the velocity slower. Not in this case, the simulation experiment found out that as the diameter increase, the velocity will also increase thus making this part of the study not valid and need further investigation.

References

Arnold SF (2006) Design of Experiments with MINITAB. Am Stat. https://doi.org/10.1198/tas.2006.s46

Athreya, S., Venkatesh, Y.D.: Application of Taguchi method for optimization of process parameters in improving the surface roughness of lathe facing operation. Int. Refereed J. Eng. Sci. **1**(3), 13–19 (2012)

Islam, M.N., Pramanik, A.: Comparison of design of experiments via traditional and taguchi method. J. Adv. Manuf. Syst. **15**(3), 151–160 (2016)

Khavekar, R., Vasudevan, H., Modi, B.: A comparative analysis of Taguchi methodology and Shainin System DoE in the optimization of injection molding process parameters. In: IOP Conference Series: Materials Science and Engineering, vol. 225, p. 012183 (2017). https://doi.org/10.1088/1757-899x/225/1/012183

Li, H., Feng, L., Du, D., Guo, X., Hua, M., Pan, X.: Fire suppression performance of a new type of composite superfine dry powder. Fire Mater (2019). https://doi.org/10.1002/fam.2750

Lodhi, B.K., Agarwal, S.: Optimization of machining parameters in WEDM of AISI D3 steel using Taguchi Technique. In: Procedia CIRP (2014). doi:https://doi.org/10.1016/j.procir.2014.03.080

Senthilkumaran, S., Meenakshisundaram, R., Balamurgan, N., SathyaPrabhu, K., Karthikeyan, V., Thirumalaikolundusubramanian, P.: Fire extinguisher: an imminent threat or an eminent danger? Am. J. Emerg. Med. (2012). https://doi.org/10.1016/j.ajem.2011.01.018

Zaman UK uz, Boesch, E., Siadat, A., Rivette, M., Baqai, A.A.: Impact of fused deposition modeling (FDM) process parameters on strength of built parts using Taguchi's design of experiments. Int. J. Adv. Manuf. Technol. **101**(5–8), 1215–1226

Chapter 29
Roles of 5S Practice in MARA Ecosystem Quality Standard (MQS)—A Case Study

Kauthar A. Rhaffor, Aida Salwani Mohamed, Rahim Jamian, and Pranesh Krishnan

Abstract The practice of 5S has been adopted by many organizations as a method to organize the workplace towards reducing waste and improving productivity. This study aims to explore the role of 5S practice in Majlis Amanah Rakyat (MARA) Ecosystem Quality Standard (MQS). MQS is a strategic MARA effort in ensuring the culture of quality and conducive working environment as practice in all MARA administration centres. In this study, a survey has been conducted through the distribution of the questionnaire to 100 employees of the selected university to identify the employees' perception on the role of 5S practice in MQS system, as well as the impact of the system to organization performance. The questionnaire was validated by a pilot test with an acceptable value of Cronbach's alpha. Data analysis was carried out using IBM SPSS and the Delphi technique conducted to get the feedback and comment from the panel of experts in this field for validation and evaluation. As the data being analysed, the structured interview with the representatives from management of the university also has been performed to support the preliminary data obtained from previous studies. The initial framework has been formed based on secondary data obtained from literatures and it has been improved according to the data gained in this study. Overall findings highlight there was a positive perception towards 5S practice among the respondents and there is also positive relationship between 5S practice in MQS system with organization performance.

K. A. Rhaffor (✉) · A. S. Mohamed
Manufacturing Section, Malaysian Spanish Institute Universiti Kuala Lumpur, Kulim Hi-Tech Park, 09000 Kulim, Kedah, Malaysia
e-mail: kauthar@unikl.edu.my

R. Jamian
Faculty of Engineering Technology, Universiti Tun Hussein Onn Malaysia (UTHM), Kampus Cawangan Pagoh, Hab Pendidikan Tinggi Pagoh, KM 1, Jalan Panchor, 84600 Pagoh, Muar, Johor, Malaysia

P. Krishnan
Intelligent Automotive Systems Research Cluster, Electrical Electronic and Automation Section, Malaysian Spanish Institute Universiti Kuala Lumpur, Kulim Hi-Tech Park, 09000 Kulim, Kedah, Malaysia
e-mail: pranesh@unikl.edu.my

M. H. Abu Bakar et al. (eds.), *Progress in Engineering Technology III*,
Advanced Structured Materials 148, https://doi.org/10.1007/978-3-030-67750-3_29

Keywords 5S practice · Majlis amanah rakyat (MARA) ecosystem quality standard (MQS)

29.1 Introduction

The Malaysian Administrative Modernization and Management Planning Unit (MAMPU) has agreed to allow the use of the Majlis Amanah Rakyat (MARA) Ecosystem Quality Standard (MQS) at all MARA administrative centres. The MQS was established to identify the critical elements and components of a conducive working environment. It will continue to improve the quality of service delivery system among MARA employees and increase the overall performance of MARA. It also standardizes the implementation, workmanship and monitoring of the MQS culture in all MARA administration centres in order to improve the productivity and quality of the delivery system. The MQS is a new management system to be implemented at all MARA centres and the system is in the introductory stage at the private university selected for this study.

The 5S practice is often viewed as an effective method to increase efficiency and productivity. The main purpose of 5S is to improve efficiency by eliminating the waste of motion looking for tools, materials, or information (Howell 2005). This practice has been implemented in the selected university for 3 years, from 2014 to 2017, and they have successfully obtained 5S Certification from Malaysia Productivity Corporation (MPC). In the selected university, the study on the effectiveness of 5S practice has been conducted by other researchers (Azwan Huzaimi et al. 2015). However, there was no study conducted on the role of 5S practice in MARA Ecosystem Quality Standard (MQS) system. Therefore, this study attempted to explore the status and relationship of 5S practice in MARA MQS system in the university. Furthermore, this study was also conducted to explore the significance of 5S practice as the positive factor contributed towards the improvement of MQS system performance. Framework of 5S Practice in MQS was established and the data collected through questionnaire was analyzed to identify the relationship between the variables involved in the framework.

The 5S practice was popularized as 'Japanese 5S' in 1980 by Hiroyuki Hirano and it is known as a method of organizing the work area. In order to improve the workplace efficiency and eliminate waste, the 5S system was always used as a lean manufacturing tool in the organizations. It also promotes a safe and efficient environment. By applying 5S practices, it will assist organization to have tidier workplaces that lead to fewer hazards and reduce risk. 5S stands for Japanese words consisting of *Seiri* (Sort), *Seiton* (Set in Order), *Seiso* (Shine), *Seiketsu* (Standardize) and *Shitsuke* (Sustain). The term was formalized by Takashi Osada in 1980 (Ho et al. (1995); Gapp et al. 2008). The word *Seiri* refers to selecting and sorting the elements into two main categories: essential and nonessential. *Seiton* consists of establishing the adequate manner for locating and identifying the essential materials so that they can be easily accessible (Ho 1999). *Seiso* seeks to maintain the workspace under clean conditions by having a regular schedule for removing dirt and dust (Osada 1991). The fourth step, *Seiketsu* indicates that everything should be easy to identify and with clearly visible labels for all operators (Becker 2001). *Shitsuke*, the final step in 5S practice consists of the crucial steps to sustain each of the five S.

29.2 Methodology

29.2.1 *Respondent*

A sample size of 30 is held by many to be the minimum number of cases for research involving statistical analysis on their data (Cohen 1997). For this study, the researchers managed to distribute a questionnaire to 100 academicians and support staff randomly selected from academic sections and administration departments including workshops and labs area.

29.2.2 *Data Collection and Instrument*

The survey approach was employed where self-administered questionnaires were distributed. Other method such as focus groups, interviews, conversations or workshops and other activities will consume more of the respondents time. Thus, the survey method using the distribution of questionnaires is the most effective technique to attain fast and accurate response for this study. The questionnaire were divided into five (5) sections; Section A: Demographic Information, Section B: 5S Practice, Section C: The Nature of MQS Readiness, Section D: Critical Success Factors of MQS System, Section E: Impact of MQS System. The questions for Section B to Section E were rated using the 4 point Likert scale (1: Strongly Agree; 2: Agree; 3: Disagree; 4: Strongly Disagree). The questionnaire was prepared in English language.

29.2.3 *Pre-Test (Validity Test)*

Pre-Test or validity test is a method to measure the level of the question on each item in the questionnaire. The questionnaire were validated by the selected experts. Three academicians who have expertise in the area of Quality Management were selected as the experts to validate the questionnaire.

Table 29.1 The value of Cronbach's alpha (α) (Sekaran 2000)

Cronbach's alpha	Internal consistency
$\alpha \geq 0.9$	Excellent
$0.9 > \alpha \geq 0.8$	Good
$0.8 > \alpha \geq 0.7$	Acceptable
$0.7 > \alpha \geq 0.6$	Questionable
$0.6 > \alpha \geq 0.5$	Poor
$0.5 > \alpha$	Unacceptable

29.2.4 Pilot Test (Reliability Test)

Reliability test is the reliability of a measurement which shows the stability and consistency of a measuring instrument makes. The determination of reliability for this interview questions were based on the Cronbach's alpha value obtained using multivariate item analysis. Cronbach's alpha is a coefficient of reliability. Table 29.1 shows the accepted values of Cronbach's alpha (Sekaran 2000). After the improvement on the structure and the content of the questionnaire have been made during pre-test, a reliability test was conducted by distributing the questionnaire randomly to 30 staffs from various departments.

29.2.5 Initial Formation of Framework

This study focused on the effectiveness of 5S practice towards the MQS system. However, the quality, safety and environmental performance also can be influenced by the impact towards MQS system performance. Therefore, based on secondary data, the initial framework has been formed. Figure 29.1 shows the relationship between the variables involved in the framework. The critical success factors of 5S practice in MQS system are listed as independent variables that will affect the dependent variables which consist of the indicators for operational performance.

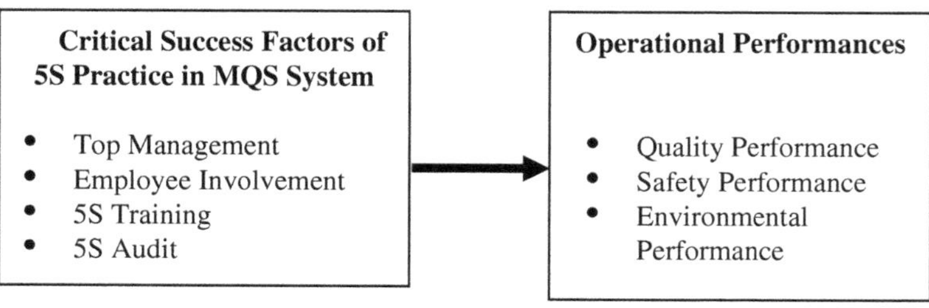

Fig. 29.1 Initial framework

Table 29.2 Strength of correlation coefficient (Davies 1971)

Correlation coefficient value (r)	Correlation strength
0.70–1.00	Very strong
0.50–0.69	Strong
0.30–0.49	Moderate
0.10–0.29	Weak
0.01–0.09	Very weak

29.2.6 Correlation Analysis

The degree of strength of the relationship between the study variables was identified using correlation analysis. The relationship between the study variables is referred to as correlation, while the correlation strength in an analysis is represented by the correlation coefficient value (r). The stronger the association of the two variables, the closer the correlation coefficient, r, will be to either $+1$ or -1 depending on whether the relationship is positive or negative, respectively. For this study, the strength of correlation is determined according to the size of correlation coefficient proposed by Davies (Davies 1971) as shown in Table 29.2. The data were analyzed using IBM SPSS.

29.2.7 Delphi Validation Analysis

Delphi technique is also known as the Delphi method, is a forecasting or estimating method based on a discussion by a group of experts. These respondents were selected based on their experience and expertise, particularly in the area of quality and safety management. The implementation procedure of the Delphi technique needs to involve multiple rounds for the purpose of verifying the topic, data collection and data analysis. In this study, Delphi technique is used to validate the established framework by the distribution of Delphi Questionnaire.

29.3 Results and Discussion

29.3.1 Pre-Test (Validity Test)

Pre-Test or validity test (face validity test) was conducted to make an adjustment or correction on the questionnaire. The questions were improved based on the comments by the experts.

Table 29.3 Internal consistency test results

Variables	Cronbach's alpha
Section B: 5S practice	0.8880
Section C: the Nature Of MQS readiness	0.8880
Section D: critical success factor of MQS system	
• Top management commitment	0.9531
• Critical success: employee involvement	0.8660
• MQS audit	0.9513
• MQS training	0.9844
Section E: impact of MQS system	
• Safety performance	0.9556
• Quality performance	0.9661
• Environment performance	0.9240

29.3.2 Pilot Test (Reliability Test)

Internal consistency test results are shown in Table 29.3. Based on the results, it shows that all variables measured were above acceptable level. Thus, no further changes needed for the questionnaire.

29.3.3 Actual Test

29.3.3.1 Respondent Information

Table 29.4 shows the respondents information. Based on the information, it can be concluded that 67% of the respondents are academicians and 71% of the respondents' academic qualification are bachelor degree and above.

29.3.4 Correlation Between Variables

Based on the results in Table 29.5, it shows that Safety and Quality Performance have moderate to strong correlations with all the critical factors studied. The Top Management Commitment has a significant positive relationship with all operating performances. It has moderate correlation with Safety and Quality performance and only weak correlation with Environmental Performance. Top management commitment is always considered as one of the most crucial factors in any initiative implemented in organizations. Previous study conducted by Kauthar et al. (2019) also identified a strong positive correlation between top management commitment and safety performance. The study conducted by Ablanedo-Rosas et al. (2010) has identified that the

Table 29.4 Demographic information

Gender	Frequency ($N = 100$)	Percentage (%)
Male	48	48
Female	52	52
Age		
21–25	3	3
26–30	23	23
31–40	57	57
41 and above	17	17
Education		
SPM	11	11
Diploma	16	16
Degree	24	24
Master	26	24
PhD	23	23
Profession		
Support staff	31	31
Academician	67	67
Others	2	2

Table 29.5 Correlation between variables

5S practice critical factors in MQS	Performance		
	Safety	Quality	Environmental
Top management commitment	0.312	0.304	0.244
Employee involvement	0.614	0.317	– 0.121
5S audit	0.352	0.303	0.199
5S training	0.464	0.304	0.225

commitment from top management will help the organization to face the challenges in implementing 5S.

The analysis has identified strong correlation strength between Employee Involvement and Safety Performance. Similar as top management commitment, employee involvement is another crucial factors need to be taken into account for any initiative to be implemented in organizations. 5S Audit and 5S Training are both very important factors to ensure the success of MQS implementation in the university. To ensure the right implementation of 5S practice, 5S Audit is considered crucial (Ho 1999). Since the workforce was trained to understand the fundamental of 5S practice, and they also familiar with the 5S Audit process, therefore those experiences have helped them to prepare for MQS system. The results shown that the respondents

Table 29.6 Benefits and barriers of MQS

Panel	Benefit of MQS	Barriers of MQS
Panel 1: senior executive for administration	This system helps to improve the 5S practice in the university	• Most of the employees do not be familiar with the system • More suitable to be adopted to government agencies
Panel 2: lecturer	• It gives positive impact to the workplace environment. • Helps to establish more organized management system	• More suitable to be adopted to government agencies

have perceived that both 5S Audit and 5S Training affected the Safety and Quality Performance of their organization.

29.3.4.1 Interview Protocol

Interview protocol was conducted with two panels from the management of the university and the details are shown in Table 29.6. This method has provided an opportunity to interact closely with the respondents and the information collected used to support the results from the survey that was conducted earlier. Both panels agreed on the benefits of MQS system in improving the workplace environment and having a more organized management system. However, both panels also found out that MQS system is more suitable to government agencies and public sectors as this system was adopted from Conducive Ecosystem for Public Sectors (EKSA). Both of the panels stated that the environment in public sectors may be slightly different from the environment in the private university, thus the employees need to be trained to be familiar with the system.

29.3.4.2 Delphi Validation Analysis

The Delphi technique was applied to validate the framework established in this study. The process involved three (3) experts panel in the selected university from different departments which are Academic Section, Management and Finance and also one expert from Quality Assurance Section. The results of the questionnaire shown that all the mean scores obtained were above 3.0 (Agree to Strongly Agree). According to Tigelaar et al. (2004), a minimum percentage of 75% for experts approval is needed in order to validate the framework. Therefore, the results from the Delphi Questionnaire confirmed that the experts has evaluated and approved the overall structure of the MQS system framework. The experts comments are shown in Table 29.7.

Table 29.7 Results of Delphi Questionnaire

Item		Mean score
Suitability of framework	Use of language and terminologies in the framework could be easily understood by the workforce	3.50
	The elements used in the framework could fulfill the needs of the university	3.67
	The framework could be implemented within the university regardless of its structural year of establishment	4.00
Effectiveness of framework	The framework is effective to improve the role of 5S Practice in MQS in the university	3.67
	The framework is effective to improve the safety, quality and environmental performance	3.00
	The framework clearly illustrated the importance of critical success factors and impact towards successful implementation of MQS in the university	3.33
	The framework clearly illustrated the benefits of MQS implementation that could be gained by absorbing and integrating the concept of 5S	3.00
Overall implementation structure	Sequences and steps of implementation are accurately illustrated	3.67
	Sequences and steps of implementation are consistent and systematically illustrated	3.00
	In overall, the framework could be easy, understand, simply and effectively used within the university	3.67

29.3.4.3 Framework of MQS

The framework proposed and presented in this section is the final MQS implementation framework which is a framework built through the process of survey, survey findings and panel validation. Analysis has been conducted in order to analyze the relationship between 5S practice and MQS system implementation in the selected university. As the data has been collected and analyzed using descriptive and inferential statistical analysis, the framework of independent variables and dependent variables has been established according to the data.

Based on Fig. 29.2, all the factors of 5S practice in MQS have a significant medium and strong relationship with Safety Performance and Quality Performance. However, the relationship between the factors and Environmental Performance was weak. Therefore Environmental Performance was eliminated from the framework.

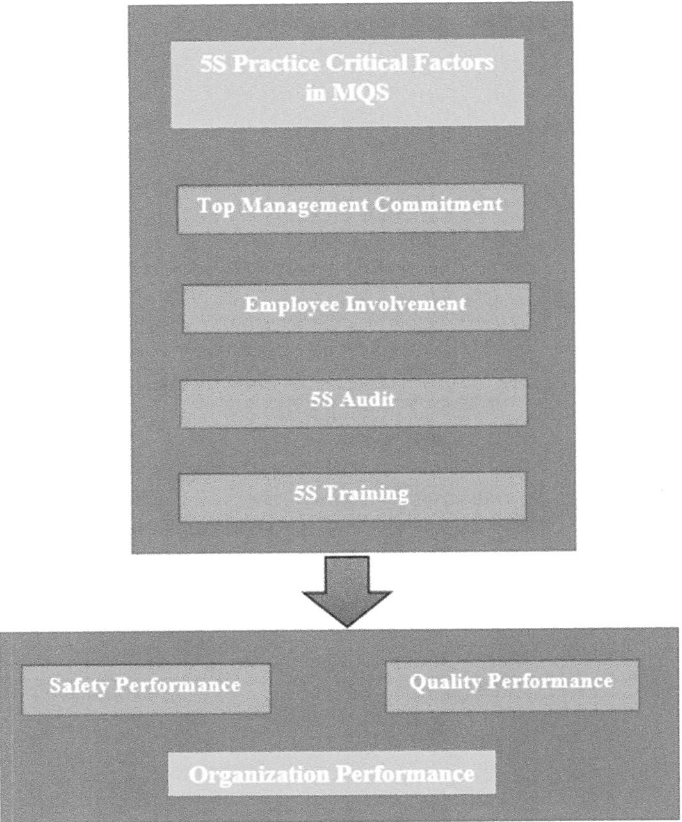

Fig. 29.2 Final framework

29.4 Conclusion and Recommendation

In this study, the status for the role of 5S practice in MQS system implementation has been explored and the relationship between the critical factors with organization performances were analyzed. Based on the findings obtained, it shows that 5S practice positively affects the implementation of MQS system in the case study. It can be concluded that the critical factors of 5S did not only give positive impact to Quality Performance, but it also helps in improving the Safety Performance of the organization. However, the workforce needs to be exposed more on the requirements for MQS system in order to fully achieve the objective of the system. With the previous success of 5S practice in the organization, it is not impossible for the organization to fulfill the requirements outlined in MQS system. This framework will be useful for other MARA centres with the background of 5S implementation to pursue MQS system certification.

References

Ablanedo-Rosas, J.H., Alidaee, B., Moreno, J.C., et al.: Quality improvement supported by the 5S, an empirical case study of mexican organisations. Int. J. Prod. Res. **48**(23), 7063–7087 (2010)

Azwan Huzaimi, M., Mohd Fauzi, Z.A., Abdul Rashid, A.H., et al.: The implementation and consequences of 5S among staff of UniKL MSI Kulim. In: Conference on Language, Education, Engineering and Technology (COLEET 2015) (2015)

Becker, J.E.: Implementing 5S to promote safety and housekeeping. Prof. Saf. **46**(8), 29–31 (2001)

Cohen, J.: Statistical Power Analysis for the Behavioral Sciences. Academic, New York (1997)

Davies, J.A.: Elementary Survey Analysis. Prentice Hall, New Jersey (1971)

Gapp, R., Fisher, R., Kobayashi, K.: Implementing 5S within a Japanese context: an integrated management system. Manag. Decis. **46**(4), 565–579 (2008)

Ho, S.K.M.: Japanese 5S—where TQM begins. TQM Mag. **11**(5), 311–320 (1999)

Howell, V.W.: 5S in the process industry. https://www.pharmamanufacturing.com/articles/2005/248.html.7. Accessed 5 Nov 2020 (2005)

Ho, S.K.M., Cicmil, S., Fung, C.K.: The Japanese 5S practice and TQM training. Training Qual. **3**(4), 19–24 (1995)

Kauthar, A.R., Nurul Hafieza, A., Rahim, J., et al.: The adoption of 5S practice and its impact on safaty management performance: a case study in a university environment. J. Occupational Saf. Health **16**(1), 9–17 (2019)

Osada, T.: The 5S's: Five Keys to a Total Quality Environment. Asian Productivity Organization, Tokyo (1991)

Sekaran, U.: Research Methods for Business: A Skill Building Approach, 3rd edn. Wiley, USA (2000)

Tigelaar, D.E.H., Dolmans, D.H.J.M., Wolfhagen, I.H.A.P., Vleuten, C.P.M.V.D.: The development and validation of a framework for teaching competencies in higher education. High. Educ. **48**, 253–268 (2004)

Chapter 30
Design and Development of a Loader Bucket Wheelbarrow

Zulkarnain Abdul Latiff, Fazidah Saad, W. Faradiana W. Maidin, and Mohamad Sazwan Faiz Mazlan

Abstract This paper briefly introduces the design and analysis of developing the loader bucket at the front of the wheelbarrow to create an assisting system for lifting load into the wheelbarrow. To operate the wheelbarrow, the user has two jobs, which is to manoeuvre the wheelbarrow and scoop the load using a shovel or lifting the load using bare hands. The process of loading and unloading the wheelbarrow could cause discomfort to the waist area for prolonged and repeating work for many hours throughout the day. Moreover, the ergonomic factor is also being considered when using a shovel or even lifting some load using bare hands. Furthermore, to overcome this problem, the wheelbarrow is designed to have its loader bucket at the front of it to do the jobs. The six-point linkages, a pneumatic actuator is assembled with the loader bucket and being controlled by an electric motor. This design has made the jobs easier, effortless, and very safe to the users. The design process starts with data collection of necessities from the users to have a clear problem statement of the current wheelbarrow. Then generations of some ideas of the loader bucket are made, and some design tools assessed them until the final conceptual design created by a CAD software. On a final note, to verify the design, the final design was analysed in terms of its strength through stress and deformation analysis by using a CAD software as well. The result of the whole assembly of loader bucket wheelbarrow after modification is acceptable with the deformation for stress is 6.297e + 07 N/m^2, the deformation for displacement is 2.062 mm and the strain is 6.247e-04.

Keywords Loader bucket · Wheelbarrow · Design process · Design analysis

Z. Abdul Latiff (✉) · F. Saad · W. F. W. Maidin · M. S. F. Mazlan
Universiti Kuala Lumpur, Malaysian Spanish Institute Kulim Hi-Tech Park, 09000 Kulim, Kedah, Malaysia
e-mail: zulkarnain@unikl.edu.my

F. Saad
e-mail: fazidah@unikl.edu.my; sazwan96@gmail.com

W. F. W. Maidin
e-mail: wfaradiana@unikl.edu.my

M. H. Abu Bakar et al. (eds.), *Progress in Engineering Technology III*,
Advanced Structured Materials 148, https://doi.org/10.1007/978-3-030-67750-3_30

30.1 Introduction

A wheelbarrow is small transporting hardware and usually with one wheel at the front, two handles for users to hold and lift it, and the load will be in between of the handle and front wheel. The wheelbarrow eases the user by having the load in the middle of the fulcrum and the effort, applying the second-class lever as in Fig. 30.1.

Nowadays, there have been numerous designs of wheelbarrow developed, aiming to ease the users and to increase productivity. There is a difference between wheelbarrow used at nursery or being used by gardeners if compared to the one that is used at the construction site. The wheelbarrow used by the gardener has a smaller and shallower tray bucket to lodge the load while the wheelbarrow used at the construction site has a deeper tray bucket. Meanwhile, wheelbarrows at the construction sites are also varying depending on their application, such as wheelbarrow with one wheel and two wheels. The wheelbarrow with one wheel is commonly used for easy manoeuvrability work through a plank or any tight corners, but for a two-wheel wheelbarrow is used for a higher amount of load when compared with a one-wheeled wheelbarrow at which the manoeuvrability is not the priority.

The existing wheelbarrow is just a simple wheelbarrow imitating the second-class lever and having a tray, wheel, and a handle, requiring a shovel to fill the tray with loads. To operate the wheelbarrow, the user has two jobs, which is to maneuverer the wheelbarrow and scoop the load using a shovel or lifting the load using bare hands. Moreover, the ergonomic factor is also needed to consider when using a shovel or even lifting some load using bare hands. It is because this process requires the user to bend their knees or waist area to lift the object or load from the ground to fill it into the tray. This process could cause discomfort to the waist area for prolonged and repeating work for many hours throughout the day. For the moment, there is still no wheelbarrow design that has a loader bucket attached to it for loading loads.

The main focus of this study is to design and develop the loader bucket in front of the wheelbarrow and creating an assisting system for lifting the loader bucket to increase the productivity of the hardware. During jobs, productivity could be increased by having this modification to ease the user. The loader bucket uses to replace the shovel in the process of scooping sand, fertilizers, and so on.

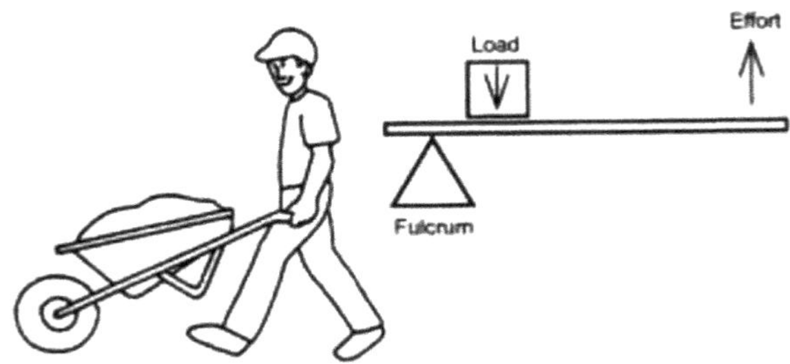

Fig. 30.1 Wheelbarrow applying second class lever

30.2 Literature

30.2.1 Optimize the Design of Wheelbarrow Loader Bucket from Wheel Loader

A wheel loader is a type of engineering machine widely used in engineering projects such as roads, railways, haven, mine, or hydroelectric projects (Yu et al. 2010). Having this equipment is an important role in speeding up the construction or building speed, reducing labour intensity and reducing cost caused by less labour intensity. The loader bucket was invented to ease the job by replacing only human to human-operated machinery and also improving project quality (Wei 2013). Figure 30.2 shows an example of ZL50 type wheel loader.

Many studies developed towards the enhancement and innovation of the wheel loader and Filla developed a simulation analysis and overall performance of the wheel loader that has high performance, efficiency, and operational in a shorter time and lower cost (Reno 2013). To establish the dynamics equations of working mechanisms of the wheel loader, Li (Guiju and Caiyuan 2018) used the Huston method and the MBDA software to predict the mechanism's security. Huang optimized the working device of loader based on the satisfactory degree theory (Zhang et al. 2020). Ning analyzed the strength of the arm, lever, and link by coupling hydraulic and mechanism (Ning et al. 2010). Cao built up a rigid, flexible coupling model of the working mechanism of the wheel loader to match the real situation (Cao et al. 2020). Dai carried out an orthogonal experiment to improve performances of the working device (Zhang and He 2018). Gao optimized the performances of parallel moving and automatic reset of the loader (Wu et al. 2012). Zu optimized the bucket angle and the force of the tilt cylinder (Zu and Wei 2013).

Fig. 30.2 ZL50 loader wheel

30.2.2 Establish the Assistance System to Load and Unload the Wheelbarrow Loader Bucket

The assistance system to load and unload the wheelbarrow loader bucket is depending on the type of system used to handle the linkages. The system can use a hydraulic servo, pneumatic cylinder, or electric actuator.

Hydraulic servo drives are controlled mainly by servo valves, proportional valves, or very often by direct control valves. The names relate to the valve's historical development, which can control four hydraulic resistances with four control edges controlled by one spool (Noskievic 2018). The main advantage of using this drive is that the separate valve control provides more degrees of freedom through the design of the control algorithms, allowing the entire drive to perform more different tasks and final functions. The control algorithm gives the final function completely.

The pneumatic cylinder is a special equipment in which the compressed air energy can be transferred to the piston rod's kinetic energy with high-velocity motion (maximum velocity up to 16 m/s). The pneumatic impact cylinder is an automatic high-speed tool that is safe and environmental friendly and is commonly used for punching, filling, riveting, running pneumatic tools, etc. This pneumatic cylinder adopts the technique of integration of the pneumatic cylinder-valve and has many superior properties such as compact structure, safety and reliability and high-frequency response (Gai et al. 2013). The advantage of using this method is the clean but strong design of the loader. Still, the disadvantages of having a complicated structure as it needs to install the air compressor, and this will result in a tedious design.

Electric actuators also referred to as electric linear actuators or electric drives are used in robotics and mechanics for the coordination of linear motion. These devices are found in three types: pneumatic, hydraulic, and electrical. The third type of actuators is more common when it comes to various amateur projects because of the relatively low prices and the simplicity of the electrical connections. The electric linear actuators linear motion is produced using the force generated by the ordinary electric motor working with both alternative (AC) and direct current (DC). However, the direct current (DC) type of motor is more often used. The mechanism of the linear actuator carries out the transformation of the torque of the motor into the linear motion of the actuator's working mechanism. Nowadays, the majority for the creation of automated linear actuators mostly chosen are the 12 V and 24 V collector motors.

30.3 Methodology

The design process used a project management guide to execute the project development, typically involving problem definition, design solutions, and evaluating ideas. The design process usually consists of a series of steps to solve a specific problem.

Usually, the design process begins by asking a few questions regarding the problem to solve.

30.3.1 Problem Definition and Data Collection from the Users

To begin developing a wheelbarrow loader bucket design, firstly identification of a problem that needs to be addressed is done by interviewing a few respondents from different working scopes but using wheelbarrow in their work. The data were collected by recorded conversations from respondents and also from some working demonstration by them in using the ordinary or traditional wheelbarrow. As a summary, most of the respondents bring a load around 30–40 kg on a trip, and their loads are such as sand, bricks, coconuts, and plant pots. It took them around 5–6 min to load the loads in the wheelbarrow bucket by using a scoop and shovel or by using bare hands, and they took about 5–6 min again to unload the loads. From the observation during the work, the respondents will use their total body strength, especially their hand muscles and waist to shovel the loads. It is not a productive way of working and harmful to the body if working in long hours. Therefore, improvement in terms of the loading and unloading mechanism is very crucial to the wheelbarrow.

To sum up, the problem faced by users while using a wheelbarrow to load and unload are:

(1) Very tiring to load and unload the wheelbarrow as the need to use a lot of strength to scoop and shovel the loads.
(2) Take a lot of time to fill up the bucket and empty the bucket afterwards.
(3) Harmful to the worker's body as they need to bend a lot during shovel and scoop the load.

30.3.2 Objective Tree Method

The objective tree method is used to build up the main objective of the design from the problems identified earlier, and this method also will detail up the objectives needed for the new design of the loader bucket wheelbarrow. The design must comply with the user's needs, and the objective tree filled with purposes of the loader bucket wheelbarrow's safety and a few features needed to comply with the data collections and interview's observations.

The objective tree of the wheelbarrow loader bucket is shown in Fig. 30.3. There are five primary objectives that need to be achieved at the end of the design.

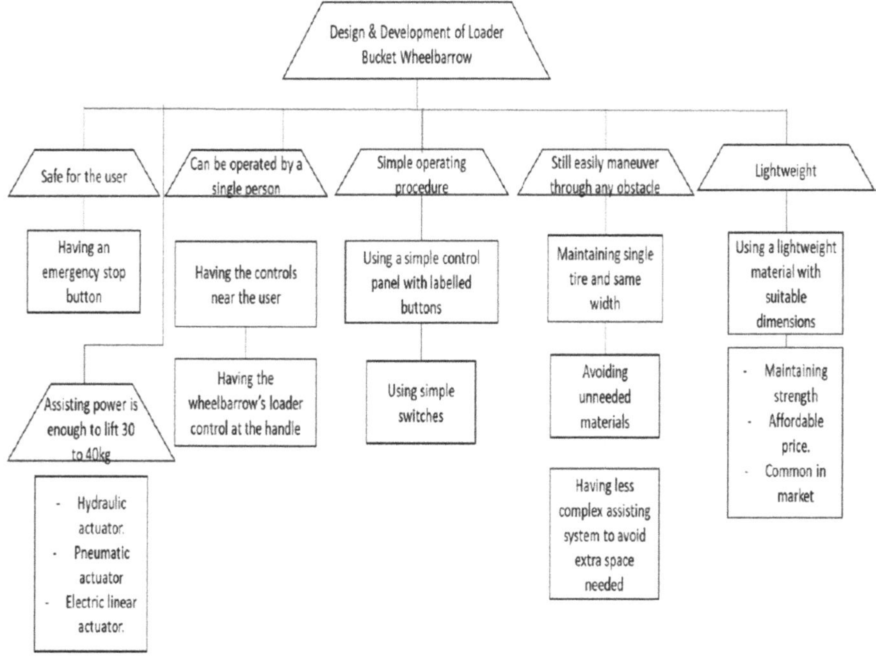

Fig. 30.3 Objective tree of wheelbarrow loader bucket design

30.3.3 Function Analysis Method

The method of function analysis offers such a means of considering essential functions and the level of addressing the problem. The essential functions are those to be fulfilled by the device, product, or system to be designed, regardless of which physical components may be used. Figure 30.4 shows the function analysis chart of the wheelbarrow loader bucket design. Here the system regulations are identified in detail to achieve the earlier objective.

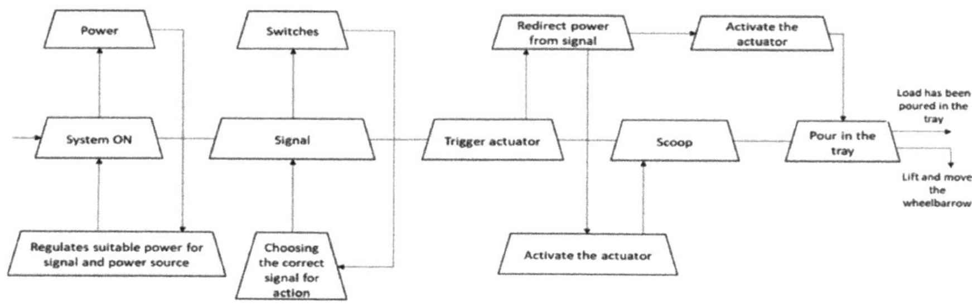

Fig. 30.4 Function analysis chart of wheelbarrow loader bucket design

Table 30.1 Performance specification of wheelbarrow loader bucket design

Physiological needs	Social needs	Technical needs
• Easy to operate the loader • Easy to maneuver the wheelbarrow • Lightweight • Have high safety features throughout the process • Ease of doing work • Reduce the effort of the worker/user while using it	• Simplicity of the design for consumers to recognize the wheelbarrow • Easy to use • Effectively reduce time and effort of the workers	• Reliable design • Max load can be loaded in the wheelbarrow bucket is 50 kg • Material used is a very lightweight but strong • Have better loading and unloading loads functionality and features

30.3.4 Performance Specification Method

The performance specification was made after the function analysis. The performance specifications consist of physiological needs, social needs, technical needs, and time needed. The performance specification method aims to make an accurate specification of the performance required of a design solution. Table 30.1 indicates the performance specification of each function of the objectives.

30.3.5 Quality Functional Deployment Method

Quality function deployment (QFD) is a process and set of tools that are used in this project to effectively define customer requirements and convert them into detailed engineering specifications and plans to produce the products that fulfil those requirements. The aim is to set targets to achieve for the engineering characteristic of the product that is essential for the product to perform and obtain the objectives in this project. At the same time, this method also is useful to fulfil the customer's requirements and evaluated the characteristics in order to identify the suitable options that relevant to the objectives and the criteria needed.

Figure 30.5 shows the QFD chart of the wheelbarrow loader bucket design. From this chart, it can be concluded that to achieve the objectives, the assisting system that needs to be used to the loader bucket is an electric actuator.

30.3.6 Brainstorming Ideas by Using Morphological Chart

The aim for this morphological chart is to generate the complete alternative design solutions of the wheelbarrow loader bucket and potentially widen and add a few features to enhance the capabilities of the product with the features that exist in the market. Table 30.2 shows the morphological chart used to identify the ideas for the designs, and the selected criteria are highlighted.

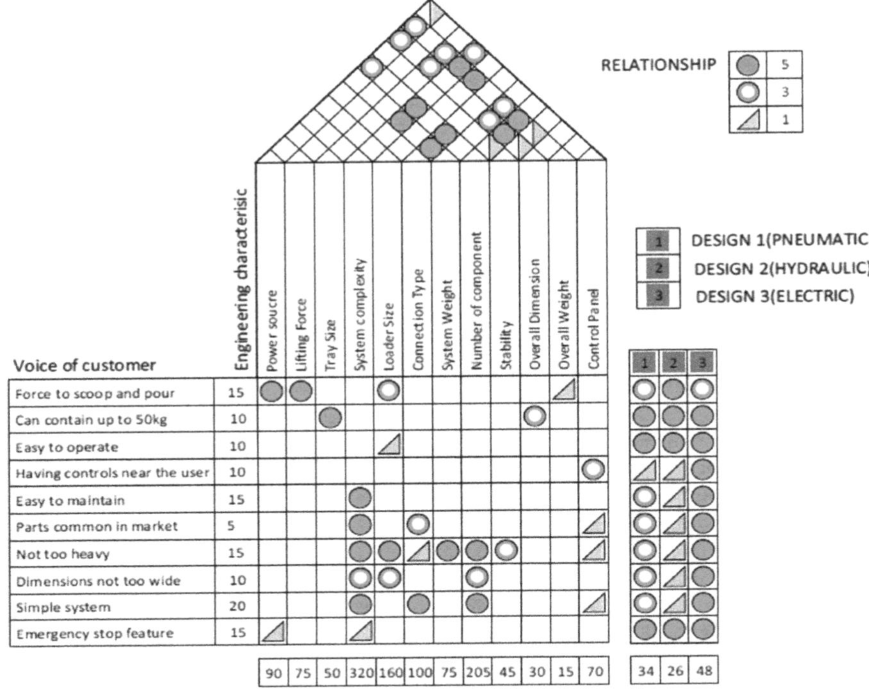

Fig. 30.5 QFD chart of wheelbarrow loader bucket design

Table 30.2 Morphological chart for wheelbarrow loader bucket design

Features	Means				
Power	Battery	AC	Generator	Generator & battery	
Lifting	Pneumatic	Hydraulic	Electric	Manual	
Loader mechanism	4 point linkage	6 point linkage			
Operator	Remote	Standing at side	Holding the handle		
Material	Aluminum	Carbon steel	Mild steel	Plastic	Fiberglass
Mounting	To tray	To chassis	To tray and chassis		
Mounting mechanism	Welding	Rivet	Bolt & Nut	Wedged shape joint (tanggam)	

30.3.7 Sketching the Ideas and Weighted Objective Analysis

Based on all the design steps taken before, three possible design sketches have been made to slightly give a picture to develop a possible design for this product. The simple 2D sketches equipped with the specifications and parts from the data collected in the previous design method process had been analyzed later by using a weighted objective analysis method. Therefore, the final design concept is available after finishing up all the designs steps.

30.4 Result and Discussion

30.4.1 Final Conceptual Design

The final conceptual design was made by using SolidWorks, Fig. 30.6a displays the final conceptual design of the wheelbarrow loader bucket, while Fig. 30.6b is the exploded view of the design. The side view of wheelbarrow bucket loader is shown in Fig. 30.7.

The design is based on the existing wheelbarrow but has a modification by using the six-point linkage based on the existing loader bucket design. The pneumatic actuator attached at the loader linkages is to make ease the loading and unloading process of the bucket. The main material used for the loader bucket is using Aluminium while the wheelbarrow body is maintains to use the existing material, which is steel.

The wheelbarrow loader bucket is attached to a loader linkage that is attached to shafts that connect to a linear actuator that acts as the arm for the bucket to move downwards or upwards. There are two linear actuators used to move the loader

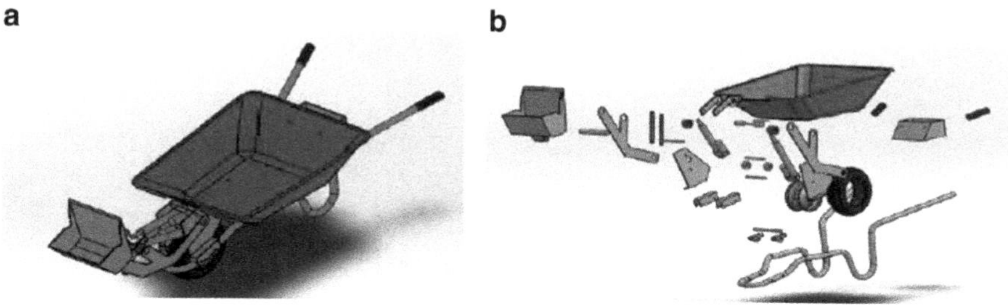

Fig. 30.6 **a** Final conceptual design of wheelbarrow loader bucket. **b** Exploded view of the design

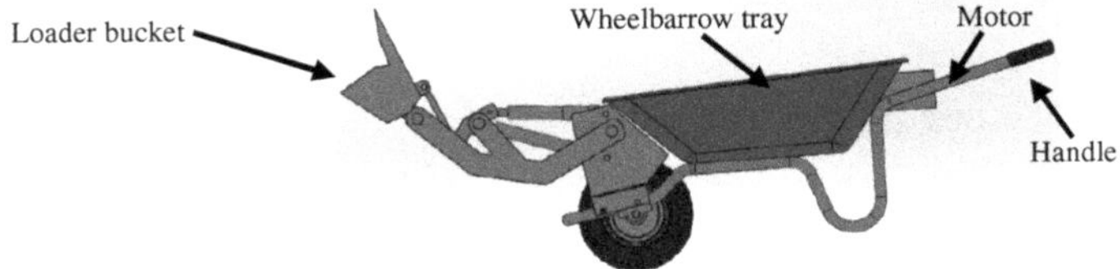

Fig. 30.7 Side vie of the wheelbarrow loader bucket

bucket, which is 100 mm linear actuator and 150 mm linear actuator depending on the length of the arm needed to move. All these actuators are controlled by a motor located at the wheelbarrow body tray and near to the handle where the user can easily manage to operate. The connection harness attached under the wheelbarrow bucket is for safety reason. All the items can be referred in Figs. 30.8 and 30.9 that shows the detail view of the loader bucket.

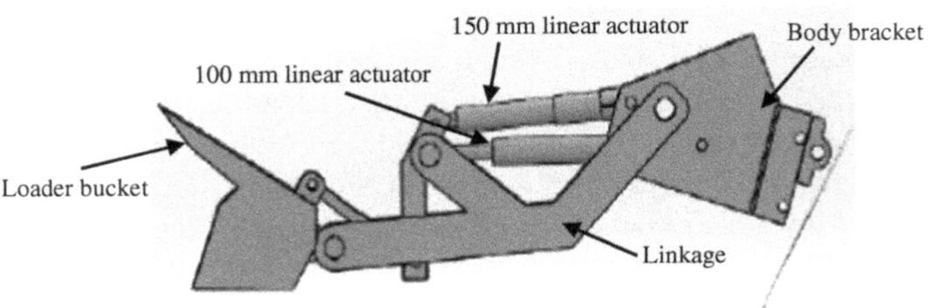

Fig. 30.8 Side view of loader bucket

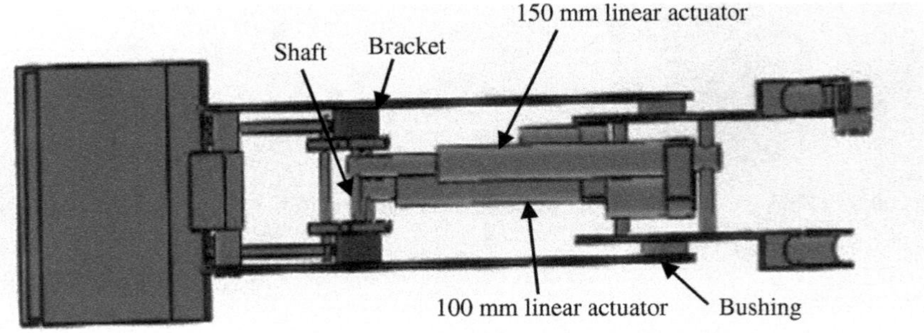

Fig. 30.9 Top view of loader bucket

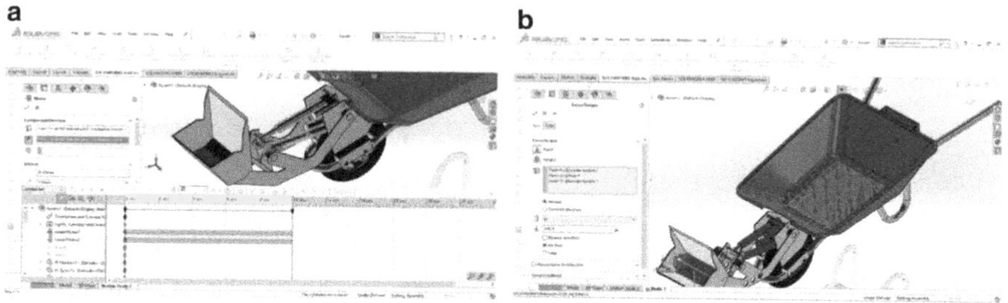

Fig. 30.10 **a** Motion study of wheelbarrow loader bucket. **b** Load applied on the loader bucket and wheelbarrow tray

30.4.2 Design Simulation and Analysis

There are a few simulations carried out using SolidWorks simulation and SolidWorks motion study. These simulations are carried out to ensure the structure design is sufficient to withstand the load applied. There are a few analysis obtained by carrying out an analysis and simulation such as stress, strain, displacement, and motor force.

30.4.2.1 Motion Study

Firstly, the motion study was done to the design. It is very crucial to study the movement of each part and to observe where parts may collide each other so that the parts can be modified to move free as intended without any problems when the fabricating process is done. The motion study is shown in Fig. 30.10a and the results show no collision happened and the design is ready for the next level of analysis which is the structural analysis.

30.4.2.2 Structural Analysis

During the structural analysis, the load of 500 N is applied to the loader bucket as in Fig. 30.10b and the deformation at certain parts is observed and analyzed to find the affected area of the developed model so that the model can be modified to have more strength and durability during the real-time use.

The stress–strain analysis results show that the highest value of stress is 6.297e + 07 $N/m^{2\cdot}$. In contrast, 6.247e-04 for a strain that occurs at the linkage between the loader bucket and body bracket that is attached to the wheelbarrow body tray and this value is still considered acceptable and can be seen in Figs. 30.11 and 30.12.

The deformation results show the highest value of deformation is 2.062 mm at the loader bucket itself during the loading of the loads and the value accepted for the design as the displacement of 2 mm is happened during the bucket is fully utilized with the maximum load of 50 kg. The results are as in Fig. 30.13.

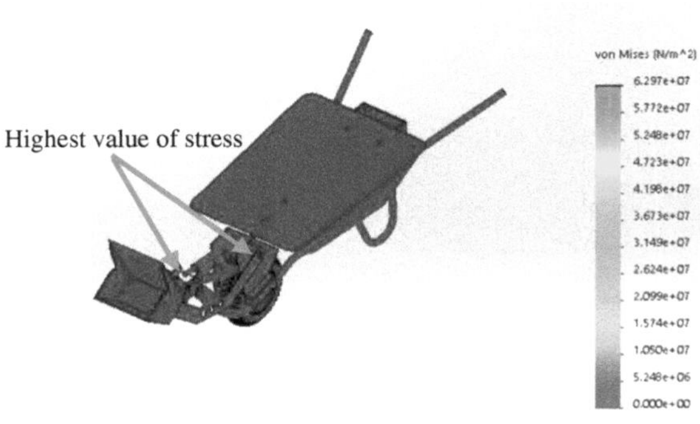

Fig. 30.11 Stress results of wheelbarrow loader bucket

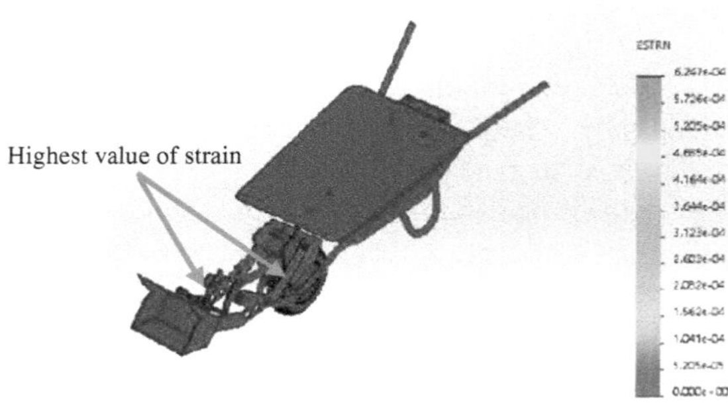

Fig. 30.12 Strain results of wheelbarrow loader bucket

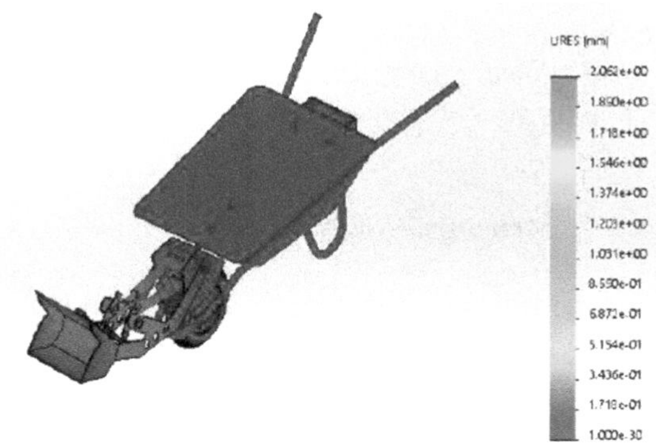

Fig. 30.13 Displacement results of wheelbarrow loader bucket

30.5 Conclusions

In conclusion, it is proved that the wheelbarrow loader bucket design is successful to load and unloaded maximum of 50 kg loads and also reduce the user's strength and effort with good safety features. The main objective of adding value to the existing wheelbarrow meant to ease the user and to increase productivity is achievable. After carefully analyzing the design from the simulation and motion study, some modifications were made to increase the structural rigidity so that the design is accepted. However, the design seems to be acknowledged, though there is still room for improvement in order to achieve a better design in the future.

References

Cao, B.W., Liu, X.H., Chen, W., Tan, P., Niu, P.F.: Intelligent operation of wheel loader based on electrohydraulic proportional control. Math Prob. Eng. (2020). https://doi.org/10.1155/2020/1730946

Gai, C.H., Wang, C.G., Xie, X.H.: Development and experimental research of two-way impact pneumatic cylinder. Appl. Mech. Mater. **331**, 236–241 (2013)

Guiju, Z., Caiyuan, X.: Dynamic simulation analysis on loader's working device. Aust. J. Mech. Eng. **16**, 2–8 (2018)

Ning, X., Shen, J., Meng, B.: Co-simulation of wheel loader working mechanism. Appl Mech. Mater. **43**, 72–77 (2010)

Noskievic, P.: Control of linear hydraulic actuator using the full hydraulic bridge. In: Proceedings of 2018 19th International Conference Research and Education in Mechatronics, REM 2018 (2018). https://doi.org/10.1109/REM.2018.8421791

Reno, F. Optimizing the trajectory of a wheel loader working in short loading cycles. In: Proc from 13th Scandinavian International Conference on Fluid Power, Linköping, Sweden vol. 292. issue no. 4, pp. 307–317 (2013)

Wei, R.S.: Modeling and motion simulation of loader working device based on ProE. Adv. Mater. Res. **722**, 410–414 (2013)

Wu, X., Xiao, G., Chen, R.: Comprehensive optimum design of working device of loader based on sensitivity. Adv. Mater. Res. **490–495**, 3027–3031 (2012)

Yu, Y., Shen, L., Li, M.: Optimum design of working device of wheel loader. In: 2010 International Conference on Mechanic Automation and Control Engineering (MACE 2010), pp. 461–465 (2010)

Zhang, Z., He, B.: Comprehensive optimum and adaptable design methodology for the working mechanism of a wheel loader. Int. J. Adv. Manuf. Technol. **94**(9), 3085–3095 (2018)

Zhang, H., Bai, G., Song, L., Zhu, S.P.: Multiobjective design optimization framework for multi-component system with complex nonuniform loading. Math Prob. Eng. (2020). https://doi.org/10.1155/2020/7695419

Zu, Y.L., Wei, J.: Simulation and optimization design of inversion six-bar linkage on loader. Appl. Mech. Mater. **416–417**, 1822–1825 (2013)

Lightning Source UK Ltd.
Milton Keynes UK
UKHW020736170522
403097UK00002B/16